Models of Tree and Stand Dynamics

Annikki Mäkelä • Harry T. Valentine

Models of Tree and Stand Dynamics

Theory, Formulation and Application

Springer

Annikki Mäkelä
Department of Forest Sciences
University of Helsinki
Helsinki, Finland

Harry T. Valentine
USDA Forest Service
Northern Research Station
Durham, NH, USA

ISBN 978-3-030-35760-3 ISBN 978-3-030-35761-0 (eBook)
https://doi.org/10.1007/978-3-030-35761-0

This Springer imprint is published by the registered company Springer Nature Switzerland AG.
The registered company address is: Gewerbestrasse 11, 6330 Cham, Switzerland

Preface

This book is designed to be a textbook for graduate-level students of forest ecology and forestry who are interested in modelling and also for mathematicians or physicists who have an interest in forest ecology. In addition, it can serve as a reference for researchers who are engaged in the modelling of tree and stand dynamics. The book is an introduction to the modelling of forest growth based on ecological theory and is specifically aimed at practical applications for forest management under environmental change. It is largely based on the work and research findings of the authors, but it also covers a wide range of literature relevant to process-based forest modelling in general.

Although the models presented in this book are based on ecological theory and research findings, they are formulated to be sufficiently simple in structure to readily lend themselves to practical application. One of our primary objectives has been to provide a bridge between so-called empirical and process-based forest models, both through mathematical description and numerical estimation. The link to empirical findings means that the models can be readily parameterized and applied using a variety of data sources and assimilation methods, while the ecological basis allows us to understand the underlying processes and further develop existing models to advance the approach as new findings become available.

The material in the book is arranged so that the reader starts from basic concepts and formulations, then moves towards more advanced theories and methods, and finally learns about parameter estimation, model testing, and practical application. To help students digest the concepts and become proficient with the methods, exercises are provided, complete with solutions and supporting R-code.

Throughout the book, we describe models using differential equations, which are the general mathematical tools for treating dynamic phenomena. It is not necessary for the reader to have in-depth knowledge about the theory of differential equations, though it is advisable to have at least some understanding of the general concepts. We provide some introductory examples of the basic concepts in Chap. 1, combined with examples that use R-code to solve and make predictions from models comprising differential equations.

Basic concepts and theory are presented in the first four chapters, including a review of traditional descriptive forest models, basic concepts of carbon balance modelling applied to trees, and theories and models of tree and forest structure. Chapter 5 provides a synthesis in the form of a core model, which is further elaborated and applied in the subsequent chapters. The more advanced theories and methods in Chaps. 6 and 7 comprise aspects of competition through tree interactions, and eco-evolutionary modelling, including optimisation and game theory, which are topical and fast-developing areas of ecological modelling in this era of climate change. Chapters 8 and 9 are devoted to parameter estimation and model calibration, demonstrating how empirical and process-based methods and related data sources can all be combined to provide reliable predictions. Chapter 10 demonstrates some practical applications and possible paths for future development of the approaches described.

Acknowledgements

The idea of writing this book first came from a publishing agent who was touring Europe to find new book titles. Without him, we probably would not have embarked on this endeavour, which has certainly represented a considerably greater investment of time than originally anticipated. However, it felt like a natural development at the time, as our collaboration on models—based on a shared and evolving philosophy—had already been going on for some years. Our first encounter was in 1986 at IIASA (International Institute for Applied Systems Analysis) in Laxenburg, Austria, where we shared an office in the kitchens (Silberkammer) of the country palace of former Empress Maria Theresa. This was a remarkable coincidence, as we knew nothing of each other though we had each just independently published a paper on the application of the pipe model theory to modelling tree growth in a carbon balance framework; but then, such encounters were very much in keeping with the mission of IIASA. It was the start of a pleasant and fruitful collaboration which has resulted in several coauthored papers applying our shared approach to modelling. In this book, we draw from this work, with the objective of consolidating and disseminating our findings and putting them into a wider context of modelling tree and stand dynamics.

Of course, many other people have also contributed to the ideas and results presented in the book, and their role is recognised below. We are indebted to all of them, but any inadequacies or errors are wholly our own responsibility.

AM: I thank Andrea Holmberg who first introduced me to biological applications of systems theory during my MSc studies at the Helsinki University of Technology and Pepe Hari whose unconventional research group at the University of Helsinki offered a stimulating environment for mathematical modelling in forest ecology. I'm indebted to John H.M. Thornley whose example on the scientific method in modelling made a profound and lasting impression on me during my 6-month visit to his lab as a PhD student and to Joe Landsberg whose work on simplified models was an inspiration.

My sincere thanks go to all my coauthors and collaborators for continuing scientific interaction, which has greatly enhanced the material in this book. Among those, special thanks are due to Frank Berninger, Roderic Dewar, Tony Ludlow, Ross McMurtrie, Eero Nikinmaa (†), and Risto Sievänen for their modelling insights; Heljä-Sisko Helmisaari, Aleksi Lehtonen, and Harri Mäkinen for fruitful collaboration towards evidence-based modelling; and Olli Tahvonen and Marcel Van Oijen for sharing methods that expanded model application. I owe special thanks to colleagues who at some point, as postgrads or postdocs in my research group, carried out empirical work on tree structure and function and/or advanced model development and application: Petteri Vanninen, Anu Kantola, Tero Kokkila, Martti Perämäki, Remko Duursma, Minna Pulkkinen, Tapio Linkosalo, Sanna Härkönen, Pauliina Schiestl-Aalto, Mikko Peltoniemi, Tuomo Kalliokoski, and Francesco Minunno. MP, TK, and FM also contributed to some of the simulations in Chaps. 8, 9, 10, and many students provided comments on the book manuscript in progress.

Throughout the years, I have also received a lot of institutional support from many sources. The Department of Forest Sciences at the University of Helsinki continues to provide a supportive work environment including excellence in ecophysiological research, a link that has been invaluable for my modelling work. I am grateful to the Natural Resources Institute Finland, the Finnish Meteorological Institute, VTT Technical Research Centre of Finland and ICOS Finland (Integrated Carbon Observation System) who have all provided access to different types of forest, climate, and flux data along with fruitful collaboration with many colleagues. I also acknowledge the project funding I have received over the years, especially from the Academy of Finland and European Commission research programmes.

Finally, I thank my husband, Tim Carter, for his enduring support and insights throughout the writing process.

HTV: My interest in forest modelling began during my senior year (1969) at Rutgers, when Prof. Benjamin Stout gave me copies of the two seminal pipe model papers. Next, at Yale, Kenneth Mitchell, Daniel Botkin, and George Furnival provided me with background on three different approaches to modelling the dynamics of forest trees and stands. I became a scientist with the Forest Service in 1974, mostly working in a unit that dealt with the gypsy moth problem and dieback and decline diseases of trees. In 1984, I decided to try to weave together the pipe model and a carbon balance formulation from John Thornley's first book, with the idea of modelling tree dieback. The Sabbatical at IIASA came soon afterwards, followed by the opportunity to concentrate on forest sampling and forest modelling research until my retirement in 2016.

I am grateful for the leadership and/or support, through the years, from David Houston, Dale Solomon, Richard Birdsey, Linda Heath, David Hollinger, and Susan McGrane. I thank all of my collaborators for their insights and contributions to our joint modelling and related sampling work. Also, I am indebted to Harold Burkhart and Ralph Amateis for supplying me with the valuable spacing trial data from the Forest Modeling Research Cooperative at Virginia Tech. I especially have benefited from frequent collaborations and discussions with Mark Ducey, George Furnival, Ed Green, Tim Gregoire, Jeff Gove, and Dave Hollinger. Most of all, I thank my wife, Joan, for 48 years of unwavering support (and first-rate proofreading).

AM&HTV: During the process of writing this book, we have benefited from the excellent suggestions and careful reviews by Ann Camp, Jeffrey Gove, Oscar García, Andrew Robinson, Roderick Dewar, Rupert Seidl, Marcel Van Oijen, and David Hollinger. We thank them all.

Helsinki, Finland
Durham, NH, USA
August 2019

Annikki Mäkelä
Harry Valentine

Contents

Symbols of the Core Model

Parameters

a_{SL}	$m^2\,kg^{-1}$	specific leaf area (leaf area per unit leaf biomass)
a_σ	–	reduction factor of rate of photosynthesis
l_b, l_s, l_t	m	specific pipe length: branch, stem, coarse root
k, k_p	–	canopy light extinction coefficient
$\mathbb{P}_0$	$Mg\,C\,m^{-2}\,yr^{-1}$	rate of gross photosynthesis (unshaded)
$r_{M_f}, r_{M_r}, r_{M_w}$	$kg\,C\,kg^{-1}\,yr^{-1}$	specific rate of maintenance respiration: leaf, fine root, wood
s_f, s_r, s_w	yr^{-1}	tissue-specific rate of litter fall: leaf, fine root, wood
T_f, T_r, T_w	yr	tissue longevity: leaf, fine root, wood
u	–	contol parameter
Y_G	$kg\,(kg\,C)^{-1}$	yield factor: biomass production per unit photosynthate
z	–	exponent
α_r	–	ratio of fine root to leaf biomass
α_x	m	empirical crown-rise parameter
β_x	–	empirical crown-rise parameter
η_b, η_s, η_t	$kg\,m^{-2}$	pipe-model ratio: branch, upper stem, coarse root
σ_f	$kg\,C\,kg^{-1}\,yr^{-1}$	leaf-specific rate of photosynthesis
$\rho_b, \rho_m, \rho_s, \rho_t$	$kg\,m^{-3}$	bulk density: branch, mature wood, juvenile wood, coarse root
ϕ_b, ϕ_s, ϕ_t	–	form factor: branch, upper stem, coarse root

Tree Variables

A_c	m^2	stem cross-sectional area at the crown base
$A(h)$	m^2	stem cross-sectional area at height h
B	m^2	tree basal area
dbh	m	stem diameter at breast height (1.37 m)
dcb	m	stem diameter at crown base
G	kg yr^{-1}	rate of whole-tree biomass production
G_f, G_r, G_w	kg yr^{-1}	rate of biomass production: leaf, fine root, wood
H	m	tree height
H_c	m	height to crown base
L_c	m	crown length
P_g, P_n	kg C yr^{-1}	rate of tree photosynthesis: gross, net
R, R_G, R_M	kg C yr^{-1}	rate of tree respiration: total, growth, maintenance
R_f, R_r, R_w	kg C yr^{-1}	rate of respiration by tree component: leaf, fine root, sapwood
r_c	–	live crown ratio
S_f, S_r, S_w	kg yr^{-1}	rate of senescence by tree component: leaf, fine root, wood
W	kg	total tree biomass
W_f, W_r, W_w	kg	biomass by tree component: leaf, fine root, total wood
W_b, W_s, W_t	kg	biomass by wood component: branch, stem, coarse root
W_p	kg	biomass of total live wood
W_q	kg	biomass of dead wood embedded in the main stem
λ_f, λ_p, λ_r	–	proportion of growth allocated to: leaves, wood, fine roots

Variables for the Mean-Tree Model

$\mathbb{L}$	–	leaf area index
N	m^{-2}	stand density
X	m	average tree spacing

Chapter 1
Introduction

This chapter introduces the purpose and intent of the book, then reviews key concepts used in dynamic ecological modelling, such as hierarchy, resolution, modelling approaches, and the concepts of state and rate variables. A simple example of growth modelling is given together with a brief introduction to solving differential equation models in R.

1.1 Setting the Stage

Models of trees and stands have existed for hundreds of years and take many forms. Some particularly realistic tree models (or 'model trees') can be seen in dioramas at the Fisher Museum, located at the Harvard Forest in Petersham, Massachusetts (https://harvardforest.fas.harvard.edu/dioramas). In most of the dioramas, model trees are positioned to form model stands that bear striking resemblance to real stands that occur in the geographical region called New England.

One particular diorama shows how the model trees are constructed. The stem system of each model tree is formed from a round bundle of fine wires. The portion of a main stem that occurs beneath the crown is formed from all the wires in a bundle, but starting at the crown base, sub-bundles of wires diverge off the main stem to become models of first-order branches, and smaller sub-bundles diverge from those branches to become models of second order branches, and so forth, until terminal shoots are modeled. Consequently, these model trees manifest *area-preserving branching*, because the total stem cross-sectional area of the branches is the same before and after a branching node. The observation that area-preserving branching pertains generally, with minor discrepancy, to real trees was first recorded

Electronic Supplementary Material The online version of this chapter (https://doi.org/10.1007/978-3-030-35761-0_1) contains supplementary material, which is available to authorized users. The videos can be accessed by scanning the related images with the SN More Media App.

A. Mäkelä, H. T. Valentine, *Models of Tree and Stand Dynamics*,
https://doi.org/10.1007/978-3-030-35761-0_1

(so far as we know) in the notebooks of Leonardo da Vinci. The wire-based model trees, which were constructed in the 1930s, also agree with the more recent pipe model theory of stem form (Shinozaki et al. 1964a,b), which subsumes the observation of area-preserving branching. In regard to stem form, the pipes of a pipe model are analogous to the wires of a museum model.

Each of the Fisher dioramas represents a static visualization of the state of a forest, but a sequential display of several dioramas depicts the ecological process of old-field succession at different points in time. A second sequence of dioramas demonstrates silvicultural practices. For a student of forestry or forest ecology, the sequential displays illuminate change under natural or managed conditions. Moreover, the model trees and stands project powerful visual evidence regarding the degree of realism that can be achieved with just two morphological (or architectural) building blocks: the pipe model and blueprints of branching patterns by species.

In contrast to static visual displays, the pioneering tree and stand simulator (TASS) of Mitchell (1975) predicts the development of trees within a stand over a stream of time. Stem growth as modelled by TASS conforms with Pressler's law of stem formation (see Assmann 1970, p. 58), and is consistent with both area-preserving branching and the pipe model. TASS provides sufficient detail for the creation of three-dimensional model stands—like those in the Fisher Museum—that afford visualization of the predicted state of a real stand at a point in time. But unlike the model stands in the Fisher dioramas, TASS also provides a description of tree and stand dynamics, and by doing so, informs us about how or why a predicted state of a stand succeeds from a previous state.

1.2 Focus of This Book

Our intent is to teach you, or refresh your memory about, how to formulate and solve mathematical models of tree and stand dynamics. The raw material for model formulation includes the observations of Leonardo and the elegant pipe model of Professor Shinozaki and his colleagues; it also includes other structural observations (data) and theories, as well as biophysical, physiological, and ecological observations, theories, and models. And, of course, our models of tree and stand dynamics are tempered by evolutionary theory, and constrained by the laws of nature.

Like all scientific disciplines, modelling of tree and stand dynamics advances incrementally with the occasional quantum leap. The formulation of a new tree or stand model ordinarily is influenced by the successes of previous tree and stand models. Different facets of, and advances in, the state of the art of dynamic tree and stand modelling are detailed in recent reviews by Mäkelä et al. (2000, 2012), Bugmann (2001), Landsberg (2003), Pretzsch et al. (2008), and Van Oijen et al. (2013), among others. Other reviews of models are contained in recent books by

Landsberg and Sands (2011), Weiskittel et al. (2011) with a 69 page bibliography, and Burkhart and Tomé (2012). Pretzsch (2009, Chap. 11) provides a very lucid review of several kinds of forest models in a historical context. A wealth of information about physiological models, and modelling in general, is provided by Thornley and Johnson (1990) and Thornley and France (2007).

Since so much background material can be gleaned elsewhere, we omit a general review of tree and forest models. The core of this book focuses on our own modelling work, supplemented with case studies of other models, and augmented with essential background material.

1.3 Dynamic Models

Trees and stands are complex life systems, and any attempt to fully describe these systems mathematically is sure to fail. Instead, the essential information about a tree or stand is summarised by one or more state variables. Since life systems are dynamic, the values of the state variables change with the passage of time. Dynamic models describe the temporal changes with difference or differential equations, or a combination of the two, one equation for each state variable. With difference equations, the values of the state variables change in discrete jumps across finite time intervals. Differential equations, by contrast, describe instantaneous rates of change.

The rate of change in a state variable may be influenced by other variables, such as environmental or biophysical variables. These are sometimes called driving or forcing variables. Other variables of interest, often called response variables, may be defined in terms of combinations of the state variables. In forestry, for example, the volume or biomass of a tree often is specified in terms of two state variables, stem diameter and tree height.

Parameters of dynamic models are components of the equations whose values do not change; beyond that, we shall not attempt any formal definition. The value of a parameter may be an estimate, a measurement, or a quantity that can be deduced from fundamental understanding. Parameters take constant values for the purposes of solving a model, but values may vary among models or be altered for different solutions of the same model. In the realm of tree and stand models, parameters may serve to translate the value of one state variable into the rate of change of another. Other parameters serve as maximum or minimum rates of change, or as asymptotes of state variables. In many of the models discussed in this book, parameters include scaling exponents, specific physiological rates, and ratios of morphological traits.

The solution of a dynamic model provides a time-course for each of the state variables. A graph of tree height versus tree age, for example, is a time-course of tree height. We describe a process of solving a model later in this chapter. Before that, however, we take up some topics essential to model formulation.

1.4 Raison D'être

A mathematical model of a dynamic system is essentially a mathematical formulation of assumptions and understanding about how the system operates. How a particular model is formulated ordinarily depends on the modelling objectives. Most commonly, tree or stand models are formulated to forecast future states of the tree or stand, as described by predicted values of the state variables. The choice of state variables naturally should accord with the objectives. Applied forecasting models may account for effects of predicted environmental or biophysical factors, prescribed courses of silvicultural treatments, availability of water and essential nutrients, and other site-related factors. Backcasting to past states is a variation on this theme.

Forecasting models known as *growth and yield models* are used widely in forestry (e.g., Pretzsch 2009; Weiskittel et al. 2011). The *growth model* of a growth and yield model is the dynamic model, and the *yield model* is the solution of the growth model: the time courses of the state variables and other variables of interest. However, one does not need a growth model to produce a yield model. Yield models, in various forms, have been advanced by foresters since the 1700s (e.g., Assmann 1970, pp. 1–2). For example, a table of stand volume versus stand age constitutes a yield model.

As currently practiced, forestry involves forest ecosystem management, usually with, but sometimes without, a profit motive. In either case, planning horizons often are measured in decades. Intelligent decision-making about management options and assessment of their likely economic and ecological consequences requires knowledge of the current state of nature, and forecasts of future states, both with and without the application of management alternatives.

In biological research, a universal goal is to increase understanding of all life systems. A dynamic model may summarise current knowledge about a life system, pinpoint gaps in that knowledge, and test current theory about how the system operates. Any critical evaluation of the model ordinarily involves comparisons of the assumptions, predicted time-courses, and possibly relationships among the time-courses with data. New hypotheses may arise from such evaluations that lead to scientific advancement and increased understanding, and perhaps the formulation of a refined model, refined analyses, new hypotheses, and so forth to an even more refined model.

Models and experiments also are used to extrapolate to new conditions. For example, free air carbon enrichment (FACE) studies monitor life systems in otherwise natural settings maintained with predicted future concentrations of atmospheric carbon dioxide (CO_2). A model formulated by Franklin et al. (2009) to maximise tree growth by optimising the tree's use of nitrogen (N) and carbon (C) has been used to interpret experimental responses to elevated atmospheric CO_2 concentrations (McCarthy et al. 2010).

1.5 Hierarchy

Many contemporary models of tree or stand dynamics have a hierarchical organization, which simply reflects the fact that this kind of organization exists throughout the biological world. Thornley and Johnson (1990, pp. 5–6) presented a hierarchical scheme for plant and crop models that we adapt to dynamic tree and stand models:

Level	Description
...	...
$i+2$	Forest landscape
$i+1$	Stand
i	Tree
$i-1$	Organ
$i-2$	Tissue
$i-3$	Cell
...	...

As noted by Thornley and Johnson (1990), each level in the hierarchy is an integration of lower levels. For example, we often express the live portion of a tree in terms of its leaf, sapwood, and fine-root biomass. In the hierarchical scheme, leaves and fine roots are organs, and sapwood (conducting xylem) is tissue that extends through the woody organs: coarse roots, central stem (trunk), and branches. The total woody biomass also includes the biomass of dead heartwood (non-conducting xylem tissue), which occurs in the main stem and large branches. In this particular scheme, the biomass of the tree, at level i, is expressed in terms of state variables at levels $i-1$ and $i-2$.

In another example, inventory data for forest stands often comprise measurements on trees that are selected under some particular sampling design (e.g., Gregoire and Valentine 2008; Mandallaz 2008). The most common tree measurements include dbh (i.e., diameter at breast height, 1.3–1.4 m), tree height, and possibly crown length. Here, the tree is summarised by the measurements of tree-level traits, and the stand, at level $i+1$, is summarised by frequency distributions of the tree-level traits.

Hierarchical language often is used to convey the focus of a model. Models with labels such as leaf-level, tree-level, and stand-level are common. Whether labeled or not, the focal level of a model is implied by the state variables, as each level in the hierarchy has its own descriptive language (Thornley and Johnson 1990, pp. 5–6). For example, leaf area or leaf biomass are leaf-level traits, whereas traits defined on the basis of horizontally projected land area, such as leaf area index or net primary productivity, apply at the stand level or higher levels.

The functional components of a system can be arranged in nested levels according to the amount of detail involved (Mäkelä 2003). Generally, the lower the hierarchical level, the smaller the spatial and temporal scales, which correspond, respectively, to faster process rates and smaller physical sizes (Thornley and Johnson 1990).

A methodological framework for dealing with hierarchical systems was developed in *systems theory* (Bertalanffy 1968; Klir 1972; Godfrey 1983; Bossel 1994) and *hierarchy theory* (Allen and Starr 1982; O'Neill et al. 1986). A focal result of these theories is that the hierarchical levels are in relative isolation from each other.

Higher levels are characterised by slower processes that appear as constraints to the level of interest, while lower levels, with faster process rates, show as effective means or steady-state properties at the level of observation. This implies that it is unnecessary to look further down than one level in the hierarchy in the search of mechanistic explanations of the system behaviour. In fact, including too many levels is not only unnecessary but even counter-productive, because models including mechanism at several levels tend to become complex and difficult to handle, hypersensitive to small changes in parameter values, and tend to lose their mechanism as noise (Landsberg 1986; Reynolds et al. 1993; Dewar et al. 2009).

The result that lower-level, higher-resolution phenomena appear as effective means to the level of interest is important, because it means that we need not focus, for example, on the day-to-day dynamics of metabolic processes in order to quantify their effects in annual-resolution models that span over the lifetime of trees. Instead, we may describe these phenomena by means of temporal constants or inputs that vary at the time step of the model. Nonetheless, we still need to devise a method for evaluating such constants or input variables. This may not be straightforward, as relevant measurements for determining such model inputs may not be at hand at our specified resolution.

Reynolds et al. (1993) suggested a method for combining scales based on the hierarchy theory. Because the levels of the hierarchy are relatively isolated, they can be modelled separately. Lower level models can be linked to the core model as summary models or response surfaces, or formulated as functional relationships or phenomenological models and tested separately at the lower level. It is important to note that these summary approaches cannot usually be derived by averaging the lower level inputs and outputs. Instead, the summary models are used to obtain the effective-mean or steady-state behaviour.

The upper-level constraints may be derived from some slowly changing environmental restrictions to model inputs (O'Neill et al. 1986). Another type of constraint is related to so-called emergent properties, which are sometimes thought about as properties derived from the lower level system, but also refer to characteristics that can only be sensibly defined at the higher level. An example of emergent behaviour that will be discussed in Chap. 7 is optimal plant function which may be regarded as a tree-level constraint to component behaviour.

The hierarchical approach produces modular models where modules at different levels are relatively independent and replaceable by alternative modules, as long as their input and output links with the rest of the system are specified (Robinson and Ek 2000). The method differs from a reductionist (mechanistic) approach because the system is also constrained from above—for example, upper-level emergent properties can be modelled as constraints to the whole system (Fig. 1.1). The

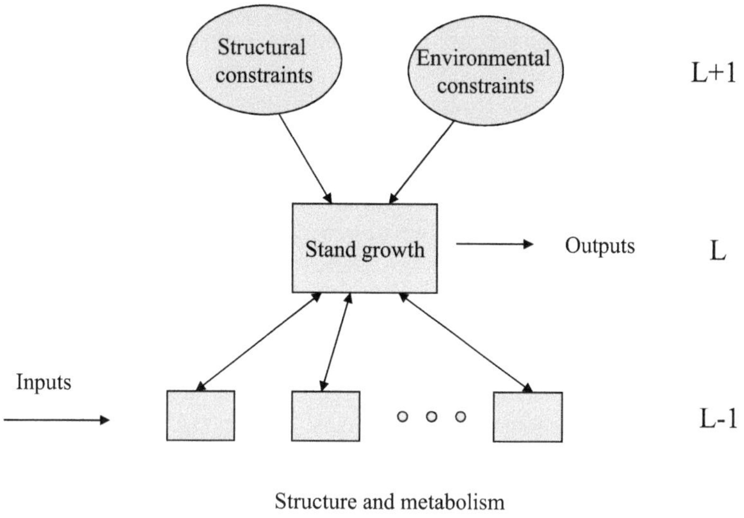

Fig. 1.1 A schematic presentation of three hierarchical levels in tree growth models. (Redrawn from Mäkelä (2003))

practical advantage of this hierarchical modularity is that each module can be developed and tested on its own before combining them all as the full model. Theoretically, it helps conceive the system as a whole, which could generally result in more robust models.

The practical significance of these concepts will be discussed especially in Chap. 8 where we consider the quantification of metabolic and structural inputs to process-based models formulated for individual trees at annual resolution.

1.6 Model Resolution

Models of stands can be organized in different ways, but most of them can be categorized as either whole-stand, mean-tree, or individual-tree models. Whole-stand models ordinarily do not distinguish individual trees, though some tree averages may be provided as response variables if the stand density—the number of trees per unit land area—is a state variable. In several models used in forestry, stand-level state variables, such as aggregate basal area, average tree height, and stand density afford the estimation of average tree diameter and biomass (e.g., Landsberg and Waring 1997). Elaborations of whole-stand models include supplemental procedures for the estimation of frequency distributions of tree diameters (e.g., Strub and Burkhart 1975; Hyink 1983).

Whole-stand models certainly are not limited to forestry applications. McMurtrie and Wolf (1983), for example, formulated a whole-stand model of carbon utilisation and allocation. The theory underpinning this model is discussed in Chap. 3, and its solution is demonstrated later in this chapter. The model is process-based, accounting for radiation interception, photosynthesis, respiration, carbon allocation, the production of new biomass, and the loss of senescent biomass.

In the mean-tree approach to stand modelling, tree-level variables summarise a mean tree. The attributes of the mean tree, supplemented by stand density, provide time-courses of the stand-level traits. The tree-level attributes may be state variables, or response variables of a dynamic model formulated at the tissue and organ levels, and possibly based on physiological processes. Moreover, a mean-tree model may employ stand-level or higher level constraints, for example, the carrying capacity per unit land area of leaf biomass (e.g., Mäkelä 1997). We speculate that mean-tree approach to stand modelling may see increased use with the advent of planted stands comprising trees of common clones.

A third kind of organization involves modelling a stand as a collection of individual model trees. Some of these models, such as TASS, use spatial information, others do not. A spatial tree-level model uses the location and size of a model tree relative to the locations and sizes of its neighbouring model trees, usually to account for crown interactions and/or shading effects (e.g., Mitchell 1975). By contrast, such detailed information is not used in a non-spatial tree-level model (e.g., Stage 1973), even though the values of the state or response variables may differ for each model tree. In some of the older modelling literature, spatial and non-spatial models, respectively, are called distance-dependent and distance-independent models (Munro 1974). The non-spatial models sort into different types according to whether the model trees are conterminous or scattered. In non-spatial gap models, for example, model trees or cohorts occupy a common patch of ground, where the patch size approximates the size of a gap created by the death of large mature tree (see, e.g., Botkin et al. 1972; Shugart 1984; Bugmann 2001). By contrast, the so-called Prognosis model of Stage (1973, et seq.) tracks the dynamics of model trees that are scattered about a stand.

1.7 Modelling Approaches

Descriptive labels of tree or stand models sometimes are modified by words such as process-based or mechanistic. As explained by Thornley and Johnson (1990, pp. 7–11), mechanistic modelling follows the reductionist approach as utilised in the basic sciences. Understanding and prediction at the ith hierarchical level is achieved through an understanding of component processes at the $(i-1)$th level and possibly lower levels.

To various degrees, stand models based on a mean tree or a collection of individual trees are mechanistic, because the behavior at the stand-level depends on processes at the tree level, and possibly lower levels. However, whereas mechanistic models generally are process-based, a process-based model is not necessarily mechanistic. The carbon budget model of McMurtrie and Wolf (1983), for example, is based on physiological processes at the stand level that translate into stand-level growth responses, so the model is process-based but not mechanistic, or at least not under the hierarchical definition.

Other descriptive modifiers of model labels include empirical and hybrid. An empirical model is basically tabular summarisation or functional interpolation of the central tendencies of observational data. Thornley and Johnson (1990, pp. 6–7) indicated that empirical models describe level *i* behavior in terms of level *i* traits, without regard to biological theory. Many of the growth and yield models used in forestry are purported to be empirical by their developers. With the exception of Chap. 2, purely or predominately empirical models of tree or stand dynamics get minor attention in this book, but the topic is well covered by Weiskittel et al. (2011) and Burkhart and Tomé (2012), among others.

Hybrid models comprise process-based and empirical components at the same hierarchical level (Mäkelä et al. 2000). However, if the model is multi-level, then it could be process-based, hybrid, or empirical at any given level, but mechanistic overall.

Yet another type of model employs optimisation to take account of the plasticity in tree structure, function, and life strategies that have evolved through natural selection (e.g., Kull 2002; Mäkelä et al. 2002). Optimisation models utilise an apparent tree-level goal or 'objective function' that is maximised with respect to one or more functional traits, usually under one or more physiological or environmental constraints (Dewar et al. 2009). The objective function serves as a proxy of fitness, and the functional traits are assumed to have sufficient plasticity to effect the optimal response. An example of an objective function at the tree level is maximum height growth rate, which is based on the assumption that the trees that grow fastest in height, while maintaining stability, are the likeliest to avoid suppression, and survive to produce offspring (Mäkelä 1985; Valentine and Mäkelä 2012). The optimisation approach to modelling is appealing because it eliminates the need for explicit dynamic equations of the plastic traits and their attendant parameters (Dewar et al. 2009). Instead, the time-courses of the plastic traits emerge from the constrained optimisation to become hypotheses for further research. We return to this subject in Chap. 7.

Process-based models that incorporate mechanism, hybridisation, or optimisation, singly or in combination, are emphasized in this book.

1.8 Growth

Everyone knows what growth is and how it happens. Nonetheless, we shall engage in a bit of review. Plant growth, and thus tree growth, results from the division and enlargement of cells (Loomis 1934). Hence, the word *growth* is descriptive of both an addition of new biomass to a tree and any consequent increase in the size of the tree. Whole tree growth is usually measured in units of either carbon or dry weight per unit of time, e.g., kg C yr^{-1} or kg DW yr^{-1}, where 'DW' signifies plant dry weight, which is equivalent in definition to biomass and dry matter.

1.8.1 Tree Growth

When measured on an annual basis, tree growth (kg C yr^{-1}) equals photosynthesis minus respiration under the implicit assumption that all labile and stored C reserves are depleted at some point during the year. The cell-based concept of growth is consistent with what ecologists call *net primary production*—the amount of new plant biomass produced by a plant or a community of plants per unit of time.

Whereas any positive change or gain in the biomass of a tree results from cell-based growth, negative change results from the loss of aggregations of cells, for example, the dispersal of seeds with accompanying pods or cones, and the shedding or sloughing of senescent organs, including fine roots, coarse roots, branches, and leaves. Other losses of biomass may result from external influences, such as silvicultural treatments, or the consumption of tree tissues and organs by herbivores.

The *change* in the biomass of a tree over a given period of time equals the gain of biomass resulting from cell-based growth minus all the losses of biomass, i.e.,

$$\text{change} = \text{gain} - \text{loss}$$

A complication, which may not be obvious at first glance, is that over a finite interval of time, the loss may include some of the gain. Consider, for example, a deciduous tree in a temperate region that grows a crop of leaves during a spring and sheds the entire crop in the fall. Over the course of a year, measured from the first day of one winter to the next, the change in leaf biomass is zero, but the biomass of the intervening leaf crop is tallied within both the gain and loss categories. Similarly, if we measure a year starting from the last day of summer, the annual change in the leaf biomass may be small or nil, even though a leaf crop was lost in the fall and a replacement crop was grown in the spring. Tracking the gains and losses over any given growth period, and accounting for the metabolic costs of the former, is fundamental to carbon-balance modelling, which is introduced in the following section.

Changes, gains, and losses are often discussed with alternative terminology. The change per unit time in either the total biomass or some component of the biomass of a tree is often called *net growth* and the growth of new biomass through cell division and enlargement is called *gross growth*. Thus, substituting net growth for change and gross growth for gain,

$$\text{net growth} = \text{gross growth} - \text{loss}$$

Some authors substitute the word *production* for gross growth so that

$$\text{net growth} = \text{production} - \text{loss}$$

1.8.2 Stand Growth

A stand comprises a population of trees, and temporal changes in population attributes, such as aggregate biomass, are usually expressed with respect to the area of land that the stand occupies. Small trees often are ignored in the calculation of stand biomass until they reach some threshold size. Upon reaching the threshold, a tree is, in effect, considered a new member of the stand, so the net growth of the stand is increased by the amount of the new member's total dry weight. By convention, trees that achieve stand membership in the current finite time interval are called *ingrowth*, and trees that die are called *mortality*. Also by convention, aggregate loss due to shedding of aboveground matter is called *litterfall*. Thus, the net growth of the aggregate biomass of a stand (Mg DW ha^{-1} yr^{-1}) is:

$$\begin{aligned}\text{net growth} &= \text{gross growth} + \text{ingrowth}\\ &\quad - \text{litterfall} - \text{mortality} - \text{other loss}\end{aligned}$$

As with individual trees, any loss over a finite interval most likely includes some of the gain from gross growth and ingrowth.

The gross growth of all trees and other plants in a stand, regardless of size, is synonymous with *net primary productivity*, or NPP. NPP, in turn, equals *gross primary productivity* (GPP) minus *autotrophic respiration* (R), where GPP is the total photosynthesis and R is the respiration of all the plants. Of course GPP and R have the same units as NPP, usually Mg C ha^{-1} yr^{-1}.

1.8.3 Instantaneous Rates of Change

Most of the models described in the book are based on instantaneous, rather than finite, rates of change. We generally use W (kg DW) to denote the aggregate biomass of the live portion of tree and we use the same symbol, but in a different font, i.e., $\mathbb{W}$ (Mg DW ha^{-1}), to denote the biomass per unit land area of the live portion of a stand. The aggregate biomass of tree organs and tissues are identified with subscripts, i.e., W_i or $\mathbb{W}_i$. For example, for a tree, some important components of biomass are: W_f, leaf; W_r, fine-root, W_t, coarse-root; and W_s, live portion of the stem. This notation is re-introduced several times in the book, and we shall add to it as needed.

The instantaneous rate of change of live biomass is denoted by $\text{d}W/\text{d}t$ (kg DW yr^{-1}) for a tree and by $\text{d}\mathbb{W}/\text{d}t$ (Mg DW ha^{-1} yr^{-1}) for the live biomass in a stand. Let G_w (kg DW yr^{-1}) be the rate of gain in tree biomass, and let S_w (kg DW yr^{-1}) be the rate of loss of senescent and any other biomass, then, by definition, the rate of change in the aggregate live biomass equals the rate of gain minus the rate of loss:

$$\frac{\text{d}W}{\text{d}t} = G_\text{w} - S_\text{w} \tag{1.1}$$

In the alternative terminology, $\mathrm{d}W/\mathrm{d}t$ is the net-growth rate of the tree and G_w is the gross-growth rate (or production rate). At the stand level, the units of change, gain, and loss are usually Mg DW ha^{-1} yr^{-1}, hence,

$$\frac{\mathrm{d}\mathbb{W}}{\mathrm{d}t} = \mathbb{G}_\mathrm{w} - \mathbb{S}_\mathrm{w} \tag{1.2}$$

where $\mathbb{G}_\mathrm{w}$ includes ingrowth and $\mathbb{S}_\mathrm{w}$ includes mortality. The use of these definitions, or identities, within a dynamic model involves specifying differential equations that provide the instantaneous rates of gain and loss in terms of state variables and environmental factors. How to do that will be addressed in Chap. 3 and later chapters. An essential point at this juncture concerns the time dimension. By specifying time, t, in units of years, we eliminate the need to model seasonal or other within-year processes that result in the waxing and waning of different components of biomass. Instead, the gains and losses are continuous and simultaneous, but scaled to an annual basis. Hence, the total gain over the course of a year—as predicted by the solutions of the instantaneous rate equations—corresponds to a total gain that may actually accrue over a shorter period of activity within the year. An analogous explanation applies to total loss. Moreover, since the rate equations for gain and loss are solved simultaneously, the gains that become losses within the year are also accounted for.

When a rate of change, gain, loss, or activity is expressed on a per unit dry-weight basis, it is called a *specific rate*. For sake of example, suppose that

$$\frac{\mathrm{d}W}{\mathrm{d}t} = aW - bW, \qquad a \geq 0, b \geq 0 \tag{1.3}$$

then

$$\frac{1}{W}\frac{\mathrm{d}W}{\mathrm{d}t} = a - b \tag{1.4}$$

where a is the specific rate of gain, b is the specific rate of loss, and $a - b$ is the specific rate of change (or net growth). We can specify the units of a and b either as kg DW (kg DW)$^{-1}$ yr^{-1} or, more simply, as yr^{-1}.

For an example of a specific rate of activity, let photosynthesis be the activity, where P (kg C yr^{-1}) is the rate of photosynthesis by a tree and W_f (kg DW) is the leaf biomass, then

$$P = \sigma W_\mathrm{f} \tag{1.5}$$

where $\sigma = P/W_\mathrm{f}$ (kg C (kg DW)$^{-1}$ yr^{-1}) is the leaf-specific rate of photosynthesis.

1.9 Carbon-Balance Models

A carbon-balance model is a mass-balance model that keeps track of the amount of carbon that enters and leaves a life system. The model may be formulated in terms of whole stands, mean trees, or individual trees. Indeed, carbon-balance models may be formulated at any of the hierarchical levels listed in Sect. 1.5, and at levels that are not listed. We may, for example, formulate a carbon balance for a leaf, a cell within a leaf, or an organelle within a cell. At the other extreme, we may formulate a carbon balance for the aggregate quantity of leaves contained within a forested landscape or a forest biome. A carbon-balance model may be formulated with an empirical, mechanistic, or optimal approach. The same can be said of nitrogen-balance, water-balance, or other kinds of mass-balance models. A mass-balance model, in essence, is a application of the mass-conservation law.

As carbon (C) is the principal currency of plant growth, the carbon balance is the basis of most of the growth models in this book. For sake of example, let us represent the biomass of an entire tree by the symbol W with units kg DW. Let [C] be the concentration of carbon in the biomass (kg C (kg DW)$^{-1}$), so the total carbon in the tree is $[\mathrm{C}]W$ (kg C). Of interest is the carbon dynamic, i.e., the annual rate of change in the tree's carbon balance,

$$\frac{\mathrm{d}\left([\mathrm{C}]W\right)}{\mathrm{d}t} = \text{rate of C gain} - \text{rate of C loss} \tag{1.6}$$

Both sides of (1.6) have units (kg C) yr^{-1}. For modelling purposes, the carbon concentration usually is assumed constant (i.e., $[\mathrm{C}] = 0.5$) among all the tissues within a tree, so $\mathrm{d}([\mathrm{C}]W)/\mathrm{d}t = [\mathrm{C}] \times \mathrm{d}W/\mathrm{d}t$. Hence, the rate of change in the dry-matter balance of the tree is

$$\frac{\mathrm{d}W}{\mathrm{d}t} = \frac{\text{rate of C gain} - \text{rate of C loss}}{[\mathrm{C}]} \tag{1.7}$$

Both sides of (1.7) have units (kg DW) yr^{-1}. At the tree level, carbon gain is achieved through photosynthesis, and loss is due to respiration and litterfall, with the latter broadly defined to include the sloughing of root matter below ground.

The whole-tree carbon-balance model is completed by expressing the rates of carbon gain and loss in terms of state variables, parameters, and driving variables. At other hierarchial levels, additional factors come into play. At the stand and higher levels, we may consider tree mortality and the consequent loss of carbon that occurs through the respiration of heterotrophic decomposer organisms.

Additional factors also come into play if we take a mechanistic approach, for example, by dividing the tree into leaves, fine roots, and wood. In effect, we track the carbon balances of the aggregate biomasses of leaves, wood, and fine roots, the sum of which equals the carbon balance of the tree. This involves modelling either the transport or allocation of carbon acquired by leaves for utilisation by wood and fine roots, and we must also account for the respiration and litterfall by tissue type.

Moreover, if we aim to formulate a carbon-balance model with the potential for practical application in forestry or forest ecology, we also need to model how the carbon balance affects the aspects tree structure that translate into rates of change in stem diameter, height, and crown ratio. Of course, we cover how to do all this in later chapters.

1.10 Solving a Model with R

We have yet to formulate or discuss any model in detail. Nonetheless, in this section, we put the cart before the horse and demonstrate how a dynamic model that comprises a system of differential equations can be solved with R, a freely available system for statistical and mathematical analysis (see www.r-project.org). There are many software options for solving models, but if you choose R, now is a good time to start learning how to use it. Jones et al. (2009) provide a well structured introduction to R, but there are many alternatives.

If you are already proficient in R, you may be interested in interfacing your R code with LaTeX, a freely available document preparation system that is especially useful for preparing mathematical and other technical documents (see www.tug.org). The manuscript of this book was coded entirely in LaTeX, except for the figures. The interface system is called 'Sweave'. Sweave affords clear documentation of a model, which may include (a) a formal introduction of the model; (b) a complete listing of the R program (script) that solves the model, punctuated by explanations of the different segments of code; and (c) the results from the model solution in tables and figures.

The electronic supplementary material of this chapter contains a file called `mw.Rnw`, which contains a mixture of LaTeX and R code for describing and solving a whole-stand carbon-balance model that was formulated by McMurtrie and Wolf (1983) (we changed some of their notation to conform with this book). This file should be copied and placed in a convenient folder (work directory) for processing. The theory underpinning this model is covered in Chap. 3.

You may need to install a package called deSolve within R before processing the `mw.Rnw` file. The deSolve package implements numerical integrators for solving sets of differential equations. This package was developed by Soetaert et al. (2010), and background material is provided by Soetaert et al. (2012). Examples of the use of this package are contained in documents within the "library folder" of the R program files.

To solve the model without processing the LaTeX: (a) enter R, (b) select or indicate the work directory (folder) that contains `mw.Rnw` and (c) enter the command: `Stangle("mw.Rnw")`. R will place a file called `mw.R` in the work directory. The model is then solved in R with the command: `source("mw.R")`. The solution is provided in the form of two graphics files, `mw_time_courses.pdf` and `mw_production.pdf`, which are placed in the work directory.

To both solve the model and process the LaTeX: Follow steps a and b, then (c) enter the command: `Sweave("mw.Rnw")`. The Sweave process implements the R script that solves the model and places a file called `mw.tex` along with the two graphics files, `mw_time_courses.pdf` and `mw_production.pdf`, in the work directory. Processing `mw.tex` with the usual (pdf)LaTeX procedures creates a document containing the following two subsections. You will need a LaTeX package called `Sweave.sty` to accomplish the last step. A copy of this package can be found among the R program files. The Sweave user manual, by Leisch and R-core (2012), can be retrieved from R's Help files.

1.10.1 Documentation: Brief Model Description

The carbon-balance model of McMurtrie and Wolf (1983) describes the rates of change in dry matter in an even-aged stand. The state variables are aggregate leaf ($\mathbb{W}_f$), fine-root ($\mathbb{W}_r$), and wood (stem, branches and coarse roots) ($\mathbb{W}_w$) dry matter (Mg DW ha^{-1}).

Rates of dry matter production depend on the rate of photosynthesis, $\mathbb{P}_g$ (Mg C ha^{-1} yr^{-1}), which is formulated as a function of leaf dry matter,

$$\mathbb{P}_g = \mathbb{P}_0 \left[1 - \exp\left(-k a_{SL} \mathbb{W}_f\right)\right] \tag{1.8}$$

where $\mathbb{P}_0$ (Mg C ha^{-1} yr^{-1}), k (ha ground (ha leaf area)$^{-1}$), and a_{SL} (ha leaf area (Mg DW)$^{-1}$) are parameters, $\mathbb{P}_0$ being the rate of photosynthesis at complete light interception, k is the canopy light extinction coefficient, and a_{SL} is the specific leaf area (leaf area per unit leaf dry matter). The product $a_{SL} \times \mathbb{W}_f$ is leaf area index—leaf area per unit land area. The net rate of photosynthesis, $\mathbb{P}_n$ (Mg C ha^{-1} yr^{-1}), equals the gross rate minus the rate of dark respiration, i.e.,

$$\mathbb{P}_n = \mathbb{P}_g - r_f \mathbb{W}_f \tag{1.9}$$

where r_f (Mg C (Mg DW)$^{-1}$ yr^{-1}) is the leaf-specific rate of dark respiration.

The dynamic equations express the rates of change in the state variables in terms of rates of gain less rates of loss:

$$\frac{d\mathbb{W}_f}{dt} = Y \eta_f \mathbb{P}_n - s_f \mathbb{W}_f \tag{1.10}$$

$$\frac{d\mathbb{W}_r}{dt} = Y(\eta_r \mathbb{P}_n - r_r \mathbb{W}_r) - s_r \mathbb{W}_r \tag{1.11}$$

$$\frac{d\mathbb{W}_w}{dt} = Y(\eta_w \mathbb{P}_n - r_w \mathbb{W}_w) - s_w \mathbb{W}_w \tag{1.12}$$

where Y (Mg DW (Mg C)$^{-1}$) is the yield of tissue dry matter per unit of photosynthate, r_w and r_r (Mg C (Mg DW)$^{-1}$ yr^{-1}) are specific rates of maintenance respiration, the η_i ($\sum_i \eta_i = 1$) are fractions of photosynthate allocated to production of leaves, fine roots, and wood, and the s_i (yr^{-1}) are specific rates of loss due to litterfall (i = f, r, w).

Documentation—R script

First, we load package deSolve:

```
> library(deSolve)
```

Next, we assign values to the parameters:

```
>      asl =  0.35
>        k =  0.6
>      P_0 = 13.6
>      r_f = 0.14
>      r_r = 0.28
>      r_w = 0.0055
>      s_f =  0.333
>      s_r =  0.5
>      s_w =  0.01
>        Y =  2.2
>    eta_f =  0.2
>    eta_r =  0.6
>    eta_w =  0.2
```

Starting values for the state variables are organized in a vector called yini:

```
>    yini <- c( W_f = 0.1,
+               W_r = 0.1,
+               W_w = 0.1 )
```

The carbon balance model is organized in a function for numerical integration, as per the instructions for using the deSolve package. We call the function mw_model:

```
>    mw_model <- function(t, y, parms) {
+      with( as.list( y ), {
+         P_n <- P_0 * (1 - exp(-k * asl * W_f)) - r_f * W_f
+        dW_f <- Y *  eta_f * P_n - s_f * W_f
+        dW_r <- Y * (eta_r * P_n - r_r * W_r)  - s_r * W_r
+        dW_w <- Y * (eta_w * P_n - r_w * W_w)  - s_w * W_w
+        return( list( c( dW_f, dW_r, dW_w ) ))
+      })
+    }
```

We solve the model from time 0 to 40 (yr) with yearly output:

```
>    m <- 40
>    times <- seq(0, m, by = 1 )
```

We invoke the differential equation solver, ode, which is contained within the package deSolve. Annual values of the state variables, as calculated by ode, are placed in a matrix called out.

```
>   out <- ode( func = mw_model, y = yini, parms = null,
+               times = times )
```

Print the last five rows of out, with results rounded to two decimal places:

```
>   round(tail(out,5),2)

     time   W_f   W_r    W_w
[1,]   36 14.43 12.92 109.20
[2,]   37 14.43 12.92 111.56
[3,]   38 14.43 12.92 113.88
[4,]   39 14.43 12.92 116.14
[5,]   40 14.43 12.92 118.36
```

The first column of the matrix out, viz., out[,1], contains the time or age, and columns 2–4, i.e., out[,2:4], respectively, contain the time-courses of leaf, fine-root, and wood dry matter. For convenience, we extract and name the time-course vectors:

```
>       t <- out[1:(m+1),1]
>   W_f_t <- out[1:(m+1),2]
>   W_r_t <- out[1:(m+1),3]
>   W_w_t <- out[1:(m+1),4]
```

We graph the time courses beginning with W_w_t to set the range (Fig. 1.2):

```
>   ylabel = expression(Dry~matter~(Mg~ha^-1))
>   pdf(file = "mw_time_courses.pdf", width = 4, height = 3.5)
>   plot(t, W_w_t, type = "l", lty = 2,
+          xlab = "Time (yr)", ylab = ylabel)
>   lines(t, W_f_t, lty = 1)
>   lines(t, W_r_t, lty = 3)
>   dev.off()
```

Calculate the rates of dry matter production at times t = 1,2,...,m+1 for: G_f, leaves; G_r, fine roots; and G_w, stems plus coarse roots.

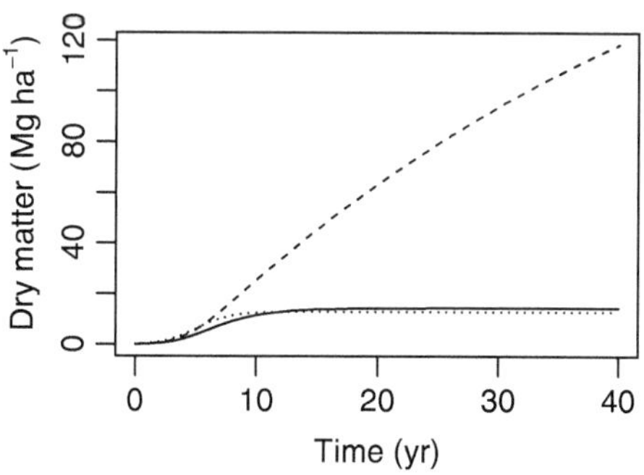

Fig. 1.2 Time-courses of leaf (*solid*), fine-root (*dot*), and wood (*dash*) dry matter

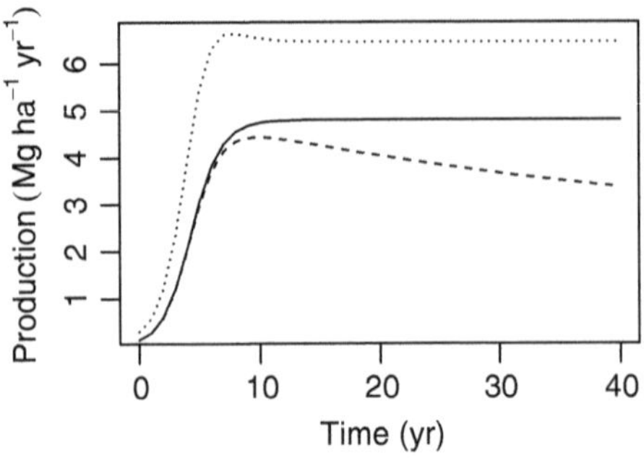

Fig. 1.3 Annual production of leaf (*solid*), fine-root (*dot*), and wood (*dash*) dry matter

```
>     Pn <- P_0 * (1 - exp(-k * asl * W_f_t)) -
+              r_f * W_f_t
>    G_f <- Y *  eta_f * Pn
>    G_r <- Y * (eta_r * Pn - r_r * W_r_t)
>    G_w <- Y * (eta_w * Pn - r_w * W_w_t)
```

Graph the annual dry matter production by component versus time (Fig. 1.3):

```
>    y2label = expression(Production~(Mg~ha^-1~yr^-1))
>    pdf(file = "mw_production.pdf", width = 4, height = 3.5)
>    plot(out[,1], G_r, type = "l", lty = 3,
+          xlab = "Time (yr)", ylab = y2label)
>    lines(out[,1], G_w, lty = 2)
>    lines(out[,1], G_f, lty = 1)
>    dev.off()
```

Note that, given the assigned parameter values, the predicted annual production (or gain) of both leaf and fine-root dry matter is similar in magnitude to the production of wood dry matter (Fig. 1.3). Indeed, the predicted annual fine-root production always exceeds wood production. The wood dry matter, however, accumulates to greatly exceed the dry matter of either leaf or fine-root dry matter (Fig. 1.2). The reason for this can be traced to the respective annual specific rates of loss, i.e., the values of s_f and s_r versus s_w.

1.11 Exercises

1.1 Retrieve the R script, `mw.R`, from the electronic supplementary material of this chapter, and place it in a convenient folder (work directory). Solve the model in R with the command: `source("mw.R")`.

1.2 In Sect. 1.8 we considered tree and stand growth dynamically as the difference between gain and loss, where gain was defined as dry matter production and loss was defined as senescence. Now consider the gain and loss of carbon of the whole ecosystem. Net Ecosystem Production (NEP) is defined as the gain of carbon of the ecosystem per unit time, and it consists of gross primary production minus total ecosystem respiration. The latter is the sum of autotroph and heterotroph respiration.

Under these premises, how would you write the conceptual dynamic equation for the state variable "total ecosystem carbon content"?

1.3 Assume that the tree stand carbon content is modelled as in Sect. 1.10.1, and consider that of the soil. To use this as a basis for calculating NEP, define soil carbon content (Mg ha^{-1}) as an additional state variable. For the sake of this exercise, assume that heterotroph respiration is proportional to soil carbon content. Now write an equation for the change of the amount of organic carbon in the soil, using assumptions given in the above text and in Exercise 1.2. Check that the change of your total ecosystem carbon content is consistent with the definition of NEP given above.

1.12 Suggested Reading

Chapter 1 of Jones et al. (2009) describes how to setup R on a computer and Chap. 2 introduces some essential features of the R calculating environment.

Fulford et al. (1997) provides a helpful introduction to modelling with differential and difference equations for biological applications.

Chapter 2
Descriptive Models

The topics discussed in this chapter span two centuries of modelling effort. We begin with sections that deal with some mathematical details of descriptive growth models of whole trees, then move to saturating functions and numerical switches that will prove useful in later chapters. We end the chapter with a simple empirical model of crown dynamics and stand density that makes use of some of the earlier material.

2.1 Descriptive Growth Models

The boundary conditions of the growth of a tree—net change in size or biomass per unit time, i.e., net growth *sensu* Sect. 1.8—are exponential growth and steady state, with exponential growth tending to occur at the earliest point in life. The term *steady state* connotes a zero growth rate, a condition that may be approached at maturation. These boundary concepts also apply to populations of trees (i.e., stands) with regard to aggregate biomass and to other population traits, for example, average tree height. Many modelling studies have been conducted at either of the two boundary conditions. Our purpose in this section, however, is to become familiar with a few simple growth models that bridge the two boundary conditions.

Several single-equation models describe the transition from an exponential growth rate,

$$\frac{\mathrm{d}W}{\mathrm{d}t} = aW \tag{2.1}$$

to steady state,

Electronic Supplementary Material The online version of this chapter (https://doi.org/10.1007/978-3-030-35761-0_2) contains supplementary material, which is available to authorized users. The videos can be accessed by scanning the related images with the SN More Media App.

A. Mäkelä, H. T. Valentine, *Models of Tree and Stand Dynamics*,
https://doi.org/10.1007/978-3-030-35761-0_2

$$\frac{dW}{dt} = 0 \tag{2.2}$$

Such models integrate to give S-shaped or sigmoid time-courses of $W(t)$, i.e., W versus time, t.

Many alternative models exist for describing the sigmoid time-courses of traits at the organ, tree, or stand level (see, e.g., Zeide 1993; García 2008; Burkhart and Tomé 2012; Panik 2013). And, for most purposes, more than one of the choices may prove efficacious in terms of goodness to fit to data. How well the underlying theory of a particular model matches the problem at hand is a different matter.

In 1942, D'Arcy Wentworth Thompson opined in his classic book, *On growth and Form* (reprinted as: Thompson 1992), that "...the characteristic form of the curve of growth in length (or any other linear dimension) is a phenomenon which we are at present little able to explain, but which presents us with a definite and attractive problem for future solution." As foreseen by Thompson, different explanations of sigmoidal time-courses have emerged from physiological observations and analyses of dynamic models. For example, several hypotheses have been put forward to explain why tree growth in general and tree height growth in particular slow following an initial (nearly) exponential growth phase (e.g. Ryan et al. 2006).

While conflicts resolve and theory advances, the use of single-equation models to describe sigmoid time-courses continues apace. We discuss the Gompertz, Verhulst (logistic), Bertalanffy, and García models. García's general model includes the Gompertz, Verhulst, and Bertalanffy models and several other models as special cases. We also distinguish between model forms that are useful for fitting data from (a) repeated measurements from a chrono-sequence and (b) a pure time-series of data.

A *time-series* of data results from the repeated measurement, over time, of the same organ, tree, or stand. Since trees are long-lived, acquisition of a time-series that spans the two boundary conditions of growth may require the attention of a sequence of investigators over several generations. One alternative is to employ a *chrono-sequence*—a sample of trees of different ages. It is best to take repeated measurements, over time, of each tree within a chrono-sequence, especially for the purpose of fitting a growth model.

2.1.1 Gompertz Model

The Gompertz model was advanced by Benjamin Gompertz, a British mathematician and actuary, to model death rate by age class in human populations (Gompertz 1825). By virtue of its shape, the Gompertz model also can be applied as a model of growth (Winsor 1932).

Thornley and Johnson (1990) suggested that the model may be appropriate where a resource necessary for the growth of some entity is saturating, but the growth machinery of that entity degrades over time according to first-order kinetics. Under this assumption,

$$\frac{dW}{dt} = \mu W \tag{2.3}$$

and

$$\frac{d\mu}{dt} = -a\mu \tag{2.4}$$

where W is the size or mass of the entity, and a is a parameter. The growth rate of the entity is governed by the exponentially decaying specific growth rate, μ, which at time $t \geq t_0$ is

$$\mu = \mu_0 \exp(-at) \tag{2.5}$$

where μ_0 is the value of μ at time $t_0 = 0$. Substituting (2.5) into (2.3) provides the growth-rate model

$$\frac{dW}{dt} = \mu_0 W \exp(-at) \tag{2.6}$$

Thornley and Johnson (1990, p. 81–82) provide a recipe for converting this time dependent form of the growth rate model to the form analysed by Winsor (1932), viz.,

$$\begin{aligned}\frac{dW}{dt} &= aW \ln\left(\frac{W_{\max}}{W}\right)\\ &= aW\,(\ln W_{\max} - \ln W)\end{aligned} \tag{2.7}$$

where $W_{\max}$, a parameter, is the asymptote of W. A near exponential growth rate is achieved when $W \ll W_{\max}$ (i.e., when W is much less than $W_{\max}$). Steady state is approached as $W \to W_{\max}$ (i.e., as W approaches its asymptote, $W_{\max}$). The maximum growth rate occurs when $W = W_{\max}/e$,

Alternatively, since $d(\ln W)/dt = (1/W)dW/dt$, we can transform the model to provide the growth rate of the logarithm of W,

$$\frac{d(\ln W)}{dt} = a\,(\ln W_{\max} - \ln W) \tag{2.8}$$

Integration of (2.8) from time t to $t + \Delta t$, i.e.,

$$\int_{W_t}^{W_{t+\Delta t}} \frac{d(\ln W)}{(\ln W_{\max} - \ln W)} = a \int_t^{t+\Delta t} dt \tag{2.9}$$

provides

$$\ln W_{t+\Delta t} = \ln W_{\max} + [\ln W_t - \ln W_{\max}] \exp(-a\Delta t) \tag{2.10}$$

or

$$W_{t+\Delta t} = W_{\max} \exp\left\{\ln\left[\frac{W_t}{W_{\max}}\right] \exp(-a\Delta t)\right\} \tag{2.11}$$

This form of the model is convenient for fitting data from remeasurements of a chrono-sequence. If the time step, Δt, is constant, then (2.10) reduces to

$$\ln W_{t+\Delta t} = \alpha + \beta \ln W_t \tag{2.12}$$

or

$$W_{t+\Delta t} = e^{\alpha} W_t^{\beta} \tag{2.13}$$

where $\beta = \exp(-a\Delta t)$ and $\alpha = \ln W_{\max}(1 - \beta)$ are parameters. Hence, given a series of measurements over a uniform interval at times $t_1, t_2, \ldots, t_n$, the Gompertz model is appropriate if the graph of $\ln W_{t_{i+1}}$ versus $\ln W_{t_i}$ approximates a straight line.

Integration of (2.8) from time $t_0 = 0$ to any time $t > t_0$ provides

$$W_t = W_{\max} \exp\left[b \exp(-at)\right] \tag{2.14}$$

where $b \equiv \ln[W_0/W_{\max}]$ is a parameter. This form of the model is convenient for fitting a time-course of data.

2.1.2 Logistic Model

The logistic model (also called the Verhulst or autocatalytic model) was used by Pierre-François Verhulst, a Belgian mathematician, to model population growth under a resource constraint (Verhulst 1838). In a population ecology context, the growth-rate model usually is written thus:

$$\frac{\mathrm{d}N}{\mathrm{d}t} = rN\left(1 - \frac{N}{K}\right) \tag{2.15}$$

where N is the population count, r is the intrinsic rate of increase (also called the Malthusian parameter), and K is the carrying capacity — the maximum population count, as constrained by a limiting natural resource within the spatial domain of the population.

Integration of (2.15) from time t_0 to any time $t > t_0$ provides the so-called logistic function:

$$N_t = \frac{K N_0}{N_0 + (K - N_0)\exp(-rt)} = \frac{K}{1 + [(K - N_0)/N_0]\exp(-rt)} \tag{2.16}$$

where N_t is the predicted population count at time t, given the starting count, N_0, at time t_0.

Applications of the logistic function, with and without elaboration, pervade many scientific disciplines, including plant science. In this regard, Thornley (1976) provided theory for the use of the logistic function to model plant growth. The model applies when the rate of change in the biomass of a plant, $\mathrm{d}W/\mathrm{d}t$, is dependent upon a single limiting substrate or natural resource, $\mathcal{S}$. The abundance of this substrate is measured in equivalent units of plant biomass.

The plant growth rate is assumed to be directly proportional to both plant biomass and substrate abundance,

$$\frac{\mathrm{d}W}{\mathrm{d}t} = \mu \mathcal{S} W \tag{2.17}$$

where μ is a parameter. Plant growth occurs solely at the expense of substrate, so

$$\frac{\mathrm{d}W}{\mathrm{d}t} = -\frac{\mathrm{d}\mathcal{S}}{\mathrm{d}t} \tag{2.18}$$

Consequently, at any time t, the residual substrate equals the maximum possible biomass of plant, W_{max}, less the current biomass, i.e.,

$$\mathcal{S} = W_{\mathrm{max}} - W \tag{2.19}$$

Substituting (2.19) into (2.17) yields a logistic model of plant growth,

$$\frac{\mathrm{d}W}{\mathrm{d}t} = \mu W (W_{\mathrm{max}} - W) = aW\left(1 - \frac{W}{W_{\mathrm{max}}}\right) \tag{2.20}$$

where $a = \mu W_{\mathrm{max}}$. Near exponential growth is achieved when $W \ll W_{\mathrm{max}}$, and steady state is approached as $W \to W_{\mathrm{max}}$. The maximum growth rate, or inflection point, occurs when $W = W_{\mathrm{max}}/2$.

The logistic model also can be arranged thus (see García 2005):

$$\frac{\mathrm{d}W^{-1}}{\mathrm{d}t} = a\left(W_{\max}^{-1} - W^{-1}\right) \tag{2.21}$$

Integrating from t to $t + \Delta t$,

$$\int_{W_t}^{W_{t+\Delta t}} \frac{\mathrm{d}W^{-1}}{W_{\max}^{-1} - W^{-1}} = a \int_{t}^{t+\Delta t} \mathrm{d}t, \tag{2.22}$$

provides a growth model over the time step, Δt,

$$W_{t+\Delta t}^{-1} = W_{\max}^{-1} - [W_{\max}^{-1} - W_t^{-1}]\exp(-a\Delta t) \tag{2.23}$$

If the time step, Δt, is constant, then we can reduce (2.23) to

$$W_{t+\Delta t}^{-1} = \alpha + \beta W_t^{-1} \tag{2.24}$$

or

$$W_{t+\Delta t} = \frac{W_t}{\alpha W_t + \beta} \tag{2.25}$$

where $\beta = \exp(-a\Delta t)$ and $\alpha = (1 - \beta)W_{\max}$ are parameters. Hence, given a series of measurements over a uniform interval at times $t_1, t_2, \ldots, t_n$, the logistic model is appropriate if the graph of $W_{t_{i+1}}^{-1}$ versus $W_{t_i}^{-1}$ approximates a straight line.

Solving (2.23) for $W_{t+\Delta t}$ provides

$$W_{t+\Delta t} = \frac{W_{\max} W_t}{W_t + [W_{\max} - W_t]\exp(-a\Delta t)} \tag{2.26}$$

which is useful when Δt is not constant. Integration of (2.20) from $t_0 = 0$ to t provides

$$W_t = \frac{W_{\max}}{1 + b\exp(-at)} \tag{2.27}$$

where $b = (W_{\max} - W_0)/W_0$ is a parameter.

Equations (2.25) are (2.26) are useful for fitting data from a remeasured chronosequence, and (2.27) is the usual choice for fitting a time-series.

2.1.3 Bertalanffy Model

This model has been described under at least five different labels, including exponential-monomolecular, Richards (after Richards 1959), Bertalanffy-Richards, and Chapman-Richards. The Bertalanffy model has been used widely in forest modelling (Pienaar and Turnbull 1973, et seq.), usually under a Bertalanffy-Richards or Chapman-Richards label.

The original Bertalanffy model may have either three or four parameters (Bertalanffy 1957), though the three-parameter variant,

$$\begin{aligned}\frac{\mathrm{d}W}{\mathrm{d}t} &= aW\left(\frac{W_{\max}^{c} - W^{c}}{cW^{c}}\right) \\ &= gW^{m} - bW, \qquad 0 < c \leq 1 \end{aligned} \tag{2.28}$$

is more common. Here a, c, and $W_{\max}$ are parameters. In the alternative formulation, $b \equiv a/c$, $g \equiv bW_{\max}^{c}$, and $m = 1 - c$. Steady-state is approached as $W \rightarrow W_{\max}$. The maximum growth rate occurs when $W = (1 - c)^{(1/c)} W_{\max}$.

Richards (1959) discovered that the model remains sigmoidal for $c < 0$ and it is equivalent to the logistic model if $c = -1$. Hence, following the suggestion of García (2008), we use the Bertalanffy label for $0 < c < 1$, the Richards label for $c < 0$, or, more generally, the Bertalanffy-Richards label for $c \neq 0$. The model is not sigmoidal if $c \geq 1$.

The principal theory underlying the original Bertalanffy model is that the growth rate of an animal, $\mathrm{d}W/\mathrm{d}t$, equals the difference between the anabolic (gW^{m}) and catabolic (bW) rates of metabolism. In plants, the rate of net growth equals the difference between the rates of gross growth and loss (see Sect. 1.8.1), where gross growth equals photosynthesis minus respiration, and loss is primarily due to the sloughing of seeds and senescent tissues (S). Respiration is often assumed to divide into two components: growth respiration, R_{G}, and maintenance respiration, R_{M}. The former is the overhead cost of growing additional biomass, the latter is the cost of maintaining the status quo. Thus, it seems reasonable to assume, in regard to tree growth, that $P - R_{\mathrm{G}}$ corresponds to the anabolic rate and $R_{\mathrm{M}} + S$ corresponds to the catabolic rate. In any event, the Bertalanffy model has proven to be an accurate descriptive model of several measures of tree growth.

The Bertalanffy model also can be written thus (after García 1983, 2008):

$$\frac{\mathrm{d}W^{c}}{\mathrm{d}t} = a\left(W_{\max}^{c} - W^{c}\right) \tag{2.29}$$

and if $c = 1$, it reduces in form to the Mitscherlich model,

$$\frac{\mathrm{d}W}{\mathrm{d}t} = a\left(W_{\max} - W\right), \tag{2.30}$$

which also has been used widely in forestry (e.g., Assmann 1970). The Mitscherlich model (also known as the monomolecular model) does not provide for a near exponential growth response for a small organism, though steady state is approached as $W \rightarrow W_{\max}$.

Integration (2.29) from time t to $t + \Delta t$,

$$\int_{W_t^c}^{W_{t+\Delta t}^c} \frac{\mathrm{d}W^c}{W_{\max}^c - W^c} = a \int_t^{t+\Delta t} \mathrm{d}t, \tag{2.31}$$

provides a growth model of W^c for the time step, Δt,

$$W_{t+\Delta t}^c = W_{\max}^c - [W_{\max}^c - W_t^c] \exp(-a\Delta t) \tag{2.32}$$

so

$$\begin{aligned} W_{t+\Delta t} &= \{W_{\max}^c - (W_{\max}^c - W_t^c) \exp(-a\Delta t)\}^{\frac{1}{c}} \\ &= W_{\max} \left\{1 - \left[1 - \frac{W_t^c}{W_{\max}^c}\right] \exp(-a\Delta t)\right\}^{\frac{1}{c}} \end{aligned} \tag{2.33}$$

If Δt is a constant time step, i.e, $t_i = t_{i-1} + \Delta t$ for $i = 1, 2, \ldots$, then it may be convenient to use

$$W_{t+\Delta t} = (\alpha + \beta W_t^c)^{\frac{1}{c}} \tag{2.34}$$

where $\beta = \exp(-a\Delta t)$ and $\alpha = W_{\max}^c (1 - \beta)$ are parameters. A graph of $W^c(t_i + \Delta t)$ versus $W^c(t_i)$, for $i = 1, 2, ..., n$, calculated with (2.34) yields a straight line of points. Hence, a graph of data, $W^c(t_i + \Delta t)$ versus $W^c(t_i)$, indicates that the Bertalanffy model is appropriate if the resultant line is approximately straight. Of course, to get this result, one must first know the value of c.

Both (2.33) and (2.34) are convenient for fitting data from a chrono-sequence. To fit data from a time-course, we integrate (2.29) from time $t_0 = 0$ to t, whence

$$W_t = W_{\max} \left[1 - k \exp(-at)\right]^{\frac{1}{c}}, \tag{2.35}$$

where $k = 1 - W_0^c / W_{\max}^c$ is a parameter (the assumption $k = 1$ is not uncommon where $W_{\max} \gg W_0$).

2.1.4 Similarities of the Classic Models

In the previous subsections, we converted the classic growth-rate models to the following forms for the purpose of integration, based on insights from García (2005):

Gompertz model

$$\frac{\mathrm{d}(\ln W)}{\mathrm{d}t} = a\,(\ln W_{\max} - \ln W)$$

Verhulst (logistic) model

$$\frac{\mathrm{d}(W^{-1})}{\mathrm{d}t} = a\left(W_{\max}^{-1} - W^{-1}\right)$$

Bertalanffy-Richards model ($c \neq 0$)

$$\frac{\mathrm{d}(W^{c})}{\mathrm{d}t} = a\left(W_{\max}^{c} - W^{c}\right)$$

Mitscherlich model

$$\frac{\mathrm{d}W}{\mathrm{d}t} = a\,(W_{\max} - W)$$

These forms are convenient because we need only remember (or look up) one integration formula. However, we also can discern that the Bertalanffy-Richards model includes the logistic ($c = -1$) and the Mitscherlich models ($c = 1$) as special cases (Fig. 2.1). And, as we shall see in the next section, the Gompertz model results in the limit as $c \to 0$ (Fig. 2.1).

2.2 García's General Model

García (2005) used the negative of the Box-Cox transformation (Box and Cox 1964) to formulate a general model that includes the Gompertz, logistic, Bertalanffy, and several other models as special cases.

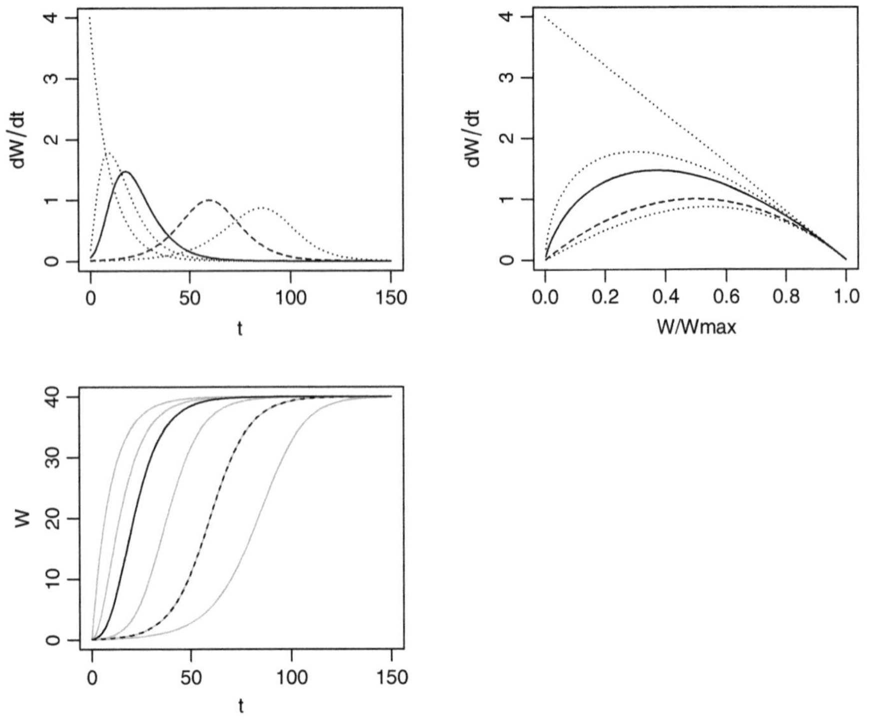

Fig. 2.1 Solutions of the Gompertz (solid), logistic (dashed), Bertalanffy-Richards model (gray) for $a = 0.1$ and $W_{\max} = 40$. Solutions of the three-parameter Bertalanffy-Richards model are provided (left to right) for $c = 1$, $c = 0.33$, $c = -0.001$, $c = -0.5$, $c = -1$, and $c = -1.5$. For $c = 1$, the Bertalanffy-Richards model is equivalent to the Mitcherlich model, which is not sigmoid. For $c = -0.001$ (i.e., $c \to 0$), the Bertalanffy-Richards model is indistinguishable from the Gompertz model; and for $c = -1$, it is equivalent to the logistic model

For our purposes, we consider the Box-Cox transformation to be a function, $T(x, \lambda)$, of a variable, $x \geq 0$, and a parameter, λ, such that

$$T(x, \lambda) = \begin{cases} \dfrac{x^{\lambda} - 1}{\lambda} & \text{if } \lambda \neq 0 \\ \ln x & \text{if } \lambda = 0 \end{cases} \tag{2.36}$$

Although it may not be obvious, the second case, $T(x, \lambda) = \ln x$, is the limit of the first case as $\lambda \to 0$. To achieve this result, write x^{λ} as $e^{\lambda \ln x}$ and apply l'Hôpital's rule: $\lim_{\lambda \to 0} f(\lambda)/g(\lambda) = \lim_{\lambda \to 0} f'(\lambda)/g'(\lambda)$. Since $f(\lambda) = e^{\lambda \ln x} - 1$ and $g(\lambda) = \lambda$, then

$$\lim_{\lambda \to 0} \frac{f'(\lambda)}{g'(\lambda)} = \lim_{\lambda \to 0} \frac{e^{\lambda \ln x} \ln x}{1} = \ln x \tag{2.37}$$

Let $B(x, \lambda)$ be the negative of (2.36), i.e., $B(x, \lambda) = -T(x, \lambda)$, then

$$B(x, \lambda) = \begin{cases} \dfrac{1 - x^{\lambda}}{\lambda} & \text{if } \lambda \neq 0 \\ -\ln x & \text{if } \lambda = 0, \end{cases} \tag{2.38}$$

García (2005) formulated both a general growth-rate model and its solution (i.e., yield model) in terms of (2.38).

Garcia's general growth-rate model of a trait, y $(0 < y \leq 1)$, is

$$\begin{aligned} \frac{\mathrm{d}y}{\mathrm{d}\tau} &= y^{1-\alpha} B(y, \alpha)^{1-\beta} \\ &= \begin{cases} y^{1-\alpha} \left(\dfrac{1 - y^{\alpha}}{\alpha} \right)^{1-\beta} & \text{if } \alpha \neq 0 \\ y\,(-\ln y)^{1-\beta} & \text{if } \alpha = 0 \end{cases} \end{aligned} \tag{2.39}$$

where α and β are parameters and τ is time. The growth-rate model integrates to a sigmoid function if $\alpha < 1$, $\beta \leq 0$, and $\alpha\beta < 1$. The asymptote is $y = 1$ and the maximum growth rate occurs when $y = y^*$, where

$$y^* = \begin{cases} \left(\dfrac{1 - \alpha}{1 - \alpha\beta} \right)^{1/\alpha} & \text{if } \alpha \neq 0 \\ \exp(\beta - 1) & \text{if } \alpha = 0 \end{cases} \tag{2.40}$$

2.2.1 Solution

Given y_0 at time τ_0, the solution of the general growth-rate model, (2.39), at time τ obtains from

$$B[B(y_{\tau}, \alpha), \beta] = B[B(y_0, \alpha), \beta] + \tau - \tau_0 \tag{2.41}$$

or, given y_{τ} at time τ, the solution at time $\tau + \Delta\tau$ obtains from

$$B[B(y_{\tau+\Delta\tau}, \alpha), \beta] = B[B(y_{\tau}, \alpha), \beta] + \Delta\tau \tag{2.42}$$

For the different possibilities of $\alpha < 1$ and $\beta \leq 0$,

$$B(B(y_{\tau}, \alpha), \beta) = \begin{cases} -\ln(-\ln y_{\tau}) & \text{if } \alpha = 0,\ \beta = 0 \\ -\ln \dfrac{1 - y_{\tau}^{\alpha}}{\alpha} & \text{if } \alpha \neq 0,\ \beta = 0 \\ \dfrac{1 - (-\ln y_{\tau})^{\beta}}{\beta} & \text{if } \alpha = 0,\ \beta \neq 0 \\ \dfrac{1 - [(1 - y_{\tau}^{\alpha})/\alpha]^{\beta}}{\beta} & \text{if } \alpha \neq 0,\ \beta \neq 0 \end{cases} \tag{2.43}$$

Substitution of these results into (2.41) provides the yield models:

$$
y_\tau = \begin{cases} \exp\{(\ln y_0)\exp[-(\tau-\tau_0)]\} & \text{if } \alpha=0,\ \beta=0 \\ \left\{1-(1-y_0^\alpha)\exp[-(\tau-\tau_0)]\right\}^{1/\alpha} & \text{if } \alpha\neq 0,\ \beta=0 \\ \exp\left\{-\left[(-\ln y_0)^\beta-\beta(\tau-\tau_0)\right]^{1/\beta}\right\} & \text{if } \alpha=0,\ \beta\neq 0 \\ \left\{1-\alpha\left[\left(\dfrac{1-y_0^\alpha}{\alpha}\right)^\beta+\beta(\tau-\tau_0)\right]^{1/\beta}\right\}^{1/\alpha} & \text{if } \alpha\neq 0,\ \beta\neq 0 \end{cases} \tag{2.44}
$$

Under the labeling convention suggested by García (2008), special cases of the general model, with $\beta = 0$, are: the Bertalanffy model, $0 < \alpha < 1$; the Richards model, $\alpha < 0$;, the Mitscherlich model (not sigmoid), $\alpha = 1$; the logistic model, $\alpha = -1$; and the Gompertz model, $\alpha = 0$. Other specific combinations of α and β provide the Korf ($\alpha = 0, \beta = -1/c$), Levakovic I ($\alpha = -1/c, \beta = -1/d$), Levakovic III ($\alpha = -1/c, \beta = -1/c$), and the Hossfeld IV and Yoshida I models ($\alpha = -1, \beta = -1/c$).

García (2005) also indicated that sigmoid models for $\alpha \leq 0$ and $0 \leq \beta < 1$ emerge from (2.44) if y is replaced with $1 - y$ and τ is replaced with $-\tau$. The inflection point, with these replacements, is $1 - y^*$, where y^* is calculated with (2.40). For example, if we perform the replacements on the third case of (2.44), then the result is a Weibull model,

$$
y_\tau = 1 - \exp\left\{-\left[(-\ln(1-y_0))^\beta + \beta(\tau-\tau_0)\right]^{1/\beta}\right\} \tag{2.45}
$$

2.2.2 *Scaling*

To scale the response of the general model in absolute, rather than relative, terms, we substitute $W/W_{\max}$ for y and at for τ in both (2.39) and (2.44). For example, for $\alpha < 1$ and $\beta = 0$, the general growth-rate model for $0 \leq W < W_{\max}$ is

$$
\frac{1}{aW_{\max}}\frac{dW}{dt} = W^{1-\alpha}W_{\max}^{\alpha-1}\left(\frac{1-W^\alpha W_{\max}^{-\alpha}}{\alpha}\right) \tag{2.46}
$$

or

$$
\frac{dW}{dt} = aW\left(\frac{W_{\max}^\alpha - W^\alpha}{\alpha W^\alpha}\right) \tag{2.47}
$$

and its solution is

$$W(t) = W_{\max}\left\{1 - \left(1 - \frac{W_0^{\alpha}}{W_{\max}^{\alpha}}\right)\exp[-a(t - t_0)]\right\}^{1/\alpha} \tag{2.48}$$

The constant a depends on the unit measure of time. The maximum growth rate occurs when $W = y^* W_{\max}$.

2.3 Saturating Responses

Trees ordinarily respond to an increasing supply of an essential growth factor or resource (e.g., light, nutrient, water) in a positive manner, for example, an increased rate of activity or growth, or an increase in yield. A saturating response occurs when some rate of activity by a tree or the yield of a stand manifests a strong response to an increase of an essential factor if the factor is scarce, but a more tempered response if the factor is already abundant. A mechanistic analysis of this topic might involve water, carbon, and nutrient budgets, transport mechanisms, and chemical conversions (Thornley and Johnson 1990). However, an empirical analysis, by definition, is intended to be descriptive rather than explanatory. In this section, we review a few functions that have proven useful for describing saturating responses.

2.3.1 Mitscherlich Model

We have already discussed a variant of the Mitscherlich model as a limiting case of the Bertalanffy growth model. However, the Mitscherlich model—named for E.A. Mitcherlich (see, e.g., Assmann (1970) and/or Harmsen (2000) for some pertinent history)—also is used to describe the effects of nutrients and other factors on crop yields. Under the assumptions of the model, if the yield of a crop equals y for a given amount of growth factor or fertilizer, x, then any increase in the yield from further additions of x will be proportional to the difference between y and the maximum yield $y_{\max}$, i.e.,

$$\frac{dy}{dx} = c(y_{\max} - y) \tag{2.49}$$

where c is a parameter with units equivalent to those of $1/x$. The yield equation is:

$$y = y_{\max} - (y_{\max} - y_0)\exp(-cx) \tag{2.50}$$

where the yield, y, equals y_0 when x equals 0. If y_0 equals 0, then the model predicts

$$y = y_{\max}\left[1 - \exp(-cx)\right] \tag{2.51}$$

The declining rate of increase in y to increasing x, as described by a Mitscherlich function, is sometimes called the "efficacy law" (e.g., Assmann 1970).

The Mitscherlich model also has been used as a model of activity. For example, the photosynthesis model, (1.8), used by McMurtrie and Wolf (1983) (see Sect. 1.10.1) is a Mitscherlich model in form. Recall that (1.8) predicts $\mathbb{P}_g$ (Mg C ha^{-1} yr^{-1})—the rate of photosynthesis of an even-aged stand—from the rate of photosynthesis at complete light interception, $\mathbb{P}_0$ (Mg C ha^{-1} yr^{-1}), and leaf area index, $\mathbb{L}$ (ha leaf (ha ground)$^{-1}$). Under the Mitscherlich assumption,

$$\frac{d\mathbb{P}_g}{d\mathbb{L}_f} = k(\mathbb{P}_0 - \mathbb{P}_g) \tag{2.52}$$

which integrates to give

$$\mathbb{P}_g = \mathbb{P}_0\left[1 - \exp\left(-k\mathbb{L}\right)\right] \tag{2.53}$$

In this application, $\mathbb{P}_0$ can be understood as the canopy photosynthesis corresponding to complete light absorption. As the actual photosynthesis, $\mathbb{P}_g$, is approaching this maximum value, the marginal gain of adding leaf area to absorb more light becomes smaller and smaller due to shading. This topic is developed more thoroughly in Chap. 3.

2.3.2 *Hyperbolas*

Rectangular and non-rectangular hyperbolas have been used as empirical models in much the same way as the Mitscherlich model. In addition, they are the most widely used models of enzyme kinetics (Thornley and Johnson 1990), and in this connection, the rectangular hyperbola is usually called the Michaelis-Menten model.

As above, let y denote a yield or a rate of activity, and let x be the factor influencing the yield or activity. The non-rectangular hyperbola may be written as a quadratic equation (after Thornley and Johnson 1990):

$$\theta y^2 - (\alpha x + y_{\max})\, y + \alpha x y_{\max} = 0 \tag{2.54}$$

where $y_{\max}$, α, and θ are parameters. Ordinarily, the value of θ is constrained in the range $0 \le \theta \le 1$ to ensure that a model makes biological sense. Solving for the lower root,

$$y = \frac{1}{2\theta}\left\{\alpha x + y_{\max} - \left[(\alpha x + y_{\max})^2 - 4\theta\alpha x y_{\max}\right]^{\frac{1}{2}}\right\} \tag{2.55}$$

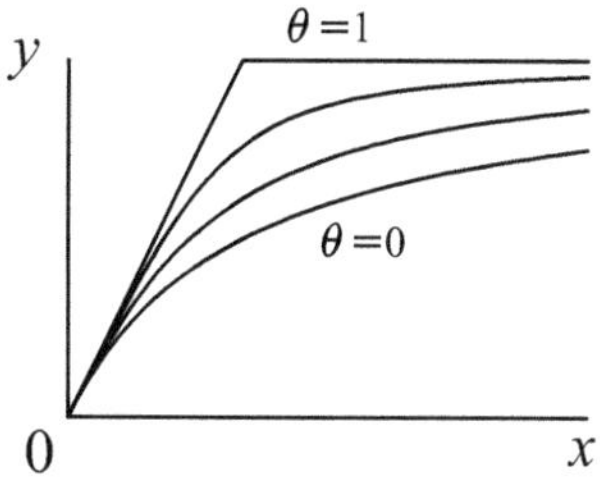

Fig. 2.2 From top to bottom are the non-rectangular hyperbolas ($\theta = 1, 0.9, 0.6$) and a rectangular hyperbola ($\theta = 0$). (Adapted from Thornley and Johnson (1990))

Thornley and Johnson (1990) demonstrate that if $\theta = 1$, then (2.54) becomes $(y - \alpha x)(y - y_{\max}) = 0$, which reduces to a segmented linear function,

$$y = \begin{cases} \alpha x & x \le y_{\max}/\alpha \\ y_{\max} & x > y_{\max}/\alpha \end{cases} \tag{2.56}$$

For $\theta = 0$, the non-rectangular hyperbola reduces to the rectangular hyperbola (see Fig. 2.2),

$$\begin{aligned} y &= \frac{\alpha x y_{\max}}{\alpha x + y_{\max}} \\ &= \frac{x y_{\max}}{x + k} \end{aligned} \tag{2.57}$$

where $k = y_{\max}/\alpha$ is the value of x that provides $y = y_{\max}/2$. The parameter, α, is the initial slope of the hyperbola, whether rectangular or non-rectangular.

Let us return to the photosynthesis example of the previous subsection, and model the rate of photosynthesis with hyperbolic models instead of the Mitscherlich model. The non-rectangular hyperbolic model for $0 < \theta < 1$ is

$$\theta \mathbb{P}_g^2 - (\alpha \mathbb{L} + \mathbb{P}_0)\mathbb{P}_g + \alpha \mathbb{L}\mathbb{P}_0 = 0 \tag{2.58}$$

where α has units Mg C (ha leaf)$^{-1}$ yr^{-1}. For $\theta = 1$,

$$\mathbb{P}_g = \begin{cases} \alpha \mathbb{L} & \text{if } \mathbb{L} \le \mathbb{P}_0/\alpha \\ \mathbb{P}_0 & \text{if } \mathbb{L} > \mathbb{P}_0/\alpha \end{cases} \tag{2.59}$$

and for $\theta = 0$,

$$\mathbb{P}_g = \frac{\alpha \mathbb{L}\mathbb{P}_0}{\alpha \mathbb{L} + \mathbb{P}_0} = \frac{\mathbb{L}\mathbb{P}_0}{\mathbb{L} + \mathbb{K}} = \frac{\alpha \mathbb{L}\mathbb{K}}{\mathbb{L} + \mathbb{K}} \tag{2.60}$$

where $\mathbb{K} = \mathbb{P}_0/\alpha$ is the leaf area index for which $\mathbb{P}_g = \mathbb{P}_0/2$.

Mäkelä et al. (2008b) used a rectangular hyperbola to model the rate of stand photosynthesis as a saturating function of leaf dry matter, i.e.,

$$\mathbb{P}_{\mathrm{g}} = \frac{\sigma_{\mathrm{fM}} \mathbb{W}_{\mathrm{f}} \mathbb{P}_0}{\sigma_{\mathrm{fM}} \mathbb{W}_{\mathrm{f}} + \mathbb{P}_0} \tag{2.61}$$

where $\mathbb{P}_{\mathrm{g}}$ and $\mathbb{P}_0$ (Mg C ha^{-1} yr^{-1}) are defined as before, σ_{fM} (Mg C (Mg leaf DW)$^{-1}$ yr^{-1}) is the light-saturated leaf-specific rate of photosynthesis, and $\mathbb{W}_{\mathrm{f}}$ (Mg DW ha^{-1}) is leaf dry matter. Substituting $\mathbb{K}_{\mathrm{f}}$ for $\mathbb{P}_0/\sigma_{\mathrm{fM}}$,

$$\mathbb{P}_{\mathrm{g}} = \frac{\mathbb{W}_{\mathrm{f}} \mathbb{P}_0}{\mathbb{W}_{\mathrm{f}} + \mathbb{K}_{\mathrm{f}}} = \frac{\sigma_{\mathrm{fM}} \mathbb{W}_{\mathrm{f}} \mathbb{K}_{\mathrm{f}}}{\mathbb{W}_{\mathrm{f}} + \mathbb{K}_{\mathrm{f}}} \tag{2.62}$$

where $\mathbb{K}_{\mathrm{f}}$ (Mg DW ha^{-1}) is the areal density of leaf dry matter that provides $\mathbb{P}_{\mathrm{g}} = \mathbb{P}_0/2$.

2.3.3 *Numerical Switch*

Application of an exponent, $n > 1$, to both x and k in (2.57), i.e.,

$$\frac{y}{y_{\max}} = \frac{x^n}{x^n + k^n} \tag{2.63}$$

provides a sigmoid function. If x is tree or stand age, the function is equivalent to a Hossfeld IV yield model, which reportedly was used to model tree size vs age as early as 1822 (Zeide 1993).

Among the varied uses of this function is that of a numerical switch, $S(x)$:

$$S(x) = \frac{y}{y_{\max}} = \frac{x^n}{x^n + k^n} = \frac{1}{1 + (k/x)^n} \tag{2.64}$$

Note that $S(x) = 0$ if $x = 0$, $S(x) = 1/2$ if $x = k$, and $S(x) \to 1$ as x increases further in value. The abruptness of the switch from 0 toward 1 increases with the value of n. Conversely, $S'(x) = 1 - S(x)$, i.e.,

$$S'(x) = \frac{k^n}{x^n + k^n} = \frac{(k/x)^n}{1 + (k/x)^n} \tag{2.65}$$

shows switch-off behavior (Fig. 2.3). An example of a numerical switch is used below in Sect. 2.4.3 to switch on self-thinning in an even-aged model stand.

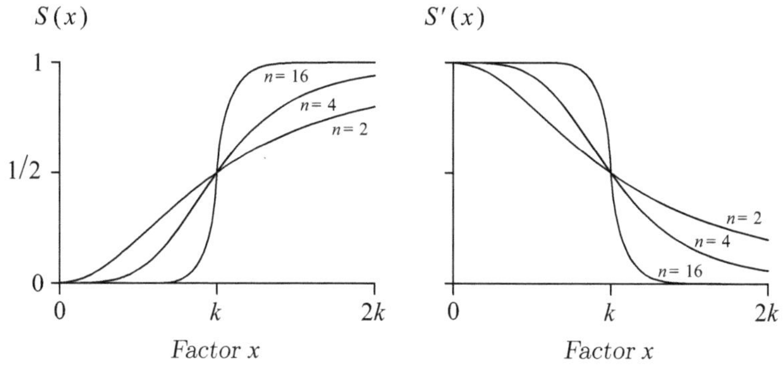

Fig. 2.3 Numerical switches: $S(x) = x^n/(x^n + k^n)$ has switch-on behavior and $S'(x) = 1 - S(x)$ has switch-off behavior for $n \geq 2$. The abruptness of the switch increases with value of n

2.4 An Empirical Crown Model

As noted in Chap. 1, empirical models are widely used in forestry and forest ecology for forecasting and summarising tree or stand dynamics. The most useful data for parameter estimation—times-series, chrono-sequences, or combinations of the two—often can be extracted from large continuously updated inventory or experimental datasets. The assortment of tree variables in such datasets, however, may be limited to diameter, height, crown ratio, and mortal state, which constrains the choice of state variables for any empirical purpose.

2.4.1 *Crown Rise*

Crown ratio, r_c, is the ratio of vertical crown length, L_c (m), to tree height, H (m), so $L_c = r_c H$ and the height to the crown base, H_c (m), is $H_c = H - L_c$. Thus, given time streams of H and either r_c or H_c, we have the raw material for modelling crown-length dynamics:

$$\frac{dL_c}{dt} = \frac{dH}{dt} - \frac{dH_c}{dt} \tag{2.66}$$

We refer to upward movement of the height of a crown base as *crown rise*. Naturally, crown rise occurs coincident with the death of the lowest live branch in a tree crown. Branch death is usually associated with external factors, principally limitation of space, light, or both. Here we present a simple empirical model of crown rise based on spatial limitation.

Let us consider an average tree within an even-aged stand and let X (m) denote the average spacing between the tree and its adjacent neighbours. From the spacing, we deduce that the stand density is $N = 1/X^2$ (trees (m^2 ground)$^{-1}$) and the average ground area per tree is $X^2 = 1/N$ (m^2).

Crown-length dynamics are known to play out in fairly predictable courses in even-aged stands, especially where the distance (or spacing) among trees is uniform, as in a plantation (Brown 1962; Beekhuis 1965; Valentine et al. 1994b, 2013; Dean et al. 2013). After establishment, crowns lengthen by height growth, with little or no rise in the base heights of the crowns until closure; then the average crown length becomes approximately constant, with crown rise keeping pace with height growth, until differentiation and self-thinning ensue.

The dynamics of the crown rise are approximated by a model formulated under three assumptions:

1. Crown rise begins in a closed stand when the average crown length, $H - H_c$, increases to the point where

$$H - H_c = \alpha_x + \beta_x X \qquad (2.67)$$

where $\alpha_x \geq 0$ (m) and β_x are parameters. Thus the crown-base height, H_c, is

$$H_c = H - \alpha_x - \beta_x X \qquad (2.68)$$

2. Between stand establishment and the onset of crown rise, the average crown length equals average tree height, i.e. $H - H_c = H$, so $H_c = 0$.
3. After a thinning or pruning in year t, crown height is constant until the stand re-closes, so $H_{c,t+1} = H_{c,t}$ until re-closure.

Combining the three assumptions into a single crown-rise model (after Valentine et al. 2013),

$$H_{c,t+1} = \max\left(0, H_{c,t}, H_{t+1} - \alpha_x - \beta_x X_{t+1}\right) \qquad (2.69)$$

This model can be fitted by non-linear regression. The model also can be elaborated to account for early weed competition and to provide estimates of crown-base height for individual trees (see Valentine et al. 2013).

The parameter β_x of the crown-rise model can be expressed in terms of angle ϕ, i.e.,

$$\beta_x = \frac{1}{2\tan\phi} \qquad (2.70)$$

as shown in Fig. 2.4.

The possible responses of the crown-rise model for mean trees are presented diagrammatically in terms of *crown-length frames* in Fig. 2.5. It is important to note that a crown-length frame only indicates the potential average crown length, given

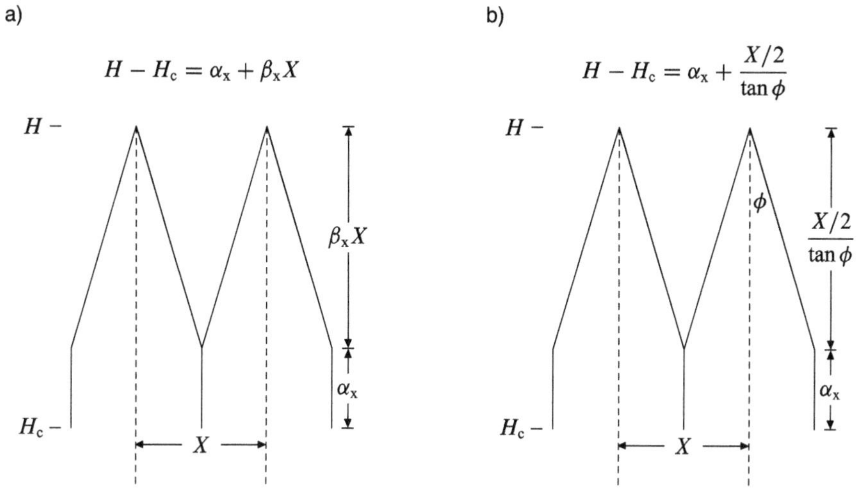

Fig. 2.4 The crown-rise model equates average crown length, $L_c \equiv H - H_c$, to $\alpha_x + \beta_x X$ during active crown rise, where X is the average spacing between adjacent trees. The vertical dashed lines in (**a**) and (**b**) represent the central stems of two identical trees, and the solid crown-length frames indicate the maximum crown length, given the spacing, X. In three dimensions, a crown-length frame forms a cone atop a cylinder. In (**b**) the parameter β_x is expressed as the tangent of the angle, ϕ

the spacing; it does not serve as a model of the shape of a crown between the tip and the base; crowns need not be cone-shaped.

The crown-rise model is generalized in Chap. 6 to deal with situations where trees have substantially different heights or where spacing is irregular, or both.

2.4.2 Height Growth

Numerical solution of the crown-rise model requires measurements or estimates of average tree height and spacing in year $t+1$ as well as crown-base in year t. Hence, let us assume that height growth is described by Bertalanffy model (see Sect. 2.1.3):

$$H_{t+1} = \left(a + bH_t^c\right)^{\frac{1}{c}} \tag{2.71}$$

It may be instructive to calculate c from the asymptotic height, $H_{\max}$. Thus, substitute $H_{\max}$ for both H_{t+1} and H_t and solve for c, i.e., $c = \log[a/(1-b)]/\log H_{\max}$ with $0 < b < 1$.

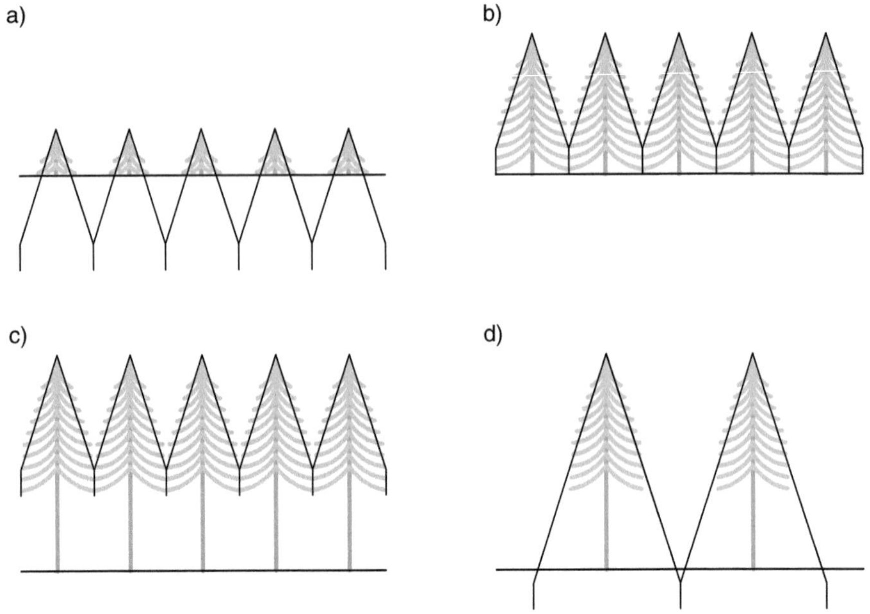

Fig. 2.5 (**a–d**): Two dimensional slices of crown-length frames indicate the maximum crown length in a row of identical model trees with uniform spacing. In the transition from (**a**) to (**b**), crown-length frames rise with height growth and crown expansion. The bases of the crown-length frames coincide with the crown bases at the onset of crown rise. The transition from (**b**) to (**c**) occurs with no alteration of spacing, so crown bases rise at the rate of height growth, and crown lengths remain constant. (**d**) Thinning of the stand increases both the average spacing and the corresponding vertical lengths of the crown-length frames. After thinning, the crown bases stay where they are while the crowns lengthen from height growth and expand from branch elongation. Crown rise would resume when the bases of the crown-length frames again rise to coincide with the crown bases. (Adapted from Valentine et al. 2013)

2.4.3 Change in Spacing and Stand Density

Let us assume that any change in average spacing results from self-thinning and that the onset of self-thinning occurs after the onset of crown rise. If there is no tree mortality between stand initiation and the onset of crown rise, then the expected tree height at the onset of crown rise is $H_{\mathrm{x}} \equiv H_{\mathrm{c},0} + \alpha_{\mathrm{x}} + \beta_{\mathrm{x}} X_0$, where $H_{\mathrm{c},0}$ and X_0 are the starting values of crown height and spacing.

Under the previously stated assumptions, crown rise occurs when $H_{t+1} \geq H_{\mathrm{x}}$. Self-thinning occurs when a stand differentiates sufficiently that the crowns of fast growing trees overtop the crowns of their slower growing neighbours. Accordingly, it seems reasonable to specify the onset of self-thinning in a model stand when the crown-base height rises to the point where $H_{\mathrm{c},t+1} > k_{\mathrm{x}} H_{\mathrm{x}}$ for $k_{\mathrm{x}} > 0$.

Tree spacing, X, does not change in the absence of self-thinning:

$$X_{t+1} = X_t \qquad \text{if } H_{c,\,t+1} \leq k_X H_X \tag{2.72}$$

To estimate how X might change after the onset of self-thinning, we subtract (2.67) at time t, i.e., $H_t - H_{c,\,t} = \alpha_X - \beta_X X_t$, from the same equation at time $t + 1$, which yields

$$(H_{t+1} - H_t) - \left(H_{c,\,t+1} - H_{c,\,t}\right) = \beta_X \left(X_{t+1} - X_t\right) \tag{2.73}$$

For clarity, we re-write this result as

$$\Delta H - \Delta H_c = \beta_X \Delta X$$

During active self-thinning, we expect the crowns of surviving trees to lengthen, so the annual change in crown-base height for the average tree, ΔH_c, is expected to be less than the change in tip height, ΔH. Therefore, let us assume that $\Delta H_c = (1 - u)\Delta H$, where u $(0 < u \leq 1)$ is a control parameter. Substituting into (2.73) and solving for X_{t+1}, provides

$$X_{t+1} = X_t + \frac{u}{\beta_X}\left(H_{t+1} - H_t\right) \tag{2.74}$$

Combining this result with (2.72),

$$X_{t+1} = \begin{cases} X_t, & \text{if } H_{c,\,t} < k_X H_X \\ X_t + \dfrac{u}{\beta_X}\left(H_{t+1} - H_t\right), & \text{if otherwise} \end{cases} \tag{2.75}$$

The first case of (2.75), where $H_c < k_X H_X$, corresponds to no change in spacing, and the second case, where $H_c \geq k_X H_X$, corresponds to fully active change, so the switch from no change to active change is perhaps unrealistically abrupt. It seems more natural for mortality and the consequent change in spacing to begin slowly as H_c closely approaches $k_X H_X$ in value and then transition to its fully active rate after H becomes greater than $k_X H_X$.

Hence, let us consider the switching function (see Sect. 2.3.3), viz.,

$$S_X = \frac{H_{c,\,t}^{n_X}}{H_{c,\,t}^{n_X} + (k_X H_X)^{n_X}} \tag{2.76}$$

S_X is bounded by 0 and 1 and equals 1/2 when H_c equals $k_X H_X$. The value of n_X regulates the abruptness of the transition from 0 to 1 as H_c transitions from less than to greater than $k_X H_X$—the larger the value of n_X, the faster the transition. Thus, utilising the switching function, (2.75) converts to

$$X_{t+1} = X_t + S_{\mathrm{X}} \frac{u}{\beta_{\mathrm{X}}} (H_{t+1} - H_t) \tag{2.77}$$

The spacing model, in turn, easily converts to a dynamic model of stand density (N, trees m^{-2}). Recalling that $X = 1/\sqrt{N}$,

$$N_{t+1} = \left[\frac{1}{\sqrt{N_t}} + S_{\mathrm{X}} \frac{u}{\beta_{\mathrm{X}}} (H_{t+1} - H_t) \right]^{-2} \tag{2.78}$$

Clutter and Jones (1980) presented a dynamic model of stand density similar to (2.78), with stand age substituting for tree height. The original Clutter-Jones model has spawned elaborations in model form (e.g., Diéguez-Aranda et al. 2005; Zhao et al. 2007) and fitting methodology (e.g., Affleck 2006). García (2009) substituted height for stand age in the Clutter-Jones model, and his resultant dynamic model of stand density during active self-thinning includes (2.78) as a special case.

2.4.4 R Script

The crown-rise model, as developed so far, comprises three difference equations that are solved on a finite timestep of one year:

$$H_{t+1} = \left(a + bH_t^c\right)^{\frac{1}{c}}$$

$$X_{t+1} = X_t + S_{\mathrm{X}} \frac{u}{\beta_{\mathrm{X}}} (H_{t+1} - H_t)$$

$$H_{\mathrm{c},\,t+1} = \max\left(0,\, H_{\mathrm{c},\,t},\, H_{t+1} - \alpha_{\mathrm{X}} - \beta_{\mathrm{X}} X_{t+1}\right)$$

The crown-rise model is driven by height growth, so the height-growth model, (2.71), is updated first, followed by the spacing model, (2.77), and finally the model of crown-base height, (2.69). Note that height growth, in this model, is not affected by crown length or spacing. This deficiency will be corrected in Chap. 5.

The following R script, called "crise_mod_runs.R," solves the model and writes the results in the form of five graphs. The script is available in the electronic supplementary material of this chapter for your convenience.

Solve the crown-rise model from year 0 to year `m` in each of `s` runs:

```
>     m <- 100
>     s <- 3
```

`T` is time in years—from year 0 to year `m`:

```
>     T <- seq( 0, m, by = 1)
```

Define time-course vectors as columns in matrices, one column of length `m+1` for each of the `n` runs:

```
>     H <- matrix( 0, nrow = m+1, ncol = s, byrow = TRUE)
>    Hc <- matrix( 0, nrow = m+1, ncol = s, byrow = TRUE)
>     X <- matrix( 0, nrow = m+1, ncol = s, byrow = TRUE)
```

Assign parameter values to the Bertalanffy height-growth model:

```
>     a <- .6
>     b <- 0.96
>     Hmax <- 30
>     c <- log( a/(1 - b) ) / log(Hmax)
```

Assign values to the parameters that control the timing and rate of change in spacing:

```
>     u <- .2
>     k_x <- 2
>     n_x <- 4
```

Assign values of the crown-rise parameters:

```
>    alpha_x <- 1.45
>     beta_x <- 1.65
```

Initialize the spacing for the `i`th of `s` runs, then solve the model:

```
>   for ( i in 1:s ){
+
+    X[1,i] <- 1.22 * i
+    H_x <- Hc[1,i] + alpha_x + beta_x * X[1,i]
+
+   for (t in 1:m){
+      H[t+1,i] <- (a + b * H[t,i]^c)^(1 / c)
+           S_x <- Hc[t,i]^n_x / (Hc[t,i]^n_x + (k_x * H_x)^n_x)
+      X[t+1,i] <- X[t,i] + S_x * (u / beta_x) * (H[t+1,i]
                    - H[t,i])
+     Hc[t+1,i] <- max(0, Hc[t,i], H[t+1,i] - alpha_x - beta_x
                    * X[t+1,i])
+     } # end t loop
+   }   # end i loop
```

Graph the results for height, crown height, crown length, and spacing from the second run (see Fig. 2.6):

```
>   pdf(file="crise_two.pdf",width=4,height=4)
>    plot(T, H[,2], type = "l", xlab = "Time (yr)", ylab = "Size",
+     ylim=c(0, 30), col = gray(.6) )
>     lines( T, Hc[,2],         lty = 2  )
>     lines( T, H[,2]-Hc[,2], lty = 4  )
>     lines( T, X[,2],          lty = 3  )
>   dev.off()
```

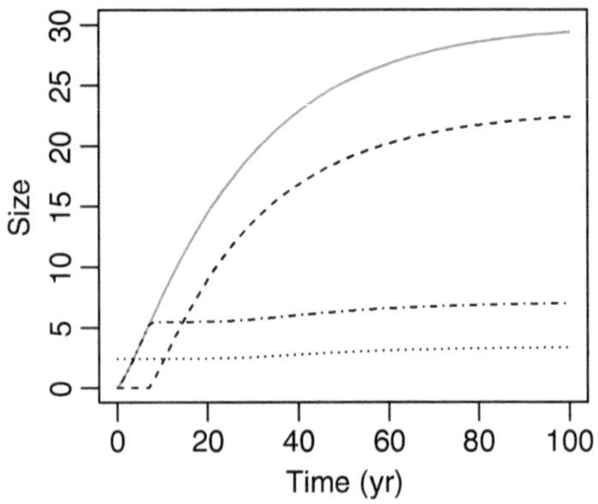

Fig. 2.6 Time-courses of state variables with an initial tree spacing of 2.44 m: tree height, solid; crown-base height, dash; crown length, dash-dot; spacing, dot

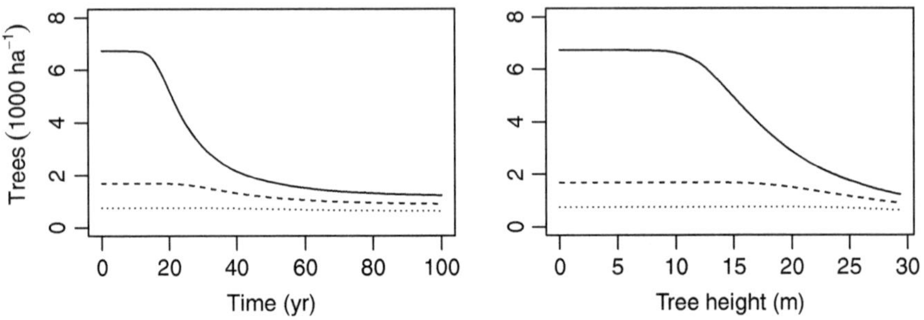

Fig. 2.7 Stand density (trees ha^{-1}) versus time (yr), left; and tree height (m), right—initial density: 6720, solid; 1680, dash; 420, dot

Graph the results for tree density—all runs (see Fig. 2.7):

```
> ylabel = expression(Trees~(1000~ha^-1))
> pdf(file = "crise_density.pdf", width=4, height=3.5)
>   plot( T, 10 / X[,1]^2, type = "l", xlab = "Time (yr)",
+              ylab = ylabel, ylim=c(0,8) )
>   lines( T, 10 / X[,2]^2, lty = 2 )
>   lines( T, 10 / X[,3]^2, lty = 3 )
> dev.off()
```

```
> pdf(file = "crise_density_height.pdf", width=4, height=3.5)
>   plot( H[,1], 10 / X[,1]^2, type = "l", xlab = "Tree height (m)",
+     ylab = " ", ylim=c(0,8) )
>   lines( H[,2], 10 / X[,2]^2, lty = 2 )
>   lines( H[,3], 10 / X[,3]^2, lty = 3 )
> dev.off()
```

Graph the results for corresponding tree spacing (see Fig. 2.8):

```
> slabel = expression(Spacing~(m))
> pdf(file = "crise_space.pdf", width=4, height=3.5)
>   plot( T, X[,1], type = "l", xlab = "Time (yr)",
+    ylab = slabel, ylim=c(0, 5) )
>   lines( T, X[,2], lty = 2 )
>   lines( T, X[,3], lty = 3 )
> dev.off()
```

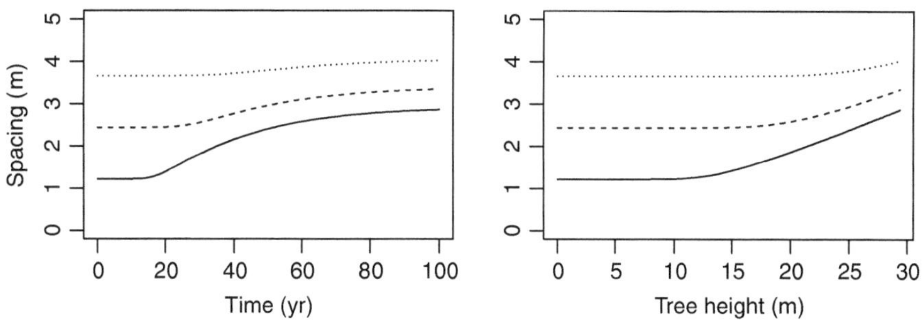

Fig. 2.8 Average tree spacing (m) versus time (yr), left; and tree height (m), right—initial spacing: 1.22, solid; 2.44, dash; 3.66, dot

```
> pdf(file = "crise_space_height.pdf", width=4, height=3.5)
>   plot( H[,1], X[,1], type = "l", xlab = "Tree height (m)",
+    ylab = " ", ylim=c(0, 5) )
>   lines( H[,2], X[,2], lty = 2 )
>   lines( H[,3], X[,3], lty = 3 )
> dev.off()
```

2.5 Exercises

2.1 Show that Eq. (2.6) is equivalent to Eq. (2.7).

2.2 Show that Eq. (2.21) is equivalent to Eq. (2.20).

2.3 The height growth equation, (2.71), can be written in the following form, where the original model parameter maximum height, $H_{\max}$, is explicitly included:

$$H_{t+1} = \left(aH_{\max} + (1-a)H_t^c\right)^{\frac{1}{c}} \tag{2.79}$$

Study the behaviour of the model by modifying its parameter values (a, c, $H_{\max}$). Start with $a = 0.6$, $c = 0.5$ and $H_{\max} = 30$ m. Recall that $0 < c < 1$ to give the Bertalanffy type of behaviour. How do you choose the parameter values if you have data of tree height against age?

2.4 Retrieve the R script, `crise_mod_runs.R`, from the electronic supplementary material of this chapter, and place it in a convenient folder (work directory). Solve the model in R with the command: `source("crise_mod_runs.R")`. Assess the effect of the control parameter, u, by changing its value.

2.5 Suppose that a model stand is thinned in some year t, doubling the spacing. What is the effect on N_t? How would you alter the script to implement a simulated thinning?

2.6 Suggested Reading

Burkhart and Tomé (2012, Chap. 6) provide details for a large number of sigmoid growth functions, and Bontemps and Duplat (2012) motivate and describe a non-asymptotic sigmoid growth function.

Jones et al. (2009, Chap. 3) provide a primer on the use of loops and branches in R scripts.

Chapter 3
Carbon Balance

This chapter introduces the basic carbon-balance approach for trees, when considered over their lifetime, at an annual (or growing-season) resolution. At this scale, the key issues of model development include: (1) realistic long-term dynamic properties, (2) responses of growth and mortality of competing individuals, and (3) responses to eco-physiological inputs.

3.1 Photosynthesis Is the Source of Growth

Tree growth is fueled principally by the products of photosynthesis. The carbon assimilated by leaves is converted into sugars that are used for the production of new structural tissues in the form of leaves, wood, and fine roots. Part of the carbon is lost in respiration. Other carbon is lost as tissues senesce and turn over, causing losses of carbon as litterfall. Taking account of all processes involving the gains and losses of carbon provides the carbon balance of the tree. Virtually all eco-physiological growth models are formulated from a carbon balance. Like all material balances, the carbon balance is an application of the physical law of conservation of mass. The carbon balance also provides a means for accounting for the impacts of environmental factors on growth through their impacts on the eco-physiological processes of photosynthesis, respiration, turnover and biomass allocation.

As already discussed in the introduction, carbon-balance models have been constructed at variable temporal scales. Here, we focus on long-term growth, considering an annual time resolution. At this scale, we need not worry about within-year carbon dynamics that may be much more complicated than the between-year dynamics are. For example, within a growing season different tissues have variable

Electronic Supplementary Material The online version of this chapter (https://doi.org/10.1007/978-3-030-35761-0_3) contains supplementary material, which is available to authorized users. The videos can be accessed by scanning the related images with the SN More Media App.

A. Mäkelä, H. T. Valentine, *Models of Tree and Stand Dynamics*,
https://doi.org/10.1007/978-3-030-35761-0_3

growth rates with internal growth rhythms during the season, and the carbon storage dynamics are important. If we focus on a longer time span, however, we may assume that all carbon assimilated during a time step is consumed in biomass production.

Carbon-balance models were introduced for agriculture in 1970s, largely by the Dutch group led by Professor C. T. deWit (e.g., de Wit 1978), and much developed in the work of J. H. M. Thornley (e.g., Thornley 1976; Thornley and Johnson 1990). Among the first carbon-balance models of tree stands was the classic model by McMurtrie and Wolf (1983), which considered a forest stand in terms of its net photosynthetic production, introducing the basic concepts required in this approach for trees. Since then, several models with different degrees of complexity have been developed, and virtually all process-based forest growth models include a carbon-balance component (Mäkelä et al. 2000). An important difference between agricultural crop models and tree or stand models is that the latter account for the accumulation of wood in trees during their lifetimes. Therefore, one of the main issues in devising a carbon-balance model for trees is how to describe the allocation of carbon to wood, such that the development of the woody structures, described in Chap. 4, can be predicted dynamically.

In the following sections, we introduce the basic carbon-balance approach for trees when considered over their lifetime, at an annual (or growing-season) time resolution. At this scale, key issues of model development include: (1) long-term dynamic properties, (2) responses of growth and mortality of competing individuals, and (3) responses to eco-physiological inputs.

3.2 Basic Carbon Balance of Trees and Stands

Consider a tree with leaf biomass W_{f}, fine-root biomass W_{r}, and woody biomass, W_{w}, all three tissues measured in units of kg DW. We may write a growth equation for each of these biomass components as follows:

$$\frac{\mathrm{d}W_i}{\mathrm{d}t} = \lambda_i G - S_i, \qquad i = \mathrm{f, r, w} \tag{3.1}$$

where G (kg DW yr^{-1}) is the rate of new biomass production, λ_i has been defined as the fraction of that production allocated to tissue i ($\lambda_{\mathrm{f}} + \lambda_{\mathrm{r}} + \lambda_{\mathrm{w}} = 1$), and the S_i (kg DW yr^{-1}) are the corresponding litter fall rates of tissues.

We assume, on an annual basis, that the new biomass production of the whole tree, G, follows from net photosynthetic production, such that:

$$G = \frac{P - R}{[\mathrm{C}]} \tag{3.2}$$

where P (kg C yr^{-1}) is the rate of *gross* photosynthetic production and R (kg C yr^{-1}) is the whole-tree respiration, so $P - R$ is the rate of *net* photosynthetic production, and [C] (kg C (kg DW)$^{-1}$) is the concentration of carbon in new biomass, which we use to convert the units of net photosynthetic production, kg C yr^{-1}, to the units of new biomass production, kg DW yr^{-1}, and vice versa

Combining (3.2) with (3.1), we have

$$\frac{dW_i}{dt} = \lambda_i \frac{P - R}{[C]} - S_i, \qquad i = f, r, w \tag{3.3}$$

This is a basic carbon-balance equation relating the annual rate of change in biomass to the processes of photosynthesis, respiration and litter fall.

In order to understand the dynamics of tree growth, as presented in (3.3), we need to express the right-hand side of the equation in terms of the state variables—W_f, W_r, and W_w—and independent inputs. The latter may consist of driving variables dependent upon time, and constant or time-dependent processes. A first step towards defining the dependence of process rates on states is based on the observation that each metabolic process is carried out by some active tissue that functions with some specific activity per unit amount of the tissue. For example, photosynthesis takes place in leaves with a leaf-area or leaf-mass based specific activity which depends on the environment to which the leaves are exposed. Here, we measure the amount of tissue in terms of biomass (dry weight), so we consider process rates in terms of active biomass multiplied by a *specific rate* that expresses the activity of the tissue per unit active biomass.

First, the rate of photosynthetic production is proportional to the leaf biomass:

$$P = \sigma_f W_f \tag{3.4}$$

where σ_f (kg C (kg DW)$^{-1}$ yr^{-1}) is the leaf-specific rate of gross photosynthetic production. It may depend on various environmental factors, such as the average light available to the tree during the growing season, the length of the growing season, and other factors. The important dependence in the above equation, however, is that any tree's rate of photosynthetic production essentially depends on the tree's total leaf biomass.

Respiration is customarily divided into growth respiration, R_G, and maintenance respiration, R_M:

$$R = R_G + R_M \tag{3.5}$$

Growth respiration represents the loss of carbon in the process of converting stored, labile sugars into permanent tissue structures; it is proportional to total growth:

$$R_G = r_G[C]G \tag{3.6}$$

where r_G is the proportional constant. This result is not directly usable for evaluating R_G in (3.3), however, because it retains the dependence of R_G on G which we need to eliminate. To do that, we use (3.3), (3.5), and (3.6), which provide

$$G = Y_G\,(P - R_M) \tag{3.7}$$

where

$$Y_G = \frac{1}{[C]}\left(\frac{1}{1+r_G}\right) \quad (3.8)$$

We call Y_G (kg DW (kg C)$^{-1}$) the *yield factor*. Equation (3.8) indicates that a fraction, $1/(1+r_G)$, of each unit of carbon allocated to growth yields $\{[C]\,(1+r_G)\}^{-1}$ dry-weight units of new biomass. The remaining fraction of carbon, $r_G/(1+r_G)$, is lost to growth respiration.

Maintenance respiration is caused by the maintenance of live tissue, and can be expressed in terms of tissue-specific respiration rates, r_{Mi} (kg C (kg DW)$^{-1}$ yr^{-1}):

$$R_{Mi} = r_{Mi}\,W_i, \qquad i = \mathrm{f, r, w} \quad (3.9)$$

Similarly, litter fall can be expressed in terms of tissue-specific rates:

$$S_i = s_i\,W_i, \qquad i = \mathrm{f, r, w} \quad (3.10)$$

where the tissue-specific litter fall rate, s_i, is in units yr^{-1}. The equation of net growth rate, (3.3), can now be written as follows:

$$\frac{dW_i}{dt} = \lambda_i Y_G \left[\sigma_f W_f - \sum_j r_{Mj} W_j\right] - s_i W_i, \qquad i = \mathrm{f, r, w} \quad (3.11)$$

where the growth rates of the components have been expressed in terms of component biomasses, specific process rates and carbon allocation coefficients.

A key question for further model development is: how do the specific rates and allocation coefficients depend on (1) external environmental factors and (2) the states of the trees and the whole stand? Before turning to these questions, however, let us first consider model (3.11) under the assumption that all its parameters are constant. In this case, the model is a linear, time-independent system of differential equations, which can be presented in matrix form as follows:

$$\frac{d}{dt}\begin{bmatrix} W_f \\ W_r \\ W_w \end{bmatrix} = \mathbf{A}\begin{bmatrix} W_f \\ W_r \\ W_w \end{bmatrix} \quad (3.12)$$

where $\mathbf{A}$ is the matrix

$$\mathbf{A} = \begin{bmatrix} Y_G\lambda_f(\sigma_f - r_{M_f}) - s_f & -Y_G\lambda_f r_{M_r} & -Y_G\lambda_f r_{M_w} \\ Y_G\lambda_r(\sigma_f - r_{M_f}) & -Y_G\lambda_r r_{M_r} - s_r & -Y_G\lambda_r r_{M_w} \\ Y_G\lambda_w(\sigma_f - r_{M_f}) & -Y_G\lambda_w r_{M_r} & -Y_G\lambda_w r_{M_w} - s_w \end{bmatrix} \quad (3.13)$$

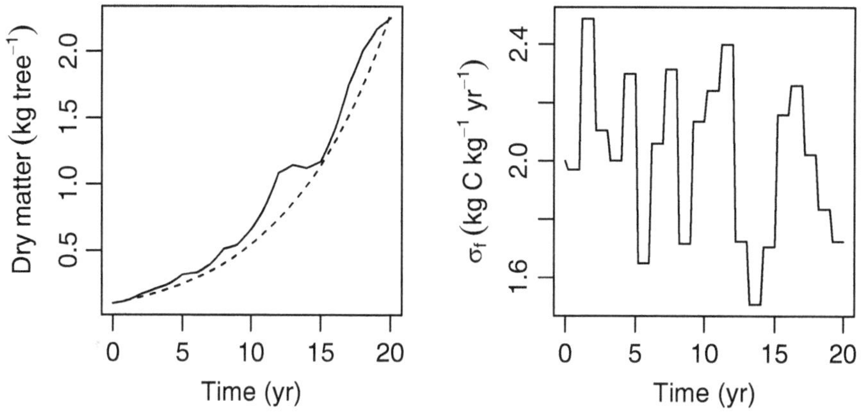

Fig. 3.1 Exponential growth in the carbon balance model (3.11). The solid line in the left panel depicts leaf biomass when the leaf-specific photosynthetic rate, σ_f, varies from year to year as shown in the right-side panel, while the dashed line is leaf biomass when all parameters are constant, with σ_f at the mean value of those shown in the right-side panel

According to the theory of linear differential equations, this equation has a solution where all the state variables are exponential functions of time (Luenberger 1979). The same qualitative behaviour is retained even if the metabolic rates and carbon allocation are allowed to vary stochastically with time (Fig. 3.1). While exponential growth may be a very good and illustrative way of analysing the growth of, e.g., small seedlings (Exercise 3.3) (Ingestad 1980), it clearly can not represent the entire lifetime of a stand, nor a tree. Indeed, the descriptive models reviewed in Chap. 2 all showed close-to-exponential behaviour for small trees, but later adopted a saturating pattern due to some "catabolic" terms in the growth equation. While the descriptive models deduced from empirical observation that some catabolic process must be included, our key question here is: what biological or physical feedbacks are there between the state of a tree or stand and the rates of the metabolic processes that lead to the observed reduction in relative growth rates as the trees grow larger?

In stands, growth reducing feedbacks have been related to resource depletion as the trees grow to occupy the site, notably mutual shading and nutrient limitation. For individual trees, it has been suggested that growth reduction may be due to, e.g., changes in the ratio of respiring to assimilating tissue, self-shading, difficulties to raise water to the crown, changes in growth allocation over time, or switching from vegetative to reproductive growth. In the following, we examine how these ideas can be incorporated in the above basic carbon balance model, and what their implications are for further model development. We begin with stand-level models, reviewing the impacts of shading and nutrient depletion, then move on to possible feedbacks in individual, isolated trees.

3.3 Stand-Level Feedbacks

3.3.1 Shading and Photosynthesis

Mutual shading by neighbours is perhaps the most obvious interaction between trees in a canopy, leading to a decline in the leaf-specific rate of photosynthesis as the amount of leaf area increases in the canopy. This is customarily described using the so-called Lambert-Beer model (Monsi and Saeki 1953), which is based on the transmission of radiation through a homogeneous medium. This leads to exponential extinction of light in the canopy, the leaves absorbing the radiation, I_{abs}, that does not reach the ground:

$$I_{abs} = I_0[1 - \exp(-k\mathbb{L})] \tag{3.14}$$

where I_0 is the radiation incident above the canopy (mol quanta m^{-2} s^{-1}), k is a canopy-specific extinction coefficient and $\mathbb{L}$ (m^2 leaf (m^2 ground)$^{-1}$) is the leaf area index of the canopy.

The light use efficiency (LUE) approach to canopy photosynthesis assumes that over a prolonged period of time, e.g., one year, canopy photosynthesis is directly proportional to the light absorbed by the canopy, such that

$$\mathbb{P}_g = \beta I_{abs,\,cum} \tag{3.15}$$

where P_g is gross photosynthetic production (kg C m^{-2} yr^{-1}) and β is light use efficiency (kg C [mol quanta]$^{-1}$), and $I_{abs,\,cum}$ is the cumulative absorbed radiation over the time period considered. We further assume that leaf area is proportional to leaf mass,

$$\mathbb{L} = a_{SL}\mathbb{W}_f \tag{3.16}$$

where a_{SL} (m^2 leaf (kg leaf)$^{-1}$)—leaf area per unit leaf biomass—is the *specific leaf area*. We may express the rate of photosynthetic production as a saturating function of leaf area as follows:

$$\mathbb{P}_g = \mathbb{P}_0[1 - \exp(-ka_{SL}\mathbb{W}_f)] \tag{3.17}$$

where $\mathbb{P}_0 = \beta I_{0,\,cum}$ represents the annual rate of gross photosynthesis assuming complete light interception.

Although the original Lambert-Beer equation was derived for homogeneous media, Duursma and Mäkelä (2007) showed that the same form of equation can be used for canopies with clustered leaves as well, provided that the extinction coefficient, k, is modified accordingly (see Sect. 4.6). The annual average extinction coefficient is also generally different from a momentary or daily average extinction coefficient, because the extinction generally depends on the incident angle of the radiation.

Inserting (3.17) into (3.4) now allows us to express the specific rate of photosynthesis, σ_f, as follows:

$$\sigma_f = \frac{\mathbb{P}_0[1 - \exp(-k a_{SL} \mathbb{W}_f)]}{\mathbb{W}_f} \tag{3.18}$$

The specific rate of photosynthesis therefore declines as the amount of leaf biomass in the canopy increases. This represents a feedback from leaf biomass to photosynthetic production which means that the system no longer manifests exponential growth but shows stable behaviour with an equilibrium (Fig. 3.2).

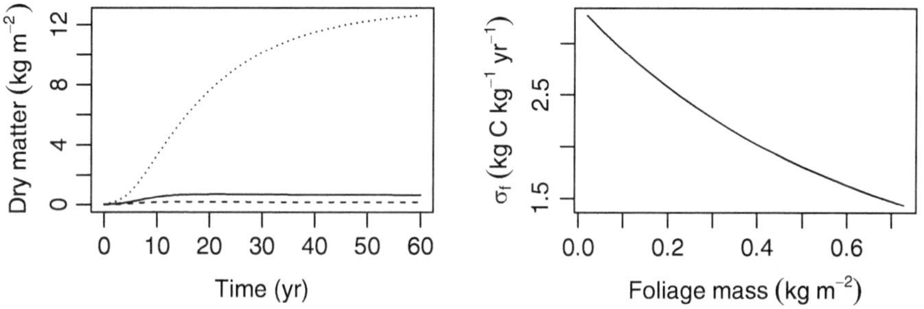

Fig. 3.2 The carbon balance model, (3.11), with the leaf-specific rate of photosynthesis, σ_f, as defined in (3.18). Left: Stand leaf mass (solid line), fine-root mass (dashed line) and woody mass (dotted line). Right: Decline of σ_f due to shading as a function of stand leaf mass. The parameter values are in Table 3.1

Table 3.1 Parameter values used in example simulations of the carbon balance model (3.11) with (3.18)

Symbol	Value	Units	Definition
$\mathbb{P}_0$	1.2	kg C m^{-2}	Maximum photosynthetic production per unit ground area
λ_f	0.2	–	Growth allocation to leaves
λ_r	0.15	–	Growth allocation to fine roots
λ_w	0.65	–	Growth allocation to wood
r_f	0.2	kg C (kg DW)$^{-1}$ yr^{-1}	Specific maintenance respiration rate of leaves
r_r	0.2	kg C (kg DW)$^{-1}$ yr^{-1}	Specific maintenance respiration rate of fine roots
r_w	0.01	kg C (kg DW)$^{-1}$ yr^{-1}	Specific maintenance respiration rate of wood
s_f	0.3	yr^{-1}	Specific turnover rate of leaves
s_r	0.9	yr^{-1}	Specific turnover rate of fine roots
s_w	0.05	yr^{-1}	Specific turnover rate of wood
Y_G	1.3	kg DW (kg C)$^{-1}$	Growth efficiency
a_{SL}	14.0	m^2 (kg DW)$^{-1}$	Specific leaf area
k	0.2	–	Light extinction coefficient

The equilibrium of the system is found when the rates of change in (3.11) vanish. However, because of the exponential in (3.17), the resulting equations cannot be solved in closed form. However, it is possible to analyse the equilibrium numerically and/or graphically.

First, we can see from (3.1) that in equilibrium,

$$\lambda_f \mathbb{G} - s_f \mathbb{W}_f = \lambda_r \mathbb{G} - s_r \mathbb{W}_r = \lambda_w \mathbb{G} - s_w \mathbb{W}_w = 0 \tag{3.19}$$

This allows us to find the equilibrium values of $\mathbb{W}_r$ and $\mathbb{W}_w$ as functions of $\mathbb{W}_f$:

$$\mathbb{W}_r = \frac{\lambda_r s_f}{\lambda_f s_r} \mathbb{W}_f \qquad \mathbb{W}_w = \frac{\lambda_w s_f}{\lambda_f s_w} \mathbb{W}_f \tag{3.20}$$

These can now be used for eliminating $\mathbb{W}_r$ and $\mathbb{W}_w$ from the differential equation for leaves. If we do that and rearrange, we can write (3.11), the equilibrium for leaf biomass, as follows:

$$\mathbb{P}_0[1 - \exp(-k a_{SL} \mathbb{W}_f)] = \frac{s_f}{\lambda_f} \left[\frac{r_f \lambda_f}{s_f} + \frac{r_r \lambda_r}{s_r} + \frac{r_w \lambda_w}{s_w} + \frac{1}{Y_G} \right] \mathbb{W}_f \tag{3.21}$$

The right-hand side of the above equation—seen as a function of $\mathbb{W}_f$—is a straight line through the origin, the slope of which is determined by the specific-rate parameters and allocation coefficients. It intersects the curve defined by the left-hand side at the origin and at a non-zero value of $\mathbb{W}_f$, which is a function of all the parameter values of the model (Fig. 3.3). The intersection points are the equilibria of the system. The non-zero equilibrium is stable, that is, it will be approached from

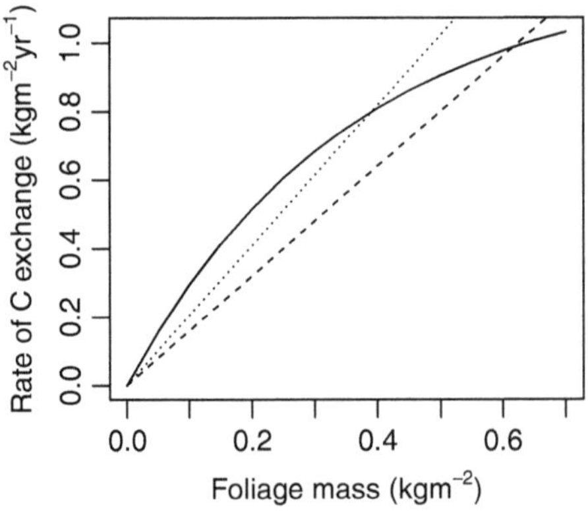

Fig. 3.3 The equilibrium for shaded leaf biomass by the carbon balance model (3.11) (3.18) can be determined graphically by finding the intersection of the photosynthesis curve (left-hand-side of (3.21)) and the straight line due to consumption in respiration and turnover (right-hand-side of (3.21)). Default maintenance respiration parameters (Table 3.1) (dashed line) and respiration parameters doubled (dotted line)

any other (positive) state of the system. The corresponding values of $\mathbb{W}_r$ and $\mathbb{W}_w$ can be found by (3.20).

We can see from Fig. 3.3 and Eq. (3.21) that if the slope of the straight line increases, the equilibrium for leaf biomass will decrease, as the intersection point is reached earlier. We can also see that the non-zero equilibrium can only be reached if the slope is below a certain value which corresponds to the tangent of the photosynthesis equation at the origin. Taking the derivative of (3.17) with respect to $\mathbb{W}_f$ and evaluating it at $\mathbb{W}_f = 0$ we find that this limiting slope is $\mathbb{P}_0 k a_{SL}$, so a non-zero equilibrium exists if

$$\mathbb{P}_0 k a_{SL} \mathbb{W}_f \leq \frac{s_f}{\lambda_f}\left[\frac{r_f \lambda_f}{s_f} + \frac{r_r \lambda_r}{s_r} + \frac{r_w \lambda_w}{s_w} + \frac{1}{Y_G}\right] \mathbb{W}_f \tag{3.22}$$

The above shows that the equilibrium $\mathbb{W}_f$ increases with increasing photosynthetic capacity (parameters $\mathbb{P}_0$, k and a_{SL}) and with increasing allocation to leaf biomass, λ_f. Increasing turnover rates of roots and wood, s_r and s_w, also increase the equilibrium of leaf biomass by reducing the biomasses of roots and wood relative to leaves and hence reducing the respiration of these components relative to production. Increasing respiration rates reduce the equilibrium of leaf biomass (Fig. 3.3).

The above model—in a slightly different form—was first presented by McMurtrie and Wolf (1983) in a paper that has become a classic in carbon-balance-based forest growth modelling. The presentation of the equilibria of the model also follows that by McMurtrie and Wolf (1983). The modification made here concerns the definition of the allocation coefficients; while McMurtrie and Wolf allocate "net photosynthesis," defined as gross photosynthesis less *leaf* respiration (see Sect. 1.10.1), the present model allocates total biomass production, i.e., gross photosynthesis less *total* respiration (see (3.3)). The present treatment is more conventional (see, e.g., Thornley 1972). Mathematically, the different formulations are equivalent, although the present form leads to slightly more complicated mathematics in the analysis of the equilibria of the model.

3.3.2 *Nitrogen Limitation*

The availability of nutrients has long been recognised to be crucial for tree and stand growth. Nitrogen (N), in particular, has been regarded as a key limiting factor. Because plant tissue requires a certain minimum nitrogen content, $[N]_{min}$ (kg N (kg DW)$^{-1}$), growth cannot exceed a maximum rate determined by the rate of availability of nitrogen for growth. If nitrogen availability is $\mathbb{U}_0$ (kg N m^{-2} yr^{-1}), then maximum production rate of biomass (kg DW m^{-2} yr^{-1}) is

$$\mathbb{G}_{max} = \frac{\mathbb{U}_0}{[N]_{min}} \tag{3.23}$$

Nitrogen limitation occurs whenever the potential rate of net carbon assimilation is greater than the availability of nitrogen.

The above idea has been included in many growth models with different practical formulations. A simple but powerful approach to the limiting role of nitrogen in tree growth is provided by the theory of *nitrogen productivity*, developed by Torsten Ingestad and Göran Ågren (Ingestad 1980; Ingestad et al. 1981; Ingestad and Ågren 1992). The principal idea is to estimate stand growth from leaf nitrogen concentration instead of from photosynthesis and respiration.

Let $[\mathrm{N}]_\mathrm{f}$ (kg N (kg DW leaf)$^{-1}$) be the nitrogen concentration in leaf biomass. Ågren (1983) defined nitrogen productivity, $\mathbb{P}_{(\mathrm{N})}$ (kg DW (kg N)$^{-1}$ yr^{-1}) as the rate of biomass production, $\mathbb{G}$ (kg DW m^{-2} yr^{-1}), per unit leaf nitrogen, $[\mathrm{N}]_\mathrm{f}\mathbb{W}_\mathrm{f}$ (kg N m^{-2}):

$$\mathbb{P}_{(\mathrm{N})} = \frac{\mathbb{G}}{[\mathrm{N}]_\mathrm{f}\mathbb{W}_\mathrm{f}} \tag{3.24}$$

so that

$$\mathbb{G} = \mathbb{P}_{(\mathrm{N})}[\mathrm{N}]_\mathrm{f}\mathbb{W}_\mathrm{f} \tag{3.25}$$

Applying (3.25) in a growth model requires the inclusion of leaf nitrogen dynamics. Denote nitrogen uptake by $\mathbb{U}_\mathrm{T}$ and the fraction of nitrogen resorbed from the senescing leaves by f_i, then

$$\frac{\mathrm{d}\left([\mathrm{N}]_\mathrm{f}\mathbb{W}_\mathrm{f}\right)}{\mathrm{d}t} = \mathbb{U}_\mathrm{T} - (1 - f_i)s_\mathrm{f}[\mathrm{N}]_\mathrm{f}\mathbb{W}_\mathrm{f} \tag{3.26}$$

where s_f is the specific rate of senescence of leaves (yr^{-1}). The uptake term in (3.26) refers only to the fraction of nitrogen uptake that is allocated to leaves. In a first approximation, we may assume that this is a constant ratio of total nitrogen uptake (McMurtrie 1991). A variety of models have been developed to describe the dynamics of nitrogen in components of forest ecosystems, including variable degrees of detail about the processes of decomposition of plant tissue in the soil and the related release of nitrogen in either mineralised or organic forms to be available to plant uptake. Here it is sufficient for our purposes to consider a constant availability of nitrogen, the rate of availability depending on the site type and the prevailing environmental conditions. Furthermore, following Ågren (1983), we assume that the trees compete for the available nitrogen with the field layer, the relative shares of nitrogen uptake depending on the amount of leaf biomass. The leaf biomass of the field layer, $\mathbb{w}$, is assumed constant. The nitrogen uptake of trees is therefore

$$\mathbb{U}_\mathrm{T} = \frac{\mathbb{W}_\mathrm{f}\mathbb{U}_0}{\mathbb{W}_\mathrm{f} + \mathbb{w}} \tag{3.27}$$

We can see from (3.26) that if $\mathbb{U}_T$ is constant, then $d([N]_f\mathbb{W}_f)/dt = 0$, and therefore leaf nitrogen has a stable equilibrium:

$$[N]_f\mathbb{W}_f = \frac{\mathbb{U}_T}{(1 - f_i)s_f} \tag{3.28}$$

On the one hand, if leaf biomass approaches an equilibrium, then the ratio of total nitrogen taken up by trees will become constant according to (3.27). On the other hand, if $\mathbb{W}_f$ does not approach equilibrium, but instead continues to increase without limit, then $\mathbb{U}_T$ would approach $\mathbb{U}_0$, which is constant. This shows that in either case, leaf nitrogen will indeed approach a stable equilibrium.

If we now apply the basic carbon-balance equation, (3.1), for leaves, roots and wood, where biomass production, $\mathbb{G}$, is calculated according to nitrogen productivity as in (3.25), and we use (3.26) and (3.27) for evaluating the dynamics of leaf nitrogen, the resulting model can be summarised as follows:

$$\frac{d\mathbb{W}_i}{dt} = \lambda_i \mathbb{P}_{(N)}[N]_f\mathbb{W}_f - s_i\mathbb{W}_i, \tag{3.29a}$$

$$\frac{d([N]_f\mathbb{W}_f)}{dt} = \frac{\mathbb{W}_f\mathbb{U}_0}{\mathbb{W}_f + w} - (1 - f_i)s_f[N]_f\mathbb{W}_f \tag{3.29b}$$

This model has a stable equilibrium that depends on the availability of nitrogen in the site, $\mathbb{U}_0$ (Fig. 3.4). It thus demonstrates in a simple way how limited availability of nitrogen leads to limited growth of the stand when all the resources are being fully used. More detailed developments of the nitrogen productivity have been provided, e.g., by Ågren (1983), McMurtrie (1991) and Ågren and Franklin (2003).

Ågren (1983) considered the reduction of nitrogen productivity as a result of shading, providing an additional feedback from leaf mass to growth, much in the same way as in the previous section:

$$\mathbb{P}_{(N)} = a - b\mathbb{W}_f \tag{3.30}$$

where a and b are species-specific parameters that can be evaluated empirically (Ågren 1983).

Ingestad et al. (1981) also presented a simple model for the nitrogen cycling between the soil and the tree stand, which was used for determining $\mathbb{U}_T$. McMurtrie (1991) considered a saturating dependence between growth and nitrogen availability. Both authors also limited nitrogen uptake if the nitrogen content of the leaves increased above a threshold. These modifications may make the results somewhat more realistic quantitatively but the basic dynamic behaviour is already included in the present simple version of the model.

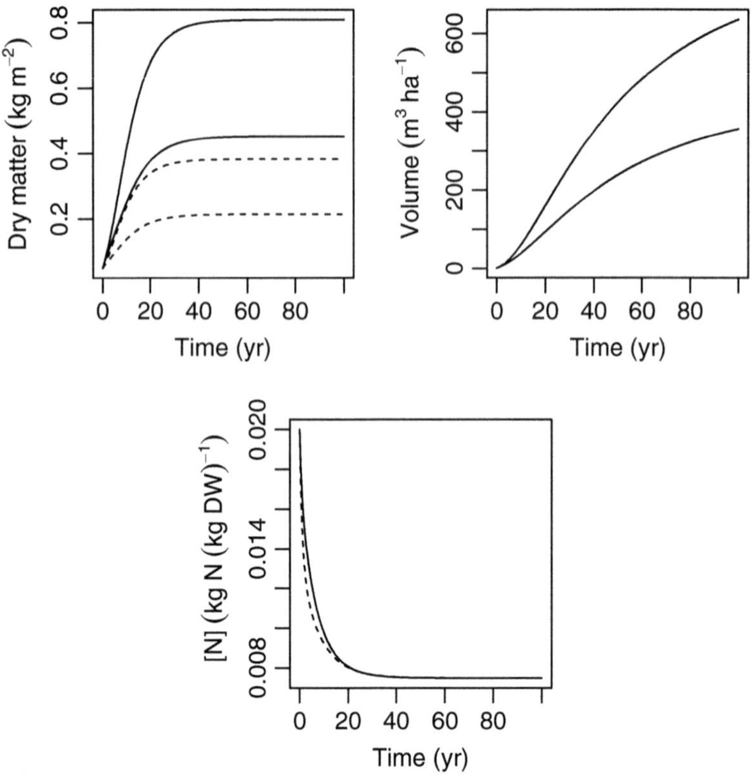

Fig. 3.4 Top: Time courses of leaf (solid line) and fine root (dashed line) biomass (left) and wood volume (right) in the nitrogen productivity model (3.29) with two different maximum nitrogen uptake rates. The higher and lower biomasses and volume are obtained with $\mathbb{U}_0 = 1.0\,\mathrm{g\,m^{-2}\,yr^{-1}}$ and $\mathbb{U}_0 = 0.6\,\mathrm{g\,m^2\,yr^{-1}}$, respectively. Bottom: Leaf nitrogen concentration in the two scenarios. Solid line: $\mathbb{U}_0 = 1.0\,\mathrm{g\,m^2\,yr^{-1}}$, dashed line: $\mathbb{U}_0 = 0.6\,\mathrm{g\,m^{-2}\,yr^{-1}}$. The other parameter values used in the simulation are summarised in Table 3.2

Table 3.2 Parameter values used in example simulations of the nitrogen productivity model (3.29)

Symbol	Value	Units	Definition
$\mathbb{P}_{(\mathrm{N})}$	200.0	kg C (kg N)$^{-1}$ m^{-2} yr^{-1}	Nitrogen productivity
λ_f	0.20	–	Growth allocation to leaves
λ_r	0.30	–	Growth allocation to fine roots
λ_w	0.50	–	Growth allocation to wood
f_i	0.50	–	Nitrogen resorption from senescent tissue $i =$ f, r, w
w	0.08	kg N m^{-2}	Nitrogen in ground vegetation
s_f	0.3	yr^{-1}	Specific turnover rate of leaves
s_r	0.95	yr^{-1}	Specific turnover rate of fine roots
s_w	0.02	yr^{-1}	Specific turnover rate of wood
ρ_w	400.0	kg DW m^{-3}	Wood density

3.4 Tree-Level Feedbacks

The above feedbacks are essentially stand-level phenomena, caused by the limitation of resources available to the trees per unit land area. In the shading feedback, the limiting factor is the amount of photosynthetically active radiation reaching the canopy. No matter how many leaves form a canopy, their collective photosynthetic production can never exceed the production corresponding to the conversion of all incoming radiation into photosynthates. In the nitrogen feedback, the limiting factor is the rate of nitrogen becoming available per unit ground area; the stand cannot grow faster than that allowed by the conversion of this nitrogen into stand biomass.

We have already seen in Sect. 3.2 that with no shading feedback, the carbon balance model (3.11) will result in exponential growth. In the case of nitrogen limitation, the uptake of nitrogen by a single tree would not be independent of the size of the tree, but would be proportional to the size of its fine-root system:

$$U_T = \sigma_r W_r \tag{3.31}$$

where σ_r is the root-specific rate of nitrogen uptake (kg N (kg C)$^{-1}$ yr^{-1}). If that were the case, (3.26) would not have a stable equilibrium, but instead would show an exponential increase of nitrogen leading to an exponential growth pattern of the whole tree.

In the following sections, we consider factors that may lead to stable behaviour of the carbon-balance model for trees growing in the absence of neighbours competing for the same resources.

3.4.1 Allocation

Above, we have assumed that total biomass production is allocated to the different components of the tree in constant proportions (λ_i). Empirical evidence suggests, however, that wood production relative to both leaf and fine-root production increases as trees grow larger, possibly levelling off later (Guillemot et al. 2015). Landsberg (1986) may have been the first to suggest that this could be modelled by utilising empirical information about allometric relationships between the different compartments of the model.

Assume that allometric relationships can be established between diameter at breast height (D) and leaf biomass on one hand, and between diameter and woody biomass on the other hand:

$$W_f = a_f D^{x_f} \tag{3.32a}$$

$$W_w = a_s D^{x_s} \tag{3.32b}$$

Above, a_f, a_s, x_f and x_s are empirical parameters. If these allometries can be assumed to hold for the entire life cycle of the tree, we may consider the temporal change of the biomasses relative to diameter as follows:

$$\frac{dW_f}{dt} = a_f x_f D^{x_f - 1} \frac{dD}{dt} \tag{3.33a}$$

$$\frac{dW_w}{dt} = a_w x_w D^{x_w - 1} \frac{dD}{dt} \tag{3.33b}$$

We observe that the original allometric equations (3.32) may be substituted into the right-hand-side of the above differential equations, yielding

$$\frac{1}{W_f} \frac{dW_f}{dt} = x_f \frac{1}{D} \frac{dD}{dt} \tag{3.34a}$$

$$\frac{1}{W_w} \frac{dW_w}{dt} = x_w \frac{1}{D} \frac{dD}{dt} \tag{3.34b}$$

This form allows us to eliminate diameter, D, from the equations, yielding a relationship between the relative growth rates of leaves and wood:

$$\frac{x_w}{W_f} \frac{dW_f}{dt} = \frac{x_f}{W_w} \frac{dW_w}{dt} \tag{3.35}$$

By conservation of mass, $G_{fw} = (1 - \lambda_r)G$ of total biomass production of leaves plus wood, after allocating the share, λ_r, to the production to fine roots. This implies that

$$\frac{dW_f}{dt} = \hat{\lambda}_f G_{fw} - s_f W_f \tag{3.36a}$$

$$\frac{dW_w}{dt} = (1 - \hat{\lambda}_f) G_{fw} - s_w W_w \tag{3.36b}$$

where $\hat{\lambda}_f$ is the share of G_{fw} allocated to leaves, thus the share of total biomass production allocated to leaves is $\lambda_f = \hat{\lambda}_f(1 - \lambda_r)$. Combining (3.35) and (3.36) allows us to solve for the allocation of growth to leaves that maintains the allometric relationships (3.32). The share of leaf growth is dependent on the current leaf and woody biomass, the allometric exponents x_f and x_w, and the turnover rates of leaves and wood, s_f and s_w:

$$\hat{\lambda}_f = \frac{x_f W_f G_{fw} + (x_w s_f - x_f s_w) W_f W_w}{G_{fw}(x_f W_f + x_w W_w)} \tag{3.37}$$

Empirically, $x_w > x_f$, i.e., the biomass of wood in standing trees increases faster with basal diameter than that of leaves. As can be seen from (3.35), this leads to an increasing growth requirement of wood relative to leaves as the tree grows larger. This further implies that the growth model accompanied with this allocation scheme becomes stable, with an equilibrium at a point where the photosynthetic production is only sufficient to compensate for respiration and turnover (Fig. 3.5, Table 3.3).

Although a gross simplification, the allometric allocation scheme clearly demonstrates the significance of allocation of biomass production in trees. A further consequence of the scheme is that, with woody biomass increasing in proportion, the respiring tissue increases in proportion to the productive, photosynthetic tissue, causing the production to become less efficient. This would not happen under exponential growth.

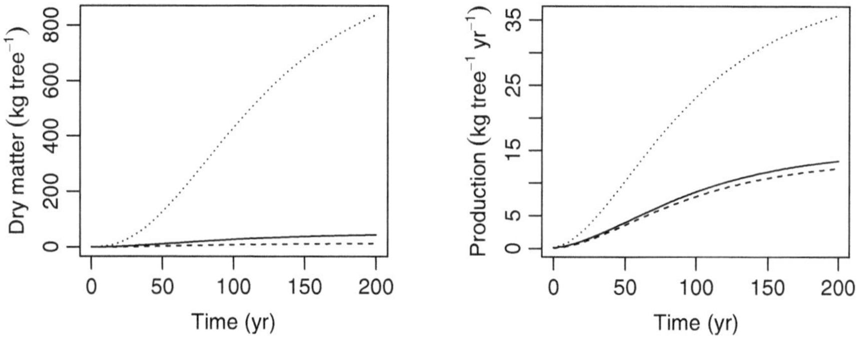

Fig. 3.5 Time courses of tree biomass components (left) and their rate of production (right) in the carbon balance model constrained by the allometries of Eq. (3.32). Leaves (solid line), fine roots (dashed line) and wood (dotted line) are shown. The parameter values used in the simulation are summarised in Table 3.3

Table 3.3 Parameter values used in example simulations of the allometric carbon balance model, comprising equations (3.36) and (3.37)

Symbol	Value	Units	Definition
σ_f	1.6	kg C (kg DW)$^{-1}$ yr^{-1}	Leaf-specific photosynthetic rate
λ_r	0.15	–	Growth allocation to fine roots
r_f	0.2	kg C (kg DW)$^{-1}$ yr^{-1}	Specific maintenance respiration rate of leaves
r_r	0.2	kg C (kg DW)$^{-1}$ yr^{-1}	Specific maintenance respiration rate of fine roots
r_w	0.01	kg C (kg DW)$^{-1}$ yr^{-1}	Specific maintenance respiration rate of wood
s_f	0.3	yr^{-1}	Specific turnover rate of leaves
s_r	0.9	yr^{-1}	Specific turnover rate of fine roots
s_w	0.04	yr^{-1}	Specific turnover rate of wood
Y_G	1.2	kg DW (kg C)$^{-1}$	Growth efficiency
x_f	1.8	–	Allometric exponent, leaf
x_w	2.6	–	Allometric exponent, wood

3.4.2 Self-Shading

In addition to shading by neighbours, tree crowns also shade themselves as the crowns grow and expand, causing a size-related reduction of photosynthetic efficiency. This can be modelled in a way similar to the shading by neighbours. Key variables include crown leaf area L_A and surface area S_A.

At a given incident angle of the sun, the amount of intercepted light is related to the crown projection area in the direction of the sun (Nilson 1999). As the sun moves across the sky during a day, or during a year, the mean proportion of light intercepted by the crown is a result of integration over all incident angles and the corresponding light intensities. For the development of a simplified model, it is helpful to utilise Cauchy's theorem (see, e.g., Lang 1991) which states that the mean projection area over all incident angles of a convex body of any shape equals one fourth of its surface area. On the basis of this, Duursma and Mäkelä (2007) proposed the following model for light interception by an individual isolated crown ($Q_{\text{abs, cum}}$), i.e.,

$$Q_{\text{abs, cum}} = Q_{0,\text{ cum}}\,\phi S_A\left[1 - \exp\left(\frac{-k_{\text{H}}L_A}{\phi S_A}\right)\right] \tag{3.38}$$

where $Q_{0,\text{ cum}}$ is the integrated photosynthetic photon flux above the crown over the time period considered (mol quanta m^{-2} s^{-1}), k_{H} is light extinction coefficient, and ϕ is an empirical parameter that depends on the distribution of sun angles and the related light intensities over the period considered. Duursma and Mäkelä (2007) showed that this model was adequately in agreement with a more detailed light interception and photosynthesis model that considered the annual intercepted light at given locations, with the time course of the solar angle and the intensities of the related direct and diffuse beams as input (Oker-Blom et al. 1989). Both ϕ and k_{H} increased with the mean zenith angle of the sun.

In analogy with the light use efficiency model for the whole stand, the crown photosynthetic production can also be modelled as proportional to the intercepted light, therefore

$$P = \beta Q_{\text{abs, cum}} \tag{3.39}$$

These, and similar models presented by others (e.g., Grace 1990; Ludlow et al. 1990), can be used to account for the self-shading in isolated or semi-isolated crowns. Because of the exponential decline of efficiency, the model potentially leads to stable behaviour with equilibrium. However, self-shading is not sufficient to explain the observed reduction in growth rate. In addition, there are factors that counteract its significance. Especially, the leaf-area to crown surface area ratio does not increase as fast as expected if leaf area was distributed in the crown volume (Duursma et al. 2010).

3.4.3 Hydraulic Limitation

Yoder et al. (1994) presented evidence that the specific rate of assimilation (i.e., net carbon exchange per unit of foliar dry matter) declines in some coniferous trees as they increase in stature. The decline is thought to result from a decrease in hydraulic conductance within stems and branches as they age and elongate, probably due to the increased chance of cavitation and number of branching internodes (Hubbard et al. 1999). Compared with younger trees, the lower hydraulic conductance of older trees may cause greater stomatal closure, which suppresses photosynthesis. In the ponderosa pines measured by Yoder et al. (1994), this effect could result in a difference of as much as 30% in daily net carbon exchange between trees of height 10 and 32 m.

This result has been included in models as a reduction in photosynthetic production proportional to tree height, H, or crown length, L_C (Landsberg and Waring 1997; Mäkelä 1997). Using the model developed in Sect. 3.2, we can relate this reduction to the leaf-specific rate of photosynthetic production, σ_f:

$$\sigma_f = \sigma_{f0}(1 - \alpha X) \tag{3.40}$$

where X is either H or L_C and α (m^{-1}) is a semi-empirical parameter (Mäkelä and Valentine 2001).

More recent evidence (Sala and Hoch 2009; Guillemot et al. 2015) suggests that the growth decline of taller trees or trees suffering from drought may be related to reduced growth rather than reduced photosynthesis. This was shown by increased concentrations of stored carbohydrates in trees suffering from water limitation. Limited availability of water may cause reduction in the sink strength of woody growth, thus reducing the growth efficiency.

3.4.4 Respiration

A standard explanation for the decline in forest growth has been the *respiration hypothesis*, suggesting that as trees get taller, the proportion of respiring tissue relative to photosynthesising tissue increases, causing a decline in the overall growth efficiency (Kira and Shidei 1967). Although this proposition seems logical in view of our empirical understanding of the development of tree form (e.g., the allometric relationships presented above), there are several issues that deserve commenting on in relation to it.

First, we have seen above that including respiration in a carbon balance model does not automatically lead to a stable model with an asymptotic equilibrium, if no genuine feedback mechanisms are included. On the other hand, if we assume the allometric relationships, (3.32), we obtain stable behaviour where the increased share of woody respiration in the carbon balance contributes to growth limitation. In this sense, it seems that it is not respiration but the allocation scheme that should be regarded as the cause of the resulting growth decline in larger trees.

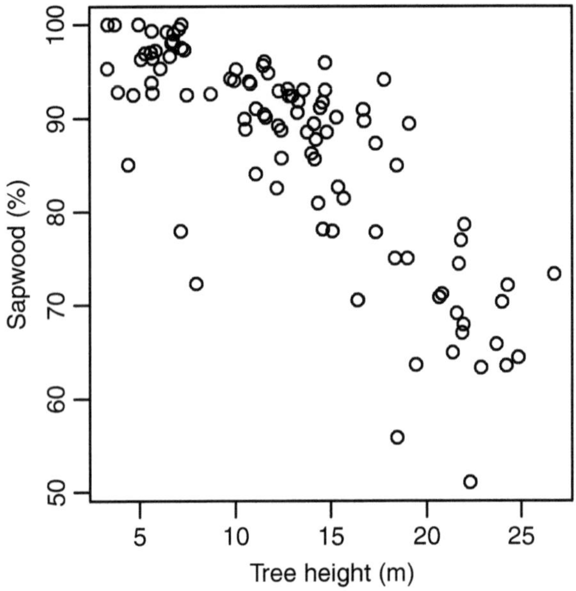

Fig. 3.6 Proportion of sapwood in stems of Scots pine (*Pinus sylvestris* L.) in southern Finland. (Data from Vanninen and Mäkelä (2000))

Secondly, up to now we have not considered the fact that not all woody tissue is respiring. There is strong empirical evidence that maintenance respiration in wood is proportional to sapwood volume or mass (Sprugel 1990; Ryan 1991). The share of sapwood in total woody biomass declines with tree size (Fig. 3.6), implying that the increase of woody respiration with tree/stand age is also less than suggested by the total amount of wood accumulated in the stand. Further empirical findings suggest that maintenance respiration is most active in the outer sapwood and possibly also in sapwood cut from the upper portion of the stem (Pruyn et al. 2002, 2005).

In general, a significant proportion of maintenance respiration is thought to be driven by protein turnover (Penning de Vries 1975). Therefore, the relative rates of maintenance respiration of different tissues should be proportional to the N concentrations of the tissues. (Ryan 1995; Pruyn et al. 2005) and others have successfully tested the hypothesis that N concentrations of different tissues explain differences in their respiration rates in a given environment. No significant trends with age or size have been observed in these relationships. The N concentration of woody tissue is less than a tenth of that in leaves.

3.4.5 Summary

The above sections show that in order to understand the feedbacks leading to stable behaviour of individual trees, it is not sufficient to consider trees in terms of carbon, nitrogen, or biomass pools, but an understanding of the dynamics of tree structure is also required. Some information about tree structure was needed in all the feedback mechanisms discussed. The allometric equations started off from empirical evidence relating basal diameter and biomass, while the self-shading model was defined in terms of crown surface area. The hydraulic limitation hypothesis was based on either tree height or crown length, and finally, realistic presentation of maintenance respiration requires that we are able to separate between total wood and sapwood.

In addition to being of crucial importance for understanding the feedbacks in material balances, and thus the dynamics of growth, tree structure is also important for developing more detailed models of tree-to-tree interactions in stands. Many shading models utilise information about crown structure and relative heights of the crowns. The following chapter is therefore devoted to a review of models of tree structure.

3.5 Problems

3.1 A simulation programme "`Cbal.R`" of the model (3.11) accompanied by the Lambert-Beer type shading as in (3.18) is provided in the electronic supplementary material of this chapter. Table 3.1 gives suggestions for parameter values that could be used in the simulation. Study the parameter values and make sure that you understand what the parameters mean. Then simulate the system with different sets of parameter values. Study the implications of varying (a) the allocation coefficients, (b) the turnover rates, and (c) the maximum production rate.

3.2 For one or two sets of parameter values, find the equilibrium of (3.11) using the graphical method indicated by (3.21) and Fig. 3.3. Check that your simulation yields the same equilibrium as indicated by the graphical solution.

3.3 Derive (3.21) and (3.22).

3.4 Simulate the model in its linear form, i.e., when all parameters in (3.11) are constant. Over a short period, e.g. 10–20 years, study the relative shares of leaves, roots, and stems in this situation. What happens to them? What can you say about the claim that "increasing respiration causes reductions to growth as trees grow larger"?

3.5 Solve the linear differential equation given in (3.12) and (3.13) by finding the eigenvalues and eigenvectors of the system. Initial state is $W_i(0) = W_{i0}$.

3.6 You will find the R code for simulating the nitrogen limitation model of (3.25) in the electronic supplementary material of this chapter, titled "`Nprod.R`". (a) Simulate the model to find how the availability of nitrogen restricts growth as the stand biomass increases. (b) Consider the hypothetical situation that an increase in root allocation (or in general, input of carbon below ground) would enable the tree to take up more nitrogen, for example, by means of feeding mycorhizza or through the priming effect of decomposition. For this, assume that root allocation increases from 0.3 to 0.34 at the cost of both leaves and wood equally, and this helps the tree to increase U. How much would U have to increase in order for this change to increase the total equilibrium biomass and total woody growth? Analyse this either through simulation or by analytical calculation of the implied changes in steady states. Discuss the ecological aspects of this.

3.7 You will find the R code for simulating the model of (3.11) combined with the allocation scheme of (3.37) in the electronic supplementary material of this chapter, titled "`allo.R`". (a) Simulate the model to find how the allocation scheme restricts growth as the trees grow larger. (b) Make an additional output to study the development of λ_{f} with time. Explain the development of λ_{f} by analysing (3.37). How do the different terms affect λ_{f} and how does this change with tree growth? (c) Compare the allocation scheme of suppressed and dominant trees by varying σ_{f}.

Chapter 4
Tree Structure

Interactions between structure, function and survival under natural selection provide a plausible explanation for the remarkable regularity observed in tree structures, regardless of the apparent wide variability of form between species and individuals. The regularity has been described and modelled in a number of ways, ranging from simple empirical observation to complicated mathematical evolutionary optimisation models. In this chapter, we introduce some of the key ideas about structural regularity that have been helpful in modelling the allocation of growth to different tree components in material balance models.

4.1 Introduction

The term structure is used in several different meanings, ranging from the form, or geometry, of an object to its material constituents. In botany, *morphology* is the science that studies plant and organ form, while the more detailed structures of organs and cells are the focus of plant *anatomy*. In all cases, structure is characterised by means of size relationships and geometry. For example, the ratio of leaf biomass to woody biomass in a tree provides a quantification of structure, and so does the description of the crown as a cone, or the definition of the crown ratio as crown length relative to tree height. More detailed descriptions of structure include the full three-dimensional spatial representation of the tree, with all growth units and their orientation (Sievänen et al. 2000).

Because the basic physiological processes are shared by almost all tree species, the variability among the species is largely due to their structural characteristics. Moreover, the acclimation of individuals to different environments is usually manifested by variable tree form. For example, the crowns of shade-tolerant tree species may grow wide and umbrella-like under shade, whereas in full light they are narrow and rigid (Chen et al. 1996; Pearcy et al. 2005). The height of the live

A. Mäkelä, H. T. Valentine, *Models of Tree and Stand Dynamics*,
https://doi.org/10.1007/978-3-030-35761-0_4

crown rises in dense canopies (Sect. 2.4), and stems become tapered and short when exposed to strong winds (Larson 1965; Morgan and Cannell 1994). We conclude that considerable variability in structure exists among species as well as among individuals of the same species that grow in different environments.

Variability notwithstanding, considerable regularity also has been observed in tree structure. As we shall see in the following sections, there appears to be constraints on structure that apply to all species and individuals. These regularities can be understood by studying the significance of structure for the functioning and, ultimately, the survival of trees. It is postulated that some structural characteristics, which are essential for the different functions of trees, have been subject to strong selective pressures, leading to forms that satisfy some important constraints relevant for survival.

An evolutionary perspective to tree structure, therefore, is concerned with how the growth and reproductive processes depend on the function of the different tree organs, and how these organs depend on the overall tree structure. From this point of view, we consider leaves and fine roots as the organs responsible for resource capture, with the woody organs providing a branching network that allows for the physiologically active organs to occupy sites where resources are available. In doing so, however, the branching network must also be equipped to transport nutrients and substrates between the fine roots and leaves. Moreover, it must provide a skeleton that is strong enough to support both itself and the physiologically active organs.

Building a woody skeleton that satisfies the requirements of resource capture, substrate transport, and mechanical stability is costly in terms of the assimilated resources. Natural selection favours forms that are efficient in their use of resources. We should therefore expect that, while satisfying all the different requirements of function, tree form has evolved to waste no resources. Following this principle we would expect, for example, that the transport pathway of substrates between leaves and fine roots be no larger than that required by the overall transport need, as related to the activities of the fine roots and leaves. In the same way, the activities of leaves and fine roots should also be balanced with each other to support the required balance of the respective substrates for growth, and to avoid any waste in growing one part relatively larger than the other, as this would not lead to increased growth rates. Similarly, tree stems would be sufficiently strong to prevent the tree from toppling over under unusually severe wind forces, but not any stronger, as this would lead to a waste of resources.

These interactions between structure, function and survival under natural selection provide a plausible explanation for the remarkable regularity observed in tree structures, regardless of the apparent wide variability of form between species and individuals. The regularity has been described and modelled in a number of ways, ranging from simple empirical observation to complicated mathematical evolutionary optimisation models. In this chapter, we introduce some of the key ideas about structural regularity that have been helpful in modelling the allocation of growth to different tree components in material balance models.

4.2 Allometry

In the context of structure, allometry is usually concerned with growth-related changes in shape of an organism or of parts of the organism. An allometric relationship between two dimensions of shape, e.g., length and width, is expressed with a power function:

$$y = kx^a \tag{4.1}$$

where y and x are dimensions of shape, a is called the scaling exponent, and k is a coefficient of proportionality. According to (4.1), y scales with the ath power of x.

If a growing organism follows the allometric law given in (4.1), then the two dimensions of shape show proportional relative growth rates. This can be seen by taking the time derivatives of both sides of (4.1), yielding

$$\frac{1}{y}\frac{\mathrm{d}y}{\mathrm{d}t} = a\frac{1}{x}\frac{\mathrm{d}x}{\mathrm{d}t} \tag{4.2}$$

If $a > 1$, the variable y will grow proportionately faster, and the ratio of y to x will increase as overall size increases. If $a < 1$, the opposite will happen.

Allometric relationships can be illustrated by means of simple geometrical objects. Consider a cube with edge length s. The volume, V, of the cube is

$$V = s^3 \tag{4.3}$$

and the area, A, of each of its sides is

$$A = s^2 \tag{4.4}$$

If the cube has uniform density ρ, then the cube's mass is $M = \rho V$, and the allometric equation for the relationship between edge length and mass is

$$M = \rho s^3 \tag{4.5}$$

From the above equations, we may deduce the relationships between mass, volume, area, and length, often called the linear dimension of the object. The exponent between volume and mass is 1, that between area and mass is 2/3, between length and mass, 1/3.

For organisms that retain their shape for most part of their development, similar allometric equations can be applied as those for the cube. These organisms are called *isometric*. However, it is more common that the shape of living organisms would change as they grow, giving rise to genuine allometry. This entails allometric exponents that are different from those of the simple geometric objects. For example, mass could increase with volume but with an exponent greater or smaller than 3, or the area of some organ of the organism could increase with an exponent either greater or smaller than 2, in pace with the linear dimension chosen as a reference.

4.2.1 Allometry of Trees

The allometric equations devised for trees usually employ dbh—the diameter at 1.3 m height on the central stem—as the reference linear dimension. The dependence of leaf, stem and branch mass as well as tree height on dbh have been analysed for many species. The allometric exponent for leaf mass is typically close to 2, that of branches in the range (2, 3) and that of stems, close to 3, while the allometric exponent for total biomass is around 2.5 (Ketterings et al. 2001; Williams et al. 2003; Wang 2006; Litton and Kauffmann 2008). Tree height sometimes changes isometrically with stem diameter; other times height growth changes allometically with stem diameter (Niklas 1995).

The geometric means of heights, diameters, and volumes of trees growing in typical stands can be illustrated, by the predictions provided by growth and yield tables. For example, Landsberg et al. (2005) used the Finnish yield tables (Koivisto 1959) to estimate the relationship between breast-height diameter and stem biomass over stand rotation in different site fertility classes (Fig. 4.1).

Allometries based on dbh are descriptive of tree form, particularly illustrating the fact that the fraction of leaf to total biomass generally decreases rapidly as tree diameter increases and that the fraction of stemwood to total biomass

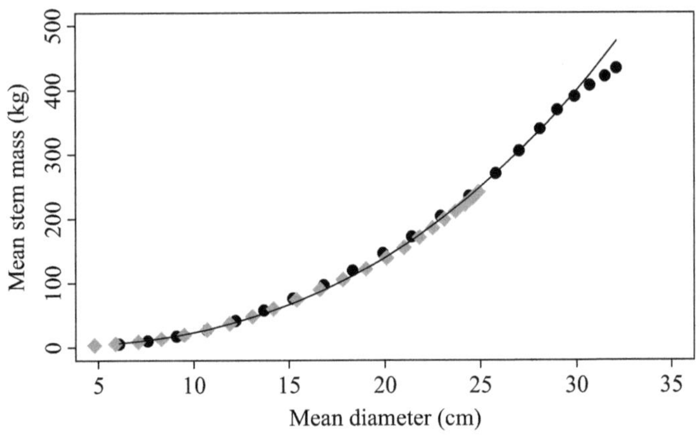

Fig. 4.1 Allometric relationships between stand-level means of stem diameter (D) and stem biomass (W_s), calculated by fitting an allometric function to data from growth and yield tables (Koivisto 1959). Stem volume was converted to stem biomass assuming that wood density was 400 kg m^{-3}. Grey diamonds: *Vaccinium* site type (VT), black circles: *Oxalis-Majanthemum* site type (OMT), fitted curve: $W_s = 0.0581 D^{2.598}$, $r^2 = 0.99$ (Landsberg et al. 2005)

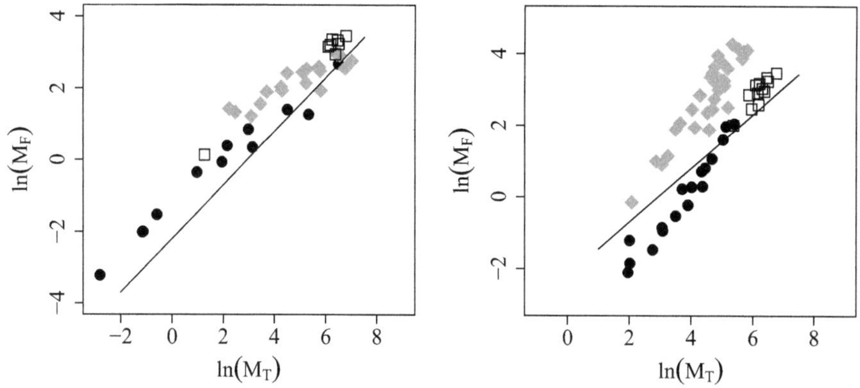

Fig. 4.2 Scaling between aboveground woody mass (M_T; kg) and leaf mass (M_F; kg) in the tree (Redrawn from Mäkelä and Valentine (2006a)). M_T includes live branches and the trunk. Key to symbols: diamonds, pine (*Pinus sylvestris*); open squares, spruce (*Picea abies*); circles, birch (*Betula pendula*). The line represents the 3/4 power law (Sect. 4.5). Left panel: Data include dominant trees from stands of variable age. Right panel: For each species, data include trees of different dominance position in one even-aged stand

simultaneously increases. Nonetheless, allometric relationships have been found to be quite variable among different data sets. Mäkelä and Valentine (2006a) found that the distribution of crown ratio in the data set was an important indicator of allometric scaling. If the trees were from a 'longitudinal data set' covering stands of different ages, it would then be characteristic that crown ratio would decrease with increasing diameter and height. By contrast, if the data came from individual stands, covering variation in size related to competition and differentiation between trees, then crown ratio would increase with increasing tree size. In these two cases, the allometric ratios obtained for the trees were distinctly different (Fig. 4.2).

The allometric relationship between tree basal diameter and tree height has also been found to vary between different data sets. Niklas (1995) presented evidence that this relationship was size-dependent in several tree species, with a decline in the allometric exponent as trees grow larger. Small trees showed almost isometric relationships between height and diameter, while in large trees the exponent was close to $2/3$.

4.3 Pipe Model

The pipe model is a more detailed approach to tree structure than the allometric models. As already noted in Chap. 1, the pipe model was presaged by Leonardo da Vinci's observation of area-preserving branching, but the pipe-model formulations we use today derive from two seminal papers that were published in 1964 by a team of ecologists at the Osaka City University, Japan (Shinozaki et al. 1964a,b). In

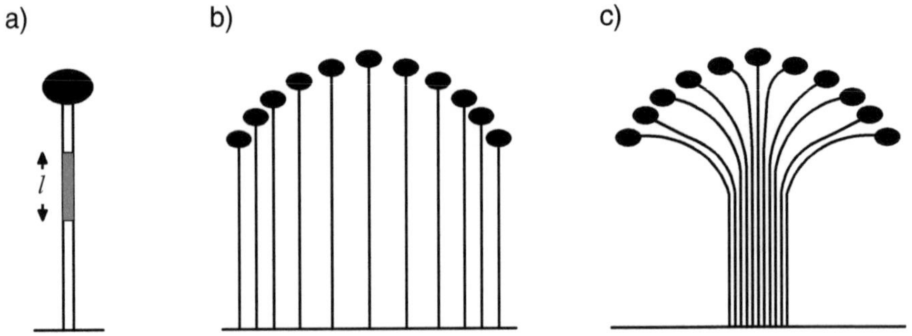

Fig. 4.3 (**a**) A single pipe with an illustration of the specific pipe length (l); (**b**) a cluster of pipes representing a stand; and (**c**) an aggregation of pipes representing a plant. (Redrawn from Shinozaki et al. (1964a))

extensive empirical studies, they found that, "the amount of leaves existing above a certain horizontal level in a plant community was always proportional to the sum of the cross-sectional areas of the stems and branches found at that level." This led them to interpret plant stems as an assemblage of *unit pipes*, each pipe supporting a fixed amount of leaves (Fig. 4.3).

The pipe model theory was formulated from data that were gathered by the stratified clip technique. Briefly, let $z_{\min}$ and $z_{\max}$, respectively, denote the base and top of a canopy formed by a tree community, and let z (m) be height above the canopy base, ($z_{\min} \leq z \leq z_{\max}$). Each stratum is defined by two parallel horizonal planes that slice through the canopy separated by a vertical distance Δz (m). Several such strata stack from the bottom to the top of the canopy. Of interest in each stratum are (a) the amount of leaf biomass per unit ground area, denoted by $\mathbb{W}_{\mathrm{f}}$ (kg DW leaf (m^2 ground)$^{-1}$), and (b) the amount of stem biomass per unit ground area, denoted by $\mathbb{W}_{\mathrm{s}}$ (kg DW stem (m^2 ground)$^{-1}$).

Dividing $\mathbb{W}_{\mathrm{f}}$ and $\mathbb{W}_{\mathrm{s}}$ by Δz within each stratum provides a continuous biomass density with respect to z from the bottom to the top of the canopy. Let $\mathbb{\Gamma}(z) = \mathbb{W}_{\mathrm{f}}/\Delta z$ be the leaf biomass per unit vertical length at z. Likewise, let $\mathbb{C}(z) = \mathbb{W}_{\mathrm{s}}/\Delta z$ be the corresponding stem biomass per unit vertical length at z. Both biomass densities have units of kg DW (m length)$^{-1}$ (m^2 ground)$^{-1}$.

From such data, Shinozaki et al. (1964a) discovered the following relationship:

$$\mathbb{C}(z) \approx \frac{1}{l} \int_{z_{\min}}^{z_{\max}} \mathbb{\Gamma}(z)\, \mathrm{d}z \tag{4.6}$$

where l [(m kg^{-1} stem) $\times$ (kg leaf)] is called the *specific pipe length*. Equation (4.6) indicates that, over a defined area of ground, the stem biomass per unit length at a given level within a tree canopy is proportional to the total leaf biomass above that level (Fig. 4.4).

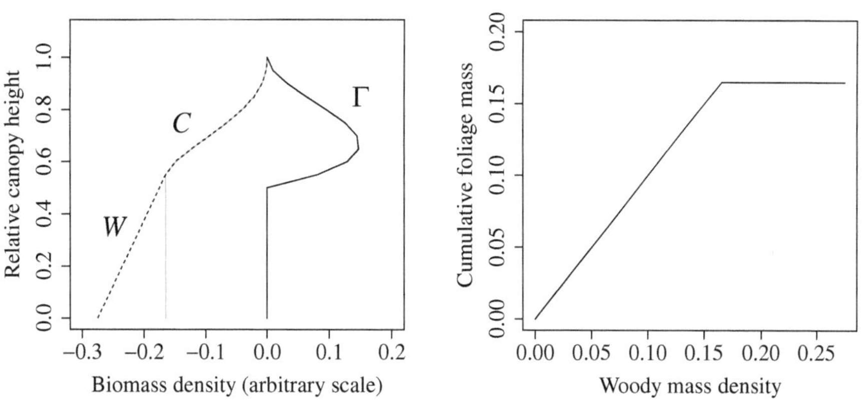

Fig. 4.4 Illustration of the relationship of different biomass densities according to the pipe model. (Redrawn from Shinozaki et al. (1964a)). Left panel: Sapwood density, $\mathbb{C}$, is the integral of foliage density, $\mathbb{\Gamma}$, while total woody density, $\mathbb{W}$, also includes disused pipes. Right panel: Above the crown base, total leaf mass above a given height is proportional to total woody density at that height. Below the crown base, total woody mass still increases downwards because of the disused pipes

The specific pipe length equals the ratio of the biomass of supported leaves to the biomass per unit length of the supporting stem(s) (Fig. 4.3a). Stem biomass per unit length can be replaced by stem volume per unit length under the assumption of uniform bulk density, and volume per unit length morphs into cross-sectional area under consideration of a vanishingly short stem length. Applying these transformations to both $\mathbb{C}(z)$ and l in (4.6) indicates that the cross-sectional area of stems at a given level within a tree canopy is proportional to the total leaf biomass above that level, and vice versa.

We suspect that the ‘vice versa method’ of estimating leaf biomass from stem cross-sectional area is the most widely used result from pipe model theory. Whether estimating leaf biomass that way is more or less accurate than estimating leaf biomass from stem weight per unit length remains to be tested against empirical data.

While the pipe model, according to Shinozaki et al. (1964a), applies to most plant genera, it has been most widely used for trees, for which the stem biomass is called woody biomass. Shinozaki et al. (1964a) made an important distinction between trees and non-woody plants, however, as they observed that, “in trees, the disused pipes, which had once supported the leaves on the already shed-off branches, remain in the trunk and bough together with the working pipes connected with the living leaves.” The term *pipe model theory of tree form* thus refers to the combination of active and disused pipes on the tree that together shape the form of stems and branches and regulate their connections to leaves.

The pipe model has had several useful applications, particularly in methods of estimating leaf biomass and other biomass components from more easily measurable quantities, such as sapwood area at either the crown base or breast height. As we shall see in the succeeding chapters of this book, the pipe model has also become an important component of tree-growth models in providing rules for biomass allocation. These are the types of applications for which the theory was intended by its developers: the assemblage of pipes is an analogy that allows us to derive a relationship between woody mass and leaf mass. However, several investigators have interpreted the term 'pipe' to mean a water pipe. Although it does seem reasonable from an evolutionary point of view to assume that the water conducting pathway should be more or less in balance with the mass of transpiring leaves, such as in the unit pipe system, this does not imply that the pipes, as such, should be interpreted as water conducting pipes or tubes. Some investigators have nevertheless done so, and furthermore, subsequently claimed that the pipe model is incorrect because such water pipes do not exist in real trees. This naïve interpretation also ignores the role of the disused pipes and the importance of both the active and disused pipes in providing mechanical support to the crown.

4.3.1 *Basic Definitions*

In many applications, the original formulation of the pipe model has been augmented to include fine roots, whereby a unit of leaf biomass is attached to an active pipe that extends through the area-preserving network of branches, stem, and coarse roots to the location of some defined amount of fine-root biomass (Mäkelä and Valentine 2006a). In this regard, it is useful to think of active pipes as strands of wood that connect leaves to fine roots. Although active pipes may differ in length, all together they account for the volume of the entire live woody portion of a tree, including the embedded vascular conduits. Similarly, disused pipes can be interpreted as strands of dead wood, which all together account for the dead wood in the interior of a tree's stem.

For many applications, it is more convenient to relate the leaf biomass distribution to a certain height in the stem than to a height of a horizontal layer in the canopy. This will allow us to more easily distinguish among stems, branches, and coarse roots, and to take account of their differences when needed.

Let $w_f(h)$ (kg DW m^{-1}) be the leaf biomass per unit vertical length at height h (m) on the stem for $0 \leq h \leq H$, where H is tree height. For convenience, we define *active pipe area* to mean 'aggregate cross-sectional area of active pipes.' Thus, we now express the pipe-model relationship between the active pipe area at h and the cumulative leaf biomass from h to H as follows:

$$A_s(h) = \frac{1}{\eta_s} \int_h^H w_f(h)\, dh \tag{4.7}$$

where $A_s(h)$ (m^2) is the active pipe area and η_s (kg DW m^{-2}) is the ratio of leaf biomass to active pipe area. This ratio is constant under the usual assumption that the bulk density of wood is uniformly constant.

Perpendicular (or excurrent) branches that ramify from the stem at height h bear the leaf biomass at h, so active pipe area of these branches $a_b(h)$ (m^2) is proportional to the leaf mass per unit vertical length at h, i.e.,

$$a_b(h) = \frac{1}{\eta_b} w_f(h) \tag{4.8}$$

where η_b (kg DW m^{-2}) is a constant. Substituting (4.8) into (4.7) provides the relationship between active pipe area at height h in the main stem and the active pipe area in the branches above h:

$$A_s(h) = \frac{\eta_b}{\eta_s} \int_h^H a_b(h)\, dh \tag{4.9}$$

This formulation allows for different specific pipe lengths for the main stem and branches. Or to put it another way, the ratio of leaf biomass to active pipe area for a branch may differ from that for the main stem.

4.3.2 Pipe Model for Tree-Level Variables

Because there are no leaves below the crown base, we may relate the active pipe area at crown base to the total leaf mass:

$$W_f = \eta_s A_c \tag{4.10}$$

Recall that A_c (m^2) is the active pipe area at crown base and W_f (kg DW) is total leaf biomass. The constant, η_s (kg DW m^{-2}), is often referred to as the *pipe model ratio*. This relationship, alone, is often referred to as the pipe model.

Integration of (4.8) over the length of the live crown yields an analogous relationship between leaf mass and the aggregate active pipe area of the first-order branches (A_b) (m^2), i.e.,

$$W_f = \eta_b A_b \tag{4.11}$$

By analogy, we can extend this analysis to coarse roots also, yielding

$$W_f = \eta_t A_t \tag{4.12}$$

where A_t (m^2) is the aggregate active pipe area of coarse roots leaving the base of the trunk and η_t (kg DW m^{-2}) is the pipe-model coefficient for coarse roots. These equations allow us to relate leaf mass to active pipe area in stems, branches, and coarse roots. They have been intensively utilised in foliage biomass estimation at both the stand and tree levels.

4.3.3 Fine Roots

Originally, the pipe model was only concerned with above-ground variables and coarse roots, though little information was provided about the latter. Subsequent studies confirmed the applicability of the pipe model with area-preserving branching to coarse root systems (e.g., Carlson and Harrington 1987; Richardson and ZuDohna 2003). Fine roots can be included in the model in the same way as leaves, which implies that fine-root biomass is proportional to active pipe area and, therefore to leaf biomass, i.e.,

$$W_r = \alpha_r W_f \tag{4.13}$$

where α_r is the coefficient of proportionality. There is convincing empirical evidence and theoretical arguments indicating that α_r is determined by the availability of nutrients to the tree (Ingestad and Ågren 1992; Vanninen and Mäkelä 1999); and it also varies among species (Helmisaari et al. 2007).

4.3.4 Biomass of Active Pipes

While leaf mass can be expressed in terms of active pipe area at the base of the crown, active stem biomass, W_s, can be evaluated on the basis of the vertical leaf biomass distribution:

$$W_s = \frac{\rho_s}{\eta_s} \int_0^H \left(\int_h^H w_f(h')\, dh' \right) dh$$

which simplifies to

$$= \rho_s \int_0^H A_s(h)\, dh \tag{4.14}$$

where ρ_s (kg DW m^{-3}) is the bulk density of stem wood. This expression can be further simplified by defining a *mean pipe length* within the stem. Since stems are vertical, we express the mean pipe length, $\hat{H}_s$ (m), in terms of height up the tree from the stem base.

Recall that A_c is the active-pipe area at the crown base, which occurs at height H_c. We assume that no leaves occur below the crown, so all the active pipes within the stem extend from the stem base at $h = 0$ to at least as high as the crown base. Above H_c, pipes branch from the main stem or terminate at leaves, so the active pipe area gradually diminishes with increasing height.

Suppose that the active stem comprises $n(0)$ active pipes at $h = 0$. Let $n(h)$ be the number of active pipes that extend upwards from the stem base to at least height h, and let a_s be the cross-sectional area of a single pipe, so that $A_c = a_s n(0)$ and $A_s(h) = a_s n(h)$. By definition, the mean pipe length, $\hat{H}_s$, is the area-weighted average of the lengths of the individual pipes, i.e.,

$$\begin{aligned} \hat{H}_s &= \frac{\int_0^H a_s n(h)\,\mathrm{d}h}{a_s n(0)} \\ &= \frac{\int_0^H A_s(h)\mathrm{d}h}{A_c} \\ &= \frac{W_s/\rho_s}{A_c} \end{aligned} \tag{4.15}$$

where W_s/ρ_s is the aggregate volume of the active pipes in the stem. Therefore, the mean pipe length within the stem equals the active pipe volume within the stem divided by the active-pipe area at the crown base. Consequently, the woody biomass represented by the active pipes within the stem is estimated by

$$W_s = \rho_s A_c \hat{H}_s \tag{4.16}$$

or substituting (4.10),

$$W_s = \frac{\rho_s \hat{H}_s}{\eta_s} W_f \tag{4.17}$$

To estimate the active pipe length of branches and coarse roots using an analogous method to the above, we would require information about the distribution of leaves in each branch along the branch axis, and the distribution of fine roots along the coarse root axis. However, the mean active pipe length of branches and coarse roots may also be defined in analogy with (4.16), yielding

$$\begin{aligned} W_b &= \rho_b A_b \hat{H}_b \\ W_t &= \rho_t A_t \hat{H}_t \end{aligned} \tag{4.18}$$

where ρ_b and ρ_t (kg DW m^3), respectively, are the bulk densities of branches and coarse roots, and $\hat{H}_b$ and $\hat{H}_t$ (m) are the corresponding mean branch and coarse-root lengths. Substituting (4.11) and (4.12) relate these components of wood mass to the total leaf mass:

$$W_i = \frac{\rho_i \hat{H}_i}{\eta_i} W_f, \qquad i = \mathrm{b}, \mathrm{t} \tag{4.19}$$

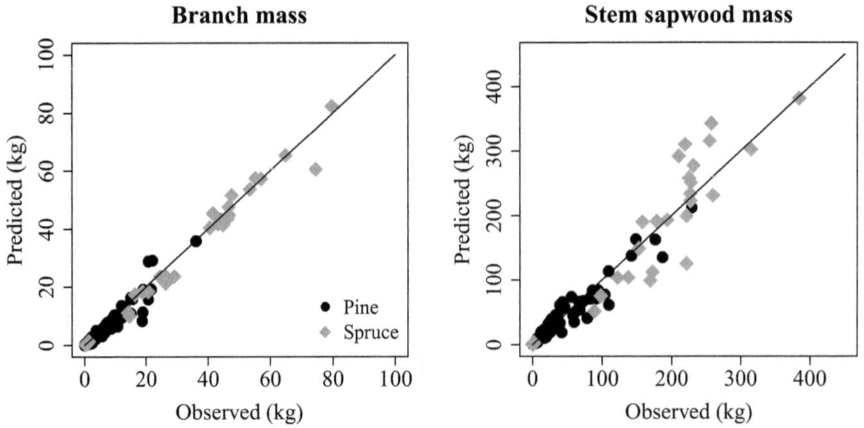

Fig. 4.5 Observed total branch biomass (left panel) and stem sapwood biomass (right panel) in *Pinus sylvestris* and *Picea abies* compared with pipe model predictions calculated with (4.16–4.20). (Data from Mäkelä and Valentine (2006a)). The form factors are $\phi_s = 0.5$, $\phi_b = 1.2$ for pine and, $\phi_s = 0.7$, $\phi_b = 0.92$ for spruce

This is convenient, because empirical evidence indicates that, in many tree species, the mean pipe lengths are directly proportional to easily measurable quantities, such as crown height, crown length and total height (Fig. 4.5). We call the coefficients of proportionality *form factors*, denoted by ϕ_i, (i = b, s, t), of the branches, the main stem above the crown base, and the coarse roots. The mean stem lengths in terms of the form factors are:

$$\begin{aligned} \hat{H}_s &= H_c + \phi_s L_c \\ &= \phi_s H + (1 - \phi_s) H_c \\ \hat{H}_b &= \phi_b L_c \\ &= \phi_b H - \phi_b H_c \\ \hat{H}_t &= \phi_t H \end{aligned} \tag{4.20}$$

4.3.5 Disused Pipes

The pipe model theory suggests that as pipes lose their connection to leaves, they become disused but are retained in the stem, giving rise to stem taper below crown (Fig. 4.6). The disuse of pipes is related to the shedding of leaves and the death of branches, and therefore occurs as a result of the crown base rising upwards, especially in dense stands. A similar phenomenon may occur in large branches.

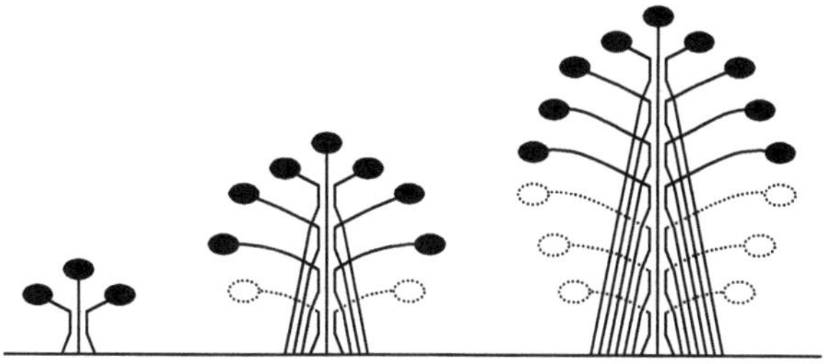

Fig. 4.6 Disused pipes are retained within the bole as branches die, giving rise to stem taper

To quantify this for any tree of interest, we would need to know the history of the tree in terms of its height growth and crown recession (Chiba et al. 1988; Osawa et al. 1991; Mäkelä 2002; Valentine et al. 2012), and dynamic models have indeed been developed to mimic the process (see Chap. 5). However, empirical evidence suggests that the tapering of the trunk can be approximated with the crown ratio, r_c (Valentine et al. 1994a):

$$A_c = r_c B \tag{4.21}$$

where B denotes basal area of the stem at the stem base. This result appears to have no direct theoretical basis, however, it is approximately consistent with both stem form as projected by the theory of constant strain (Morgan and Cannell 1994) and the dynamic development of trees according to the pipe model, given realistic patterns of crown rise (Chiba et al. 1988; Osawa et al. 1991; Mäkelä 2002; Valentine and Mäkelä 2005). The possibility of disused pipes within the crown will be omitted here but has been dealt with by Valentine et al. (2012).

4.3.6 Biomass Estimation Using the Pipe Model

For the estimation of total forest carbon required, e.g., in greenhouse gas inventories, it has become important to be able to derive biomass estimates of tree components from standard forest inventory variables. These usually comprise diameter and height and increasingly also crown height of sample and/or tally trees. Empirical biomass equations have been developed in several countries for this purpose, using extensive species-specific data sets (e.g., Zianis et al. 2005; Repola 2009; Neumann et al. 2016). The biomass equations based on the pipe model presented above offer an alternative method for biomass estimation. While empirical equations rely on statistical fitting of large data sets, the strength of the pipe-model equations derives from their well-defined parameters that can be estimated from designated measurements. Provided that the defined structure is sufficiently general, this could constitute an efficient and general method for biomass estimation.

The main idea of biomass estimation using the pipe model is that the biomass of all components apart from the stem below crown can be derived from active pipe area and mean active pipe length using the equations in the preceding sections. The biomass of the bole below crown, on the other hand, contains disused pipes and must be determined on the basis of its diameter distribution over its entire length. In reality, cross-sectional areas at crown base and in branches contain variable proportions of disused pipes (e.g., Berninger et al. 2005; Valentine et al. 2012). Nevertheless, the biomass equations may still be applicable, only the parameters must be modified accordingly.

Equation (4.10), which estimates leaf biomass, is by far the best known and most widely utilised pipe model equation for biomass estimation (Waring et al. 1982; Lehnebach et al. 2018). Lehtonen et al. (2020) tested the assumption that the pipe model ratio is a species-specific constant in a large empirical data set in Finland, and compared the leaf mass predictions with different empirical models (Marklund 1988; Repola 2009). Both Marklund (1988) and Repola (2009) provided models with different sets of independent variables, a more simple model using just diameter and height, and another also incorporating crown length. The study concluded that for Scots pine, the pipe model provided the most accurate results of all the tested models in the whole data set, whereas for Norway spruce the model of Marklund (1988) including crown length (with parameters estimated from the data) was slightly more accurate. They also found correlations between the pipe model ratio and some environmental variables in the geographically extensive data set. The ratio was somewhat larger in sites that were of higher quality according to the Finnish site classification system (Cajander 1949), indicating both higher soil moisture and more ample nitrogen availability. The pipe model ratio also increased temporarily after fertilisation. Interestingly, both species showed large pipe model ratios in trees growing on very barren sites such as rocks. In pine, increasing temperature sum and increasing annual precipitation also somewhat increased the ratio.

Hu et al. (2020) tested the accuracy of pipe-model based branch and stem biomass equations in a large data set on Scots pine and Norway spruce, collected from a subset of Finnish national inventory plots (Lehtonen 2005). The data set covered a wide range of tree sizes, ages, and competitive positions. It contained measurements of crown base area, total basal area of branches as well as branch length for a representative set of sample branches. This allowed them to evaluate an estimate of total branch mass, W_{bt}, in the form of (4.18) and (4.20), where branch active pipe area was replaced with total branch area, that is,

$$W_{\mathrm{bt}} = \phi_{\mathrm{bt}} \rho_{\mathrm{b}} A_{\mathrm{bt}} \bar{H}_{\mathrm{b}} \tag{4.22}$$

where A_{bt} is total branch cross-sectional area, $\bar{H}_{\mathrm{b}}$ is basal-area weighted mean branch length, and ϕ_{bt} is form factor for total branch mass. However, for practical applications, A_{bt} or $\bar{H}_{\mathrm{b}}$ are usually not available. In analogy to active pipes, Hu et al. (2020) approximated total branch area with total area at crown base, A_{ct}, and mean branch length with a power function of crown length, yielding the following estimate for total branch biomass:

$$W_{\mathrm{bt}} = \tilde{\phi}_{\mathrm{bt}} \rho_{\mathrm{b}} A_{\mathrm{ct}} L_{\mathrm{c}}^{x} \tag{4.23}$$

where several parameters have been embedded in $\tilde{\phi}_{\rm bt}$. Further, inserting (4.21),

$$W_{\rm bt} = \tilde{\phi}_{\rm bt}\rho_{\rm b}r_{\rm c}BL_{\rm c}^{x} \tag{4.24}$$

Further, to estimate stem mass, Hu et al. (2020) used the simple approximation that the mean basal area of the stem below the crown is a weighted average of those at crown base and breast height. The biomass of the stem below the crown ($W_{\rm sh}$) could thus be expressed as follows:

$$W_{\rm sh} = \phi_{\rm bct}\rho_{\rm s}H_{\rm c}(B + A_{\rm c}) = \phi_{\rm bct}\rho_{\rm s}HB(1 - r_{\rm c}^2) \tag{4.25}$$

where (4.21) and the definition of $H_{\rm c}$ as $(1 - r_{\rm c})H$ have been utilised. $\phi_{\rm bct}$ depends on the weights of crown base and breast height areas in the mean basal area of the bole. Total stem biomass is the sum of ($W_{\rm sh}$) and stem biomass inside the crown, approximated with (4.20), where active pipe area has been replaced by total crown base area:

$$W_{\rm stot} = \phi_{\rm bct}\rho_{\rm s}HB(1 - r_{\rm c}^2) + \rho_{\rm s}\phi_{\rm st}r_{\rm c}^2HB = \rho_{\rm s}HB[\phi_{\rm bct}(1 - r_{\rm c}^2) + \phi_{\rm st}r_{\rm c}^2] \tag{4.26}$$

Again, (4.21) and the definition of crown length as $r_{\rm c}H$ have been utilised.

The above equations allow us to estimate branch and stem biomass from tree height, basal area and crown ratio, provided that wood density and the relevant form factors are known. Obviously, for the models to be useful, the form factors need to be reasonably constant and invariant between different data sets that may vary with respect to, e.g., stand structure, age and environment. Hu et al. (2020) found that branch mass could be rather accurately estimated with (4.22) for both pine and spruce. A fairly strong correlation was retained when the branch basal area and mean length were replaced by the more generally available independent variables of (4.24), although more scatter was introduced in the results. The accuracy of the estimate was similar to empirical equations by Repola (2009) (Fig. 4.7). It is noteworthy that the correlation between crown base and breast height basal area (4.21) was very tight for both species, although the relationship was not exactly one-to-one (Fig. 4.8).

Similar to branches, the stem biomass could also be fairly accurately estimated with (4.26) (Fig. 4.9). Three parameters were first estimated from the data, namely the form factors of the stem inside and below the crown, $\phi_{\rm st}$ and $\phi_{\rm bct}$, and stem-wood density, $\rho_{\rm s}$ (over bark) (Table 4.1). This was possible because the data included detailed information about stem form and biomass. A strong linear correlation was found between the estimates and measured values. The results were comparable with empirical model predictions based on 10 (5 for stem wood and 5 for bark) parameter values (Repola 2009).

The results suggest that the pipe model structure is indeed quite general and could be used at least as a good indication of the order of magnitude of biomasses. The key explanation for the generality of the model is the fact that it separates the

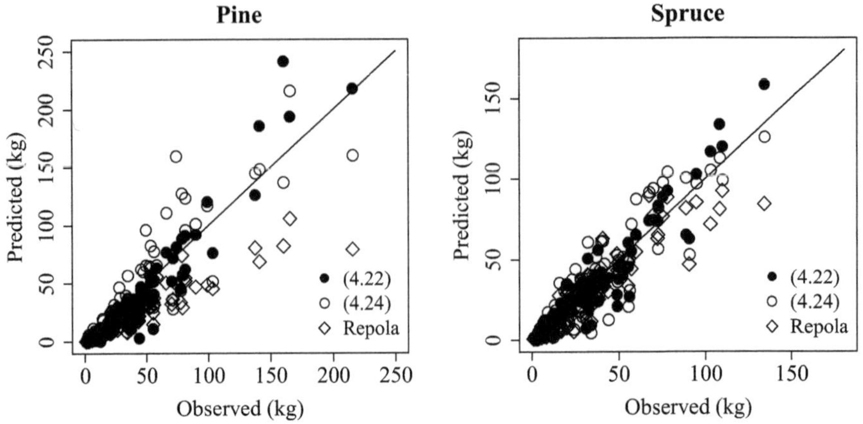

Fig. 4.7 Branch mass in trees from forest inventory plots in southern Finland (Lehtonen 2005). The predictions were made using (4.22) and (4.24) with parameters as given in Table 4.1. The parameters were fitted to the data. With these parameter values, the predictions were unbiased and their accuracy measured as r^2 of the relationship between predicted and observed branch mass was 0.89 and 0.91 (4.22), and 0.81 and 0.85 (4.24) for pine and spruce, respectively. The empirical equations came from Repola (2009) and had r^2 values 0.87 and 0.88 for pine and spruce, respectively

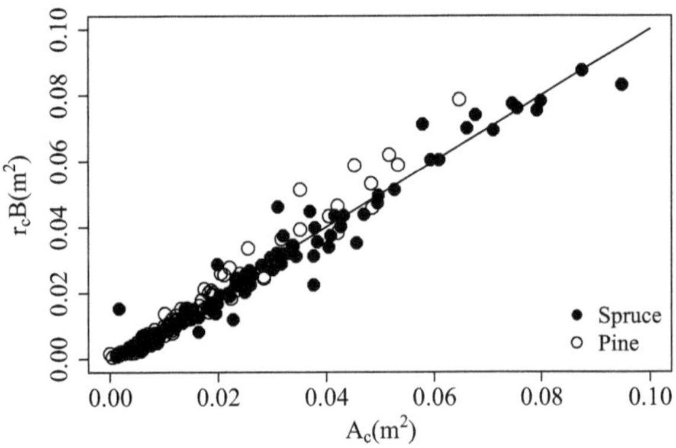

Fig. 4.8 A test of the proposition that stem cross-sectional area at crown base, A_c, is approximated by (4.21), i.e., basal area, B, multiplied with crown ratio, r_c. (Data are from national inventory plots in southern Finland (Lehtonen 2005))

crown and bole below crown. This leads to fairly stable form factors that are the key parameters of the model. This way it seems to avoid the risk of very large bias which can be obtained when empirically fitted equations are extended beyond model calibration data (Neumann et al. 2016). On the other hand, quite large scatter still seems to remain in the leaf and branch equations. Combining the general form of the pipe model with environmental factors, as suggested by the study of Lehtonen et al. (2020), could make the model more readily applicable over larger geographical regions and to a variety of stand structures.

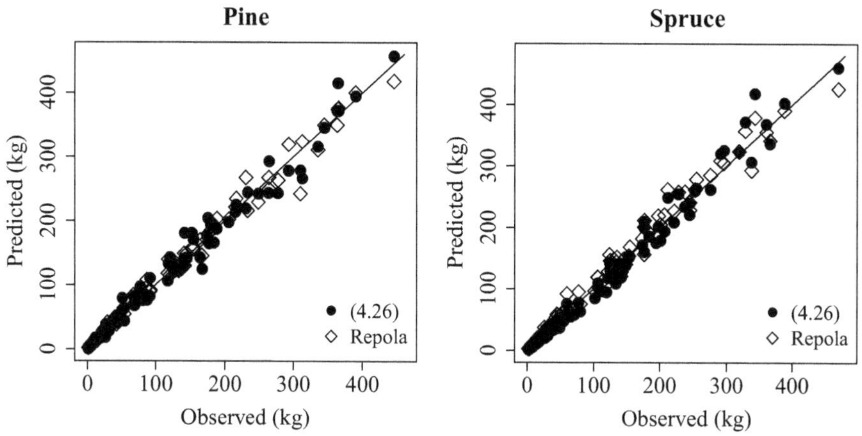

Fig. 4.9 Stem mass in trees from forest inventory plots in southern Finland (Lehtonen 2005) predicted by (4.26) with fitted parameter values listed in Table 4.1. The predictions were unbiased and their accuracy measured as r^2 of the relationship between predicted and observed stem mass was 0.99 for both pine and spruce. The empirical equations came from Repola (2009) and also had r^2 values 0.99 for both pine and spruce

Table 4.1 Parameter values for estimating biomass with equations: (4.22)–(4.26). Branch wood density ρ_b was taken from previous literature. All other parameter values were estimated from tree measurements at national inventory plots in southern Finland (Hu et al. 2020). All stem and branch measurements were over bark

Parameter	Pine	Spruce
ρ_s	370	335
ρ_b	400	590
η_s	450	600
ϕ_{bt}	1.40	1.28
$\tilde{\phi}_{bt}$	1.00	0.53
ψ_{st}	0.40	0.42
ϕ_{bct}	0.53	0.55
x	0.8	0.5

4.4 Height-to-Diameter Ratios: Greenhill Scaling

4.4.1 *Vertical Biomass Density*

The pipe model provides a sound basis for relating total leaf mass or area to basal area in trees. However, it does not provide any information about or constraints for the height-to-diameter relationship of stems or crowns. As far as the pipe model is concerned, the pipes connecting fine roots to leaves could be of any length, and the pipe model could still be satisfied. In order to complete the description of tree structure, we must understand what regulates the vertical density of biomass in relation to horizontal measures such as basal area. This is needed for predicting length-to-diameter ratios of stems, branches and roots, foliage densities in crowns, or fine-root densities in the soil.

Two main theoretical approaches have been applied to describing or predicting the vertical densities of above-ground biomass. These relate to (1) mechanical stability and (2) maximum photosynthetic gain. In the mechanical stability approach it is assumed that trees will grow as tall as possible while at the same time avoiding the risk of breakage under their own weight or toppling over by windthrow. The stability analysis results in an estimate of the minimum diameter that can be sustained for a given height. If it is assumed that trees favour height growth to survive in competition for light with their neighbours, then this minimum diameter is the one that realises the most efficient and least wasteful strategy to achieve this goal.

The approach of maximum photosynthetic gain analyses crown shapes and foliage distributions that would provide the best display of foliage to light from the point of view of photosynthetic production. At the same time, growing the branches to support the foliage will consume growth resources and establish a carbon cost for the tree. Foliage density can therefore be seen as a tradeoff between maximising photosynthesis and minimising the construction cost. As we will see in Chap. 7, solving this tradeoff in the pipe model and carbon balance framework can provide meaningful constraints both for foliage density and height-to-diameter ratios.

In this Section, however, we focus on the mechanical stability approach, introducing perhaps the first ever theoretical investigation on mechanical constraints to height, by A.G. Greenhill. The results provide a first approximation of height-to-diameter ratios in trees, pivotal in the subsequent analysis of fractal trees in Sect. 4.5.

4.4.2 Greenhill Scaling

In 1881, A.G. Greenhill published a paper entitled, "Determination of the greatest height consistent with stability that a vertical pole or mast can be made, and of the greatest height to which a tree of given proportions can grow." The paper is notable because, to our knowledge, it reports the earliest mechanistic model of tree structure and, further, Greenhill's result concerning the limiting allometry of tree height versus diameter (or radius) is a cornerstone of some modern models of plant structure, including the quarter-power scaling model of plant vascular systems (see West et al. 1999).

Given a basal diameter, (D, m), homogeneous bulk density (ρ_W, kg m^{-3}) and Young's elastic modulus (E, kg m^{-2}), the Greenhill model predicts the critical height (H_{max}, m) at which a mast (or other entity) with the shape of column, cone, or paraboloid would become unstable and suffer Euler buckling. Short of this critical height, a slight deflection over the length of the entity is righted owing to the entity's resilience to elastic deformation. At the critical height the resilience is overcome by the weight of the object. For each profile, the critical height scales with the 2/3 power of the basal diameter. i.e.,

$$H_{\max} = c\left(\frac{E}{\rho_{\mathrm{w}}}\right)^{1/3} D^{2/3} \tag{4.27}$$

where c is a dimensionless constant whose value depends on the profile of the entity and other considerations. Inverting the model, the minimum basal diameter scales with the 3/2 power of height.

In regard to trees, Greenhill concluded that the power relationship between height and diameter accounts for "the slender proportions of young trees, compared with the stunted appearance of very large trees." Greenhill derived his results without consideration of wind or other environmental forces. In a more realistic setting, the constant, c, depends on whether a lateral force is applied at the centre of mass or over the full extent of the entity (Niklas 1994). King and Loucks (1978) indicated that, for a tree, c depends on the ratio of crown mass to trunk mass. Naturally, wind and other environmental forces have selected for trees with safety margins; trees are shorter than the critical height given the diameter, or thicker than minimum diameter given the height (e.g., McMahon 1973), though how much shorter or how much thicker remains an open question (Horn 2000).

Greenhill's result is now sometimes called *Greenhill scaling* and it has been extended to describe structural characteristics of branches. For this purpose, we define a branch as the part of a tree that falls when you saw through a stem just beyond any branching node. Let ℓ and d, respectively, denote the length and basal diameter of the branch. If Greenhill scaling holds generally, that is, if $\ell \propto d^{2/3}$ regardless of where you choose to use your saw, then the branching system is said to be *self similar*.

The Greenhill model is based on assumptions of homogeneity and *elastic similarity*. King and Loucks (1978) indicated that, by definition, elastically similar branches bend through similar arcs under their own weight, or as Niklas (1994, p. 165) put it, deflection at the free end of a branch remains constant relative to the branch length. Under the homogeneity assumption, branch mass (W) is proportional to volume (V) and $V \propto d^2\ell$. Under the added assumption of elastic similarity, $\ell \propto d^{2/3}$, so it follows that $d \propto W^{3/8}$ and $\ell \propto W^{1/4}$.

McMahon and Kronaurer (1976) examined crowns of five tree species for agreement with either elastic similarity or an alternative model, *stress similarity* ($\ell \propto d^{1/2}$, $d \propto W^{2/5}$, and $\ell \propto W^{1/5}$). They concluded that branches tend to be self similar with regard to structural parameters, so that any patch of the structure (i.e., any branch sawn off any node) is a model of the whole tree. Moreover, their estimates of the structural parameters were in better agreement with elastic similarity than with stress similarity. Niklas (1994), however, noted that statistical evaluations often fail to distinguish significant differences between elastic similarity and stress similarity models.

Empirical noise notwithstanding, Greenhill scaling, elastic similarity, self similarity, and related allometric relationships have gained familiarity throughout the ecological and forestry modelling communities through their incorporation into the quarter-power scaling model of West et al. (1999).

4.5 Fractal Trees

Another important structural characteristic of trees, not covered by the allometric models, the pipe model, or the Greenhill model, is the spatial structure of the crown, including branching patterns and leaf biomass distributions. These are important for both biomass allocation of trees, and their ability to absorb light. Some interesting insights into the regularities of crown architecture have been achieved by considering the crowns as fractals or fractal-like structures (Mandelbrot 1983; West et al. 1997b; Zeide 1998).

Fractal geometry (Mandelbrot 1983) is a generalisation of the conventional Euclidean geometry to include structures with non-integer dimensions. If the fractal dimension of an object is greater than 1 but less than 2, the object is something between a line and a surface. Similarly, fractal dimension between 2 and 3 characterises something that is a hybrid between surface and volume. The applicability of fractal geometry to crown architecture is based on the observation that a tree crown is neither a three-dimensional solid nor a two-dimensional photosynthetic surface (Zeide 1998).

In order to understand fractal structures, let us first consider the concept of dimension in Euclidean geometry. This can be illustrated by a cube divided into smaller cubes of equal size. If the side of the smaller cube is a fraction r of the original cube, and we have N smaller cubes, then

$$N = \frac{1}{r^3} \tag{4.28}$$

regardless of the size of the original cube and the number of cubes cut out. Similarly, if we divide a square into N identical smaller squares with their side being a fraction r of the original square, we have

$$N = \frac{1}{r^2} \tag{4.29}$$

Finally, if it was just a rod divided into N smaller rods of relative length r, then

$$N = \frac{1}{r} \tag{4.30}$$

We may therefore conclude that the dimension, D, of a Euclidean object is

$$D = -\frac{\ln N}{\ln r} \tag{4.31}$$

This constitutes a definition of dimension which can be generalised to fractals (Mandelbrot 1983). In the case of fractals, we find the dimension by constructing a *cover* of an object. This is constructed by dividing a Euclidean object of appropriate dimension (line, square, cube) and covering the fractal object, into a number of equal

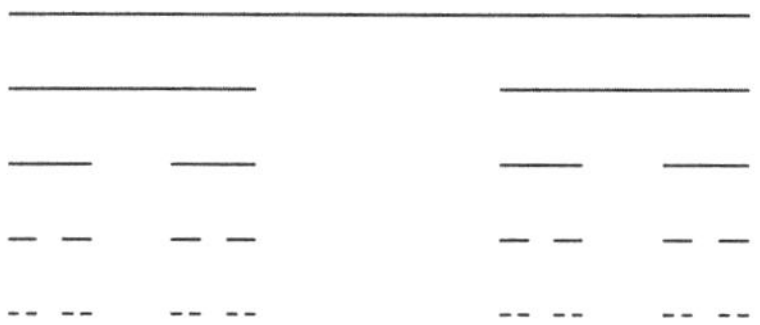

Fig. 4.10 First four iterations in the construction of the Cantor set from a line segment (*top*). The segment is divided into three equal parts, and the middle segment is removed. This procedure is repeated infinitely to construct the Cantor set

Fig. 4.11 First four steps in the construction of the Koch curve from a line segment (*top*). The segment is divided into three equal parts, and an equilateral triangle is placed in the middle. This procedure is iterated to construct the Koch curve

size smaller objects, such that their linear dimension is a fraction r of the original. We will need either a larger or a smaller number, N, of these objects, than for the corresponding objects in Euclidean space, to actually cover the fractal object. The dimension, D, of the fractal object is determined by N and r according to (4.31).

Fractal dimension can be best understood by looking at the construction of a fractal from a Euclidean object. We may either take an object of some dimension and make a construction that has a higher dimension than the original object, or we may devise a method that reduces the original dimension. An example of the former is the Koch curve with dimension 1.2619, and one of the latter is the Cantor set with dimension 0.631 (Bar-Yam 2012), both constructed from a one-dimensional line (Figs. 4.10 and 4.11). Below, we consider the construction of Menger's sponge which was given as an example of fractals by Zeide (1998) to illustrate the fractal nature of tree crowns.

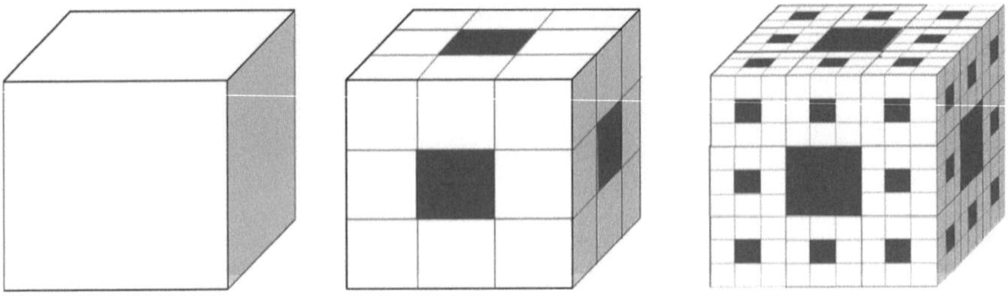

Fig. 4.12 First three steps in the construction of Menger's sponge

4.5.1 Menger's Sponge

Take a cube and divide it into 27 equal smaller cubes, the side of which thus becomes 1/3 of the original side. If you look at one face of the original cube, you can see the outer face of nine of the smaller cubes. Now remove the cube in the middle of this face, and similarly the cubes in the middle of the other five faces as well. Finally, remove the central cube. This means that there are 20 smaller cubes left, while 7 have been removed. Now repeat this procedure for all the 20 cubes that are left: divide them into 27 smaller cubes, remove 7, and so on. This means that while the structure that is being created still has its original outer dimensions, its weight has been considerably reduced, and it is no longer a three-dimensional object, but a rather light-weight sponge-like fractal (Fig. 4.12).

The dimension of this fractal can be calculated as follows. In order to contain the whole object within a cover of cubes, we need 27 smaller cubes of radius 1/3, but only 20 of these are filled. The fractal dimension of Menger's sponge is, therefore

$$D = -\frac{\ln 20}{\ln 1/3} = 2.7268 \tag{4.32}$$

Zeide (1998) devised a method for measuring the fractal dimension of crowns which is a further development of the box-counting method, utilising the idea of a cover. He found that the fractal dimension of the species studied was approximately 2.5. In other words, the suggestion of the crowns being similar to Menger's sponge was corroborated.

4.5.2 Branching Patterns and Fractal Foliage

If the foliage is distributed in the crown in a certain way, then how should the branching structure be organised to support this foliage? This problem has been analysed with fractal-like tree structures that show a regular branching pattern (Mandelbrot 1983; West et al. 1997b).

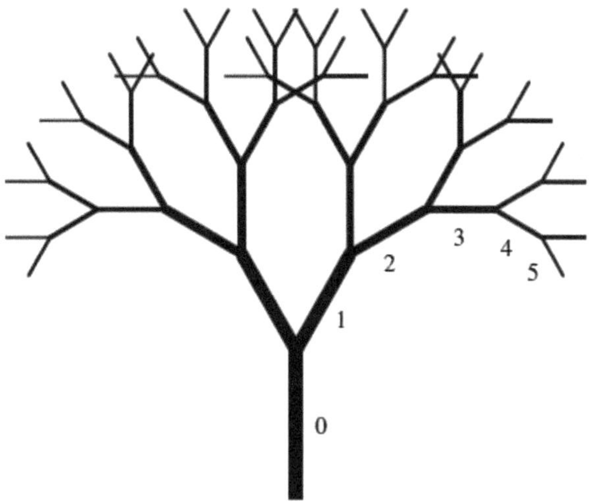

Fig. 4.13 Schematic presentation of the branching pattern

A fractal-like tree can be constructed from an initial segment or 'stem' by generating n new segments that are reduced in length by γ, and arranged in appropriate directions from the parent segment (Fig. 4.13). At each level, k, the branch length of the daughter branch, l_{k+1}, is obtained from that of the parent branch as $l_{k+1} = \gamma l_k$. A tree is constructed by repeating this infinitely many times.

A real tree, however, often does not resemble a proper fractal, because the first segment—the central stem beneath the crown base—is not similar to the rest of the tree (Bar-Yam 2012). The tips of the branches, by contrast, can be considered a fractal construction (Mandelbrot 1983). Each branch tip is identified with a unit of leaf mass, with the total assemblage of tips representing the total leaf mass, W_f. The dimension, z, of such a network can be obtained in analogy with (4.31) as follows

$$z = -\frac{\ln n}{\ln \gamma} \tag{4.33}$$

so that

$$\gamma = n^{-1/z} \tag{4.34}$$

(Mandelbrot 1983). This assumes, however, that the branching network is arranged appropriately in space. If $z = 3$, the crown is said to be *volume filling*.

In addition to the reduction in branch length in each level, the branching structure can also show a regular pattern of branch radius, r. In an *area-preserving* branching network, the sum of the basal area of the daughter branches equals the basal area of the parent branch. In such a case, we require that

$$n r_{k+1}^2 = r_k^2 \tag{4.35}$$

In a *tapering* network, the sum of the basal areas of the daughter branches is smaller than the basal area of the parent branch. This can be quantified by introducing a parameter, $a \geq 1$, such that

$$n^a r_{k+1}^2 = r_k^2 \tag{4.36}$$

This can be expressed in a form analogous to (4.34) as follows:

$$\beta = n^{-a/2} \tag{4.37}$$

where β has been defined as the ratio of the radius of the daughter branch to that of the parent branch: $\beta = r_{k+1}/r_k$.

A further regularity in the branching structure can be defined as a scaling relationship between the branch segment radius and length:

$$l^k = r_k^\alpha \tag{4.38}$$

Above, we have used the notation introduced by West et al. (1997b) in their famous analysis of biological networks as fractal-like objects. They defined a fractal-like rather than fractal network, assuming a finite number, N, of branching levels with a final element of fixed size. This construction was further used to derive allometric scaling rules for the whole network, assuming certain fixed values for the parameters. The resultant *quarter-power scaling rules* have since been widely cited in the scientific literature. The theory was first derived for networks of blood circulation, then also applied to trees and other biological networks. In tree crowns, the final smallest element of the branching network was identified as the *petiole*.

Below, we consider tree crowns as fractals and derive scaling rules consistent with the above notation, following the analysis by Mäkelä and Valentine (2006a).

4.5.3 Allometry and Fractals

Assuming that the tree crown obeys the branching network structure described above, we may express the total length of the network from level k onwards, denoted by L_k, as

$$L_k = \sum_{i=k}^{\infty} l_i \tag{4.39}$$

Because $l_{i+1} = \gamma l_i$, the sum of the infinite series is $L_k = l_k/(1 - \gamma)$. Because l_k scales as r_k^α this also means that $L_k \propto r_k^\alpha$. If $\alpha = 2/3$, the fractal trees follow Greenhill scaling, which accords with the assumption of elastic similarity (Greenhill 1881).

Denote the mass (or area) of the leaves attached to the branch tips generated by a segment at level k by W_{f_k}. Because the fractal dimension of the crown is z, it follows by definition that $W_{\mathrm{f}_k} \propto L_k^z$, and, therefore, the total leaf mass in the branching network is

$$W_\mathrm{f} \propto n_k L_k^z = L_0^z \tag{4.40}$$

Because $L_k = l_k/(1-\gamma) = \gamma^k l_0/(1-\gamma) = \gamma^k L_0$, we require that $n_k \gamma^{kz} L_0^z = L_0^z$, which obtains if and only if $\gamma = n^{-1/z}$. This is the assumption we made about the fractal dimension of foliage, in connection with (4.34), showing that the structural assumptions are internally consistent.

We may now express the woody mass of the network in terms of the sum of volumes of all the elements of the network, starting from the stem:

$$W_\mathrm{c} = \rho_\mathrm{w} \pi \sum_{k=0}^{\infty} n^k l_k r_k^2 \tag{4.41}$$

where W_c is the woody biomass and ρ_w is wood density in the fractal crown network. Inserting the rules $l_{k+1} = \gamma l_k$ and $r_{k+1} = \beta r_k$ provides

$$W_\mathrm{c} = \rho_\mathrm{w} \pi l_0 r_0^2 \sum_{k=0}^{\infty} n^k \gamma^k \beta^{2k} = \rho_\mathrm{w} \pi l_0 r_0^2 \sum_{k=0}^{\infty} \delta^k \tag{4.42}$$

where $\delta = n\gamma\beta^2$. Further, inserting (4.34) and (4.37) yields $\delta = n^{-[a-1+(1/z)]k}$. Since δ is always less than unity for $a \geq 1$, the series in (4.42) converges and therefore

$$W_\mathrm{c} = \rho_\mathrm{w} \pi l_0 r_0^2 \frac{1}{1-\delta} \tag{4.43}$$

We can now utilise the assumptions made about the shape of the branch segments and the fractal dimension of the foliage. We already argued above that $l_0 \propto L_0$, which is the total length of the network, and that $r_0 \propto L_0^{1/\alpha}$. Consequently,

$$W_\mathrm{c} \propto L_0^{1+(2/\alpha)} \tag{4.44}$$

On the other hand, $l_k = l_0 n^{-k/z} \propto r_k^\alpha = r_0^\alpha n^{-ka\alpha/2}$, so we require $\alpha = 2/za$. From (4.40), we obtain $L_0 \propto W_\mathrm{f}^{1/z}$, which provides the scaling relationship between W_c and W_f, i.e.,

$$W_\mathrm{f} \propto W_\mathrm{c}^{z/(za+1)} \tag{4.45}$$

This allometric relationship follows from our assumptions about the regular internal structure of the branching network in the crown. In other words, fractal crown structure implies a certain allometry, with the allometric exponent related to the fractal dimension of foliage and the tapering of the network. If the network is volume filling ($z = 3$) and area preserving ($a = 1$), then the allometric exponent takes on the value $3/4$, a situation referred to as *quarter-power scaling*. This was originally analysed for fractal trees by Mandelbrot (1983). West et al. (1997a,b) developed the idea for the fractal-like objects where the above relationship only holds approximately, as the number of branching levels is finite. However, a reasonably small number of branching levels is sufficient to closely approach the limiting value. A subsequent series of papers have taken the theory further, developing its implications, e.g., on water transport in trees, metabolic scaling, and community dynamics. Parts of the theory are controversial (Kozlowski and Konarzewski 2004; Mäkelä and Valentine 2006b).

4.5.4 Allometry in Pipe Model Trees with Fractal Foliage

The derivations in the previous section show that fractal or fractal-like crown structure leads to allometric relationships between foliage mass, crown woody mass and crown dimensions. It therefore provides a possible explanation for observed allometric relationships, such as (4.45), in trees (West et al. 1997b; Mäkelä and Valentine 2006a; Duursma et al. 2010). Furthermore, the distribution of foliage has been found, by empirical analysis, to follow a fractal-like structure, albeit with fractal dimension less than 3 (Zeide 1998), and the area-preserving character of the branching network (i.e., the pipe model) has been found to be reasonably well satisfied throughout tree crowns (Patrick Bentley et al. 2013). However, the actual ramification patterns have been found to be much more variable and irregular than in the fractal model (Borchert and Slade 1981; Patrick Bentley et al. 2013).

In this section, we show that the allometric relationships predicted by the fractal model are more general, and can be derived without specification of the exact branching structure, provided that certain structural regularities are met at a higher level of hierarchy. For this, we consider trees with pipe model structure and foliage distribution following the fractal model, i.e., we assume that foliage mass is proportional to a power, z, of the linear crown dimension, l. On the basis of these assumptions, we can specify equations for foliage mass, W_f, and woody mass in the crown, W_c:

$$\begin{aligned} W_f &\propto A \\ W_f &\propto l^z \\ W_c &= \rho_w A \hat{H} \end{aligned} \tag{4.46}$$

where ρ_w is wood density in the crown and $\hat{H}$ denotes the mean pipe length in the whole crown. In analogy with (4.16) and (4.19), we define a form factor, ϕ, for the whole crown as follows:

$$W_c = \rho \phi A l \tag{4.47}$$

which implies that $\hat{H} = \phi l$.

If the crown form factor ϕ can be assumed constant, then mean pipe length is proportional to crown size ($\hat{H} \propto l$), and both A and l in (4.47) can be replaced by functions of W_f, i.e., (4.46), to yield

$$W_f \propto W_c^{z/(z+1)} \tag{4.48}$$

This is a special case of (4.45) with no tapering, i.e., $a = 1$, which indicates that essentially the same allometric structure implied by the fractal model can be derived with no assumptions about the detailed branching structure of the crown, so long as the mean pipe length of the crown can be assumed proportional to the linear crown dimension. Implicitly, this assumption is made in the fractal model, as total leaf mass is assumed to scale with the total length of the branching network (see (4.40)). In real trees, crown shapes and branching structures vary, so mean pipe length relative to crown dimensions depends on things like branching angle and straightness or bending of the branches. Nevertheless, consistent species-specific form factors have been observed for both branches and whole crowns (see Fig. 4.5). This suggests that different branching patterns can be effectively simplified to a higher level of aggregation, using the concepts of mean pipe length and form factor.

4.6 Models of Crown Geometry

In the preceeding sections, we have reviewed different ways of deriving meaningful plant allometries that can serve as a basis for carbon allocation. However, these models do not explicitly describe the spatial geometry of trees or crowns. How are foliage and branches distributed in space? What angles do branches take to fill space? The geometry of foliage and branch distributions is focal for understanding how trees and canopies capture light and hence photosynthesize. Static models of crown and canopy architecture have been developed for this purpose. The dynamics of three-dimensional branching structures are the focus of the so-called structural-functional models that classify, compute and visualize the development of different tree architectures.

4.6.1 Models of Foliage Distribution for Light Interception

The simplest model of foliage distribution is the horizontally homogenous canopy assumption. This was originally introduced to allow for simple calculations of light interception (Monsi and Saeki 1953), resulting in what is known as the Lambert-Beer law. Here, leaf area density at distance z from the top of the canopy, $l(z)$ (m^2 leaf area (m^3 canopy)$^{-1}$), is assumed to be constant on the horizontal plane defined by z. Further, it is assumed that the leaves are randomly oriented, such that at any plane, the ratio of projected leaf to total leaf area is constant and independent of the angle of projection. This constant is called the extinction coefficient and denoted by k.

Assume that each time a light ray hits a leaf the light energy will be absorbed by the leaf (Fig. 4.14). The probability that a light ray hits a leaf at any distance from the top is given by kl, which is the density of projected leaf area in the canopy. Hence, if light at intensity I_0 enters the upper layer of the canopy, it will be reduced by the leaf area as follows:

$$\frac{\mathrm{d}I}{\mathrm{d}z} = -klI, \qquad I(0) = I_0 \tag{4.49}$$

This is a differential equation for light intensity, which has the following solution:

$$I(z) = I_0 \mathrm{e}^{-klz} \tag{4.50}$$

where $I(z)$ is the light intensity reaching depth z in the canopy. The difference of the incident light and the remaining light is the light energy absorbed by the canopy. We note that lz is leaf area per unit ground area in the whole canopy above depth z. Denoting this by $L(z)$ we have

$$I_{\text{abs}}(z) = I_0 \left(1 - \mathrm{e}^{-kL(z)}\right) \tag{4.51}$$

This is the Lambert-Beer law of light interception in homogeneous canopies.

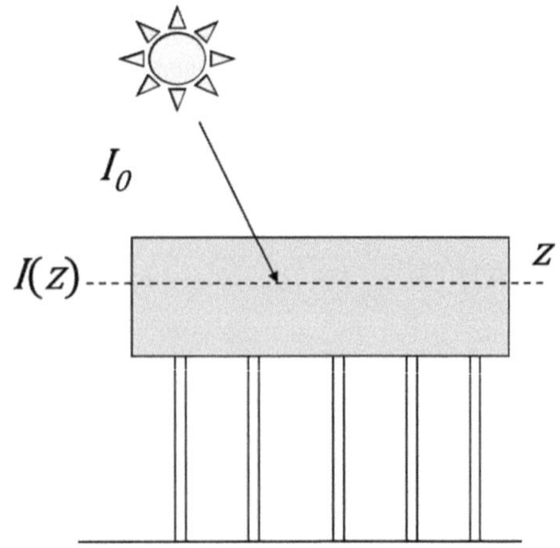

Fig. 4.14 Light interception in a homogeneous canopy

Real stands are not fully homogeneous; instead, the foliage is clumped into crowns and surrounded by gaps. Perhaps the simplest description for this is a canopy consisting of identical, randomly distributed crowns with a specified shape, such as ellipsoid or cone. This is analogous with the randomly distributed leaves in the homogeneous canopy, so it is intuitively easy to understand that the mutual shading between such crowns obeys the Lambert-Beer law such as the randomly distributed leaves in the homogeneous case (Oker-Blom and Kellomäki 1982). However, in this model the crowns themselves consist of leaves that are again randomly distributed within the crown volume, and hence have similar light interception properties as the homogeneous canopy.

Oker-Blom et al. (1989) presented a simulation model that incorporated these ideas for a canopy at a sub-daily time step. The model can be used for integrating the instantaneous light interception and photosynthesis over specified time periods with seasonal courses of the solar zenith angle affecting the direction of incident direct and diffuse light. Duursma and Mäkelä (2007) derived a simplified summary model for the long-term light interception in such canopies. It was based on the ideas that (1) for between-tree shading, the Lambert-Beer model can be applied, and that (2) for self-shading, the average light interception by a crown is related to its surface area. The latter assumption derives from the fact that the instantaneous light interception by a crown is related to its projection area in the direction of the sun (Nilson 1999). When integrating this over all angles, the mean projection is one fourth of the crown surface area (Cauchy's theorem, see Lang (1991)). For this reason, crown surface area should be a good variable for estimating time-averaged light interception.

From the above assumptions Duursma and Mäkelä (2007) were able to show that the light intercepted by the canopy could be expressed as in (4.51), provided that the extinction coefficient was modified to account for the structure of the individual crowns. They termed this the *effective extinction coefficient*, k_{eff}:

$$k_{\mathrm{eff}} = \frac{\phi S_A}{L_A}\left[1 - \mathrm{e}^{-kL_A/(\phi S_A)}\right] \tag{4.52}$$

where L_A is leaf area of the crown, S_A is crown surface area, ϕ is a parameter, and k is the homogeneous canopy extinction coefficient.

Duursma and Mäkelä (2007) showed that the above summary model was a good approximation of the more detailed model presented by Oker-Blom et al. (1989). The important implication of this result is that the surface area density of foliage in crowns is a key determinant of the efficiency of foliage in light interception. If the same amount of foliage is distributed in a large number of crowns or a small number of crowns of equal size, the light interception will be greater in the former. If the number and leaf area of crowns is kept constant but the dimensions of the crown are varied, then the bigger crowns will intercept more light. In this way the model also accounts for stand density effects on light interception.

Duursma et al. (2011) took this approach a step further in a study that was concerned with the light interception of crowns described as realistic three-dimensional structures. Crown structure was measured, digitised and input to a three-dimensional structural model (see next section), and its light interception properties were analysed with a ray-tracing method. The study concluded that, again, the Lambert-Beer model, (4.51), was adequate, provided that the extinction coefficient was modified to represent the actual crown structure. In addition to leaf area per crown surface area, in this case, an additional modification was required that accounted for the degree of clustering of the leaves in the crown. This was quantified as leaf dispersion, defined as a measure of the randomness of the leaf distribution. Leaf dispersion was found to vary between species with leaf size. Within species, leaf dispersion decreased significantly as the crowns grew in size.

4.6.2 Crown Architecture Models

Crown architecture models aim at describing the dynamic development of the crown through reiteration of basic structures, so-called growth units. A growth unit consists of a branch segment that grows within a certain period of time and carries buds that can potentially lead to the generation of new branch segments either simultaneously with the current segment or later (Prusinkiewicz and Lindenmayer 1990). The spatial orientation of all the branch segments and their subsequent growth and shedding leads to the development of the branching network, the state of which can be measured at any time.

The crown architecture models and the fractal model apply similar mathematical formulations to describe the repetitive structure of branching networks, but they also have some important differences. A fractal model of crown may resemble a real crown at a given point in time, but the fractal model cannot be made dynamic in any realistic way: Increasing the size of the fractal crown model would involve stretching the internodes. A dynamic model based on architectural rules accomplishes a fractal-like appearance by simulating the production of new shoots, the cambial expansion of subordinate internodes, and the death and self-pruning of networks of internodes that represent dead branches. The dynamic process of self-pruning redefines the internodal lengths within the crown. We use fractal models because they are easy to define and analyse; we use architectural models to gain insight into the processes of tree development.

Hallé et al. (1978) laid the ground for crown architecture studies in their comprehensive descriptive study of crown development in trees based on observations on a large number of tropical species. They proposed that a classification into 23 different crown architecture patterns could cover the great diversity of the branching systems of trees. The defined categories describe the basic growth patterns of crowns, including aspects of growth dynamics such as timing of bud set and new branch growth. However, the authors observed that while the basic pattern was often reiterated for some time from the seedling stage onwards, developmental and random processes set in later which caused irregularity and deviation of the number and size of branches.

Honda (1971) presented the first quantitative simulation model of branch growth. He observed that all three-dimensional branching structures could be described by the number of branches at the point of ramification, length of the branch segment, and branch angle. If rules are set for these quantities at each point of ramification, different three-dimensional crown structures of any size can be generated through simulation. In early simulation models of crown structure, the assumption was made that the branching parameters could be set to constant values characteristic of each type of branching. Borchert and Slade (1981) questioned this assumption and demonstrated that crowns growing with constant branching parameters (and without branch shedding) would soon become too large to be contained in the physical space available. They suggested on the basis of this theoretical calculation and empirical evidence on cottonwood that the ramification ratios must reduce radically as crown size increases.

In accordance with the conclusion of Borchert and Slade (1981), Honda et al. (1982) demonstrated that realistic crown structures could be generated by simulation if the ramification rules were set to reduce the number of daughter shoots at higher branching orders. This kind of behaviour may have evolved so as to allow trees to grow taller in height and thus survive in competition with their neighbours (Tomlinson 1983). Mechanistically, it is probably related to the hydraulic architecture of trees whereby lateral branches show lower water conductivity than the main axis, allowing more water to flow towards the top of the tree to sustain height growth even under unfavorable conditions (Zimmermann 1978).

Crown architecture models have developed greatly since the early model presented by Honda, largely owing to the pioneering work of researchers in France (de Reffye et al. 1995), Canada (Prusinkiewicz and Lindenmayer 1990) and Finland (Sievänen et al. 2000). These three-dimensional structural-functional models assume similar branching rules as Honda, however, they also allow stochastic variability and environmental interaction of the branching parameters (Fig. 4.15). For example, in the stochastic approach by de Reffye et al. (1995) a bud may either produce a daughter branch, a flower, remain dormant, or die at repeated time

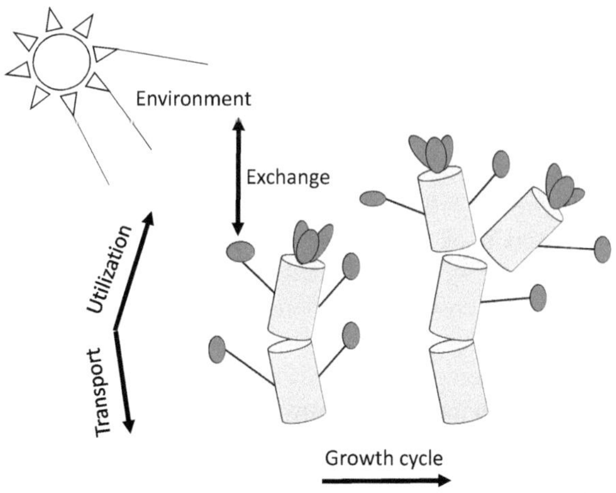

Fig. 4.15 Growth units in the LIGNUM model (Redrawn from Sievänen et al. (2000)). The above-ground structure of a plant consists of three elementary units, segment, leaf and bud. The functional processes included are exchange of elements with the environment and transport and utilisation of elements within the structure. The processes result in growth, whereby new elementary units are constructed within a growth cycle

intervals. Natural-like crowns can now be readily produced by these architectural models, and they offer an efficient work bench for analysing the relationships between structure, physiological function and environmental drivers (Nikinmaa et al. 2003, 2014).

4.7 Summary

In this chapter, we have reviewed empirical results and mathematical formulations concerning regularities in tree structure. Although we started off with a discussion about the evolutionary significance of tree structure and its interactions with tree functioning, we have only briefly touched upon the interactions of structure and function and their possible consequences for the natural selection of structural traits in this chapter. We will return to this topic in Chap. 7 dealing with evolutionary optimisation. For the time being, the important notion is that there are regularities in structure that can be utilised in modelling carbon allocation and resource capture.

Empirical allometries have been widely used in biomass estimation and modelling, however, the allometries are often dependent on the type of data at hand. Caution should be excercised particularly against applying allometries derived from spatial data to predictions of temporal development. The pipe model has been found to be a fairly general description of tree structure, with the same parameters largely applicable to spatial and temporal data sets. However, it requires additional information about height and length variables as these are not inherent in the pipe model. That foliage mass and therefore the basal area of the stem and branches in the crown are allometrically related to crown dimensions, is helpful for understanding the height-to-diameter ratios within the crown. This relationship has been observed to be fairly invariable among trees of different age and position. However, it does not directly translate to a similar relationship for the main stem below the crown. Competition for light, adaptation to frequent fires, and mechanical stability, together are important factors that have probably controlled the evolution of crown rise and stem form below the crown.

The fractal model and the dynamic models of crown architecture both include assumptions about ramification rules and the length of branch units at consecutive branching levels. Combined with the idea of branches as bundles of unit pipes carrying units of foliage, these approaches also predict length-to-diameter ratios for branches, crowns and stems. While the search for branching rules that allow us to predict crown architecture realistically is an important field of ecology *per se*, in this book our focus is on more aggregate characterisation of tree structure. To this end, we demonstrated in Sect. 4.5.4 that the allometry between foliage and wood in the crown can be derived without detailed knowledge about the branching patterns, using the concept of mean pipe length. In Sect. 4.6.1 we showed that the light interception in crowns with variable foliage distribution patterns can be described on the basis of crown dimensions combined with a few simple parameters that determining the effective light extinction coefficient. These examples illustrate that, in agreement with Borchert and Slade (1981), the properties of the crown can be adequately described at a coarser level of hierarchy, i.e., without necessarily having to deduce them from information obtained via the reductionist analysis of the branching network.

4.8 Exercises

4.1 Consider a cube of wood with edge length 10 cm and density 400 kg m^{-3}. (a) Calculate the volume, mass and surface area of the cube. (b) Now consider another cube with edge length 30 cm. Determine the ratio of volume, mass and surface area of this cube relative to those of the first cube. (c) In general, if the ratio of edge lengths of two cubes is k, what are the ratios of their volume, mass and surface area?

4.2 Consider a tree crown of length 2 m. Its total leaf mass is 2 kg. This tree grows to double its crown length while retaining its shape. What is the new foliage mass if: (a) leaf mass is evenly distributed within the crown volume, (b) leaf mass is evenly distributed across the crown surface, or (c) leaf mass is related to crown length through an allometric scaling exponent $z = 2.4$?

4.3 Assuming case (c) of Exercise 4.2, how would: (a) total leaf mass per total crown volume, and (b) total leaf mass per total crown surface area develop in a growing crown?

4.4 When crowns are exposed to light, they capture light energy falling on their surfaces. Consider the distribution leaves in crowns from this perspective.

4.5 Derive equation (4.2).

4.6 Consider two pine trees both with basal diameter 10 cm but one with crown ratio $c_r = 0.8$ and the other with $c_r = 0.25$. Work out the total leaf mass of each tree, assuming that the pipe model ratio is 500 kg m^{-2} (kg leaf dry weight per m^2 stem cross-sectional area at crown base).

Table 4.2 Results from selected long-term thinning experiments in southern Finland. The species are Norway spruce (*Picea abies*) and Scots pine (*Pinus sylvestris*). OMT: *Oxalis myrtillus* type, MT: *Myrtillus* type, CT: *Calluna* type (Cajander 1949)

Species	Site type	$\mathbb{B}$	H	c_r	Age	$\mathbb{A}_c$ (m^2ha^{-1})	$\mathbb{V}$ (m^3ha^{-1})
Spruce	OMT	55	25	0.44	61	27	694
Spruce	OMT	33	20	0.61	43	22	320
Pine	MT	28	26	0.35	70	8	335
Pine	CT	29	13	0.33	82	11	190

4.7 Above, the first tree has height 5 m and the second tree has height 10 m. Work out the woody biomasses of the trees using parameters given in Table 4.1.

4.8 Some even-aged Norway spruce and Scots pine stands have been measured in southern Finland for stand basal area (B, m^2 ha^{-1}), mean height (H, m), mean crown ratio (c_r) and stand age (Table 4.2). Use equations in Sect. 4.3 and parameters in Table 4.1 to estimate the total leaf, branch, and stem biomass in the stand. Also estimate stemwood volume. Compare estimates of cross-sectional area at crown base (A_c) and total stand volume (V) with measurements (Table 4.2).

4.9 Koch curve and Cantor set are defined in Figs. 4.10 and 4.11. Determine the fractal dimension of these sets by constructing a cover.

Chapter 5
Combining the Carbon Balance and Structure into a Core Model

In this chapter we combine material introduced in Chaps. 2, 3, 4 to formulate a dynamic model of a tree's size and biomass. The basic ingredients of the formulation are the carbon-balance model described Chap. 3, the pipe model and crown allometry described in Chap. 4, and the crown-rise model that was introduced in Chap. 2. The combination of carbon balance and structure under spatial constraints provides allocation rules by which the carbon available for growth is allocated among the components of biomass: leaf, branch, stem, coarse root, and fine root. The combination also provides the rates of loss owing to senescence or deactivation, which are integral to the tree-level model.

5.1 Combining the Carbon Balance and Structure

Trees use carbon-based photosynthates to grow new organs and tissues and to maintain the functions of already existing organs and tissues. Both growth and maintenance are accompanied by respiratory loss of some of the carbon. As modeled by Eq. (3.7) in Chap. 3, the production rate of a tree, G (kg DW yr^{-1}), is proportional to the rate of photosynthesis, P (kg C yr^{-1}), minus the rate of maintenance respiration, R_M (kg C yr^{-1}), i.e.,

$$G = Y_G(P - R_M)$$

where Y_G (kg DW (kg C)$^{-1}$) is the efficiency by which labile carbon compounds are converted to biomass, accounting for the growth respiration.

Electronic Supplementary Material The online version of this chapter (https://doi.org/10.1007/978-3-030-35761-0_5) contains supplementary material, which is available to authorized users. The videos can be accessed by scanning the related images with the SN More Media App.

A. Mäkelä, H. T. Valentine, *Models of Tree and Stand Dynamics*,
https://doi.org/10.1007/978-3-030-35761-0_5

The production rate of a real tree is the sum of the production rates of all of the live components of biomass. In concert with prior chapters, we assume that a model tree comprises five components of live or active biomass: f, leaves; b, branches; s, the main stem; t, coarse roots; and r, fine roots. Hence,

$$G = \sum_i G_i \qquad i = \mathrm{f, b, s, t, r} \tag{5.1}$$

New biomass is added to the ith component at rate

$$\begin{aligned} G_i &= \lambda_i G \\ &= \lambda_i Y_\mathrm{G}(P - R_\mathrm{M}) \end{aligned} \tag{5.2}$$

where $\lambda_i = G_i/G$ is the proportion of the total production allocated to the ith component of biomass. Conservation of mass requires that $\sum_i \lambda_i = 1$.

Of course, the production of new biomass is countered somewhat by loss (see Sect. 1.8). The net growth rate of the ith live component of a model tree, $\mathrm{d}W_i/\mathrm{d}t$ (kg DW yr^{-1}), equals the rate of production, G_i, minus the rate of loss of live biomass owing to senescence or deactivation, S_i, i.e.,

$$\frac{\mathrm{d}W_i}{\mathrm{d}t} = \lambda_i Y_\mathrm{G}(P - R_\mathrm{M}) - S_i \tag{5.3}$$

The deactivation of the live wood in the main stem and large branches creates a sixth component of the model tree, W_q (kg), the inactive (or disused) xylem that remains embedded within the tree.

Both the allocation fractions, λ_i, and the loss rates, S_i, are expected to vary over the lifespan of a tree. A structural framework based on the pipe model (Sect. 4.3) and crown allometry (Sect. 4.5.4), as constrained by the crown-rise model (Sect. 2.4), provides a basis for relating the components of biomass to each other at a given point in time. The structural framework, however, is not sufficient to elicit how the allocation fractions and loss rates change over time with respect to tree size and mass; for that, we need to merge the carbon-balance model, (5.3), with the structural model.

5.1.1 Results from the Pipe Model

We recall from Chap. 4 that leaf biomass, W_f (kg) is proportional to active-pipe area at the crown base, A_c (m^2):

$$W_\mathrm{f} = \eta_\mathrm{s} A_\mathrm{c} \tag{5.4}$$

where $\eta_s = W_f/A_c$ (kg leaf m^{-2}) is the so-called pipe model ratio. We also defined α_r of (4.13) as the ratio of fine root biomass to leaf biomass so that

$$\begin{aligned} W_r &= \alpha_r W_f \\ &= \alpha_r \eta_s A_c \end{aligned} \tag{5.5}$$

We can also use the mean and specific pipe lengths to relate the active-pipe biomass, by woody component, to the leaf biomass, i.e.,

$$W_i = \frac{\hat{H}_i}{l_i} W_f, \qquad i = \text{b, s, t} \tag{5.6}$$

The specific pipe length, l_i, $i = $ b, s, t [(m kg^{-1} stem) × (kg leaf)] is the ratio of the biomass of supported leaves to the biomass per unit length of the supporting stems (see Sect. 4.3). The mean pipe length, $\hat{H}_i$ (m), is the mean length of an active pipe in the ith woody component. Each mean pipe length, $\hat{H}_i$, $i =$ b, s, t, is defined in terms of a constant form factor, ϕ_i, and either the tree height, H, the crown base height, H_c, or both (see (4.20)).

The total active-pipe biomass of the model tree, W_p (kg), is

$$\begin{aligned} W_p &= W_b + W_s + W_t \\ &= (\beta_1 H + \beta_2 H_c) W_f \end{aligned} \tag{5.7}$$

where β_1 and β_2 [kg wood m^{-1} (kg leaf)$^{-1}$] follow from (4.20) and (5.6), i.e.,

$$\beta_1 = \frac{\phi_b}{l_b} + \frac{\phi_s}{l_s} + \frac{\phi_t}{l_t} \quad \text{and} \quad \beta_2 = -\frac{\phi_b}{l_b} + \frac{1 - \phi_s}{l_s} \tag{5.8}$$

We note that the specific pipe length, l_i, equals the ratio η_i/ρ_i where ρ_i (kg DW m^{-3}) is the bulk density of ith component of active wood. We assume that the active wood has homogenous density within the branches, coarse roots, and the main stem above the crown base, so l_i, η_i, and ρ_i are constants. By contrast, as explained later in Sect. 5.3, we provide for both juvenile and mature wood in the main stem below the crown. Whereas the main stem above the crown base contains only homogeneous juvenile wood, the below-crown segment contains both juvenile wood and denser mature wood, with the relative proportion of the latter increasing from the crown base toward the stem base. However, to remain consistent with the fundamental equation of the pipe model (see (4.6)), we assume that specific pipe length, l_s, remains constant above and below the crown base. Consequently, l_s equals the ratio η_s/ρ_s both above and below the crown base. However, the wood density $\rho_s(h)$ varies with depth h below the crown base, so the pipe-model ratio, $\eta_s(h)$, varies proportionately, i.e., $\eta_s(h) = \rho_s(h) l_s$.

Despite the minor complication within the main stem, the assumptions underlying (5.4)–(5.7), enable us to calculate all the live components of biomass from the tip height, H, the crown-base height, H_c, and either the total leaf biomass, W_f, or the active-pipe area at the crown base, A_c.

5.2 Model of Tree Dynamics

The seminal pipe model of Shinozaki et al. (1964a,b) is a static model of tree structure. Dynamic versions of the pipe model combined with carbon balance were originally formulated by Valentine (1985) and Mäkelä (1986) and subsequently elaborated for the purpose of modelling tree and stand dynamics. The relationships among tree components that obtain from the static pipe model are preserved by the dynamic models.

Before getting into the derivation of a model of tree dynamics, we need to identify a mechanism that drives active pipes to deactivate. We assume that pipes deactivate coincident with the death of branches at the crown base. This results in the upward movement of the crown base, which we call crown rise. Major factors associated with crown rise are diminished light and spatial crowding, but spatial considerations, alone, often provide an accurate estimate of maximum crown length (e.g., Valentine et al. 2013). Hence, we shall adapt the crown-rise model that was introduced in Sect. 2.4 to predict the deactivation of active pipes and the consequent loss of branch biomass.

5.2.1 *Production and Loss*

The pipe model, augmented by a simple relationship between the cross-sectional area of active wood at the crown base, A_c, and crown length, $L_c \equiv H - H_c$, provides a basis to formulate (a) the rate of production, G_i, of the ith biomass component in terms of height growth rate, $\mathrm{d}H/\mathrm{d}t$, and (b) the rate of loss, S_i, in terms of the rate of crown rise, $\mathrm{d}H_c/\mathrm{d}t$. Together, the rates of production and loss provide the net growth rates of the different components of the model tree in terms of the rate of change in crown length.

The rate of change of active-pipe area at the crown base ($\mathrm{m}^2\ \mathrm{yr}^{-1}$) is, by definition,

$$\frac{\mathrm{d}A_c}{\mathrm{d}t} = G_{A_c} - S_{A_c} \tag{5.9}$$

where G_{A_c} is the production rate of new active-pipe area and S_{A_c} is the rate of loss of old active-pipe area to the deactivation, which we assume is in association with branch death at the crown base.

Mäkelä (1997) used the pipe model combined with results from the optimal crown model of Mäkelä and Sievänen (1992) to justify an assumption that active area at the crown base is proportional to the zth power of crown length. The fractal crown model (see Sect. 4.5.3) is also theoretically consistent with this assumption. Hence, we assume

$$\begin{aligned} A_c &= \beta_A L_c^z \\ &= \beta_A (H - H_c)^z \end{aligned} \tag{5.10}$$

where β_A ($m^2\ m^{-z}$) and z are parameters.

The time derivative of this allometric model,

$$\begin{aligned} \frac{dA_c}{dt} &= \left(\frac{zA_c}{L_c}\right)\frac{dH}{dt} - \left(\frac{zA_c}{L_c}\right)\frac{dH_c}{dt} \\ &= G_{A_c} - S_{A_c}, \end{aligned} \tag{5.11}$$

decomposes, respectively, into rates of production and deactivation of active-pipe area (Valentine and Mäkelä 2005), i.e.,

$$G_{A_c} = \left(\frac{zA_c}{L_c}\right)\frac{dH}{dt} \tag{5.12}$$

$$S_{A_c} = \left(\frac{zA_c}{L_c}\right)\frac{dH_c}{dt} \tag{5.13}$$

Note that subtraction of (5.13) from (5.12) gives (5.11). With the decomposition, the production rate of new active area is associated with the rate of height growth, and the deactivation rate of (old) active area is associated with the rate of crown rise, i.e., the rate at which the live crown base rises owing to branch death.

Under the assumptions of the pipe model, leaf biomass, W_f, is proportional to the active-pipe area, i.e., $W_f = \eta_s A_c$, and, therefore, the net growth rate of leaf biomass (kg DW yr^{-1}) is:

$$\begin{aligned} \frac{dW_f}{dt} &= G_f - S_f \\ &= \eta_s (G_{A_c} - S_{A_c}) \end{aligned} \tag{5.14}$$

Ordinarily, however, the production rate of leaf biomass, G_f, is greater than $\eta_s G_{A_c}$ and, accordingly, the rate of leaf senescence is greater than $\eta_s S_{A_c}$. In a process called turnover, senescent leaves attached to elongating active pipes are assumed to be shed and replaced after T_f years, the normal leaf life span. Therefore, the rate of loss of leaf biomass to normal aging, W_f/T_f (kg DW yr^{-1}), is also the production rate of the replacement biomass, and the latter involves an expenditure of photosynthate.

The overall production rate of leaf biomass includes the rate of production of new leaves attached to new pipes, $\eta_s G_{A_c}$, plus the rate of replacement of the senescent leaves, i.e.,

$$
\begin{aligned}
G_f &= \eta_s G_{A_c} + \frac{W_f}{T_f} \\
&= \eta_s \left(\frac{z A_c}{L_c}\right) \frac{dH}{dt} + \frac{W_f}{T_f} \\
&= \left(\frac{z W_f}{L_c}\right) \frac{dH}{dt} + \frac{W_f}{T_f}
\end{aligned} \tag{5.15}
$$

Similarly, the overall rate of loss equals the sum of the rate of leaf loss due to the deactivation of branches, $\eta_s S_{A_c}$, plus the rate of leaf loss due to aging, W_f/T_f. Hence,

$$
S_f = \left(\frac{z W_f}{L_c}\right) \frac{dH_c}{dt} + \frac{W_f}{T_f} \tag{5.16}
$$

An analogous turnover process pertains to fine roots, with a normal life span of T_r years, so

$$
G_r = \left(\frac{z W_r}{L_c}\right) \frac{dH}{dt} + \frac{W_r}{T_r} \tag{5.17}
$$

$$
S_r = \left(\frac{z W_r}{L_c}\right) \frac{dH_c}{dt} + \frac{W_r}{T_r} \tag{5.18}
$$

The rates of production and deactivation of active-pipe biomass obtain from the time derivative of (5.7),

$$
\frac{dW_p}{dt} = \frac{W_p}{W_f} \frac{dW_f}{dt} + W_f \left(\beta_1 \frac{dH}{dt} + \beta_2 \frac{dH_c}{dt}\right) \tag{5.19}
$$

Using the right-hand sides of (5.15) and (5.16) to substitute $G_f - S_f$ for dW_f/dt and then decomposing provides

$$
G_p = \left(\frac{z W_p}{L_c} + \beta_1 W_f\right) \frac{dH}{dt} \tag{5.20}
$$

$$
S_p = \left(\frac{z W_p}{L_c} - \beta_2 W_f\right) \frac{dH_c}{dt} \tag{5.21}
$$

The biomass of inactive wood within the main stem, W_q (kg), accumulates within the stem of the model tree. The rate of increase of deactivated wood in the main stem equals the senescence rate of active stem wood, S_s, i.e.,

$$\frac{dW_q}{dt} = S_s$$
$$= \left(\frac{zW_s}{L_c} - \beta_{2_s} W_f\right) \frac{dH_c}{dt} \tag{5.22}$$

where, from the components of (5.7) and (5.8),

$$W_s = \left(\beta_{1_s} H + \beta_{2_s} H_c\right) W_f \tag{5.23}$$

and

$$\beta_{1_s} = \frac{\phi_s}{l_s} \quad \text{and} \quad \beta_{2_s} = \frac{1 - \phi_s}{l_s} \tag{5.24}$$

Summing up, the total mass of wood in the main stem of the model tree is $W_s + W_q$, and total mass of wood in the whole model tree is $W_p + W_q$.

5.2.2 *Height Growth Rate and Allocation Fractions*

The rate of biomass production of the model tree, G (kg DW yr^{-1}), can now be expressed as a function of height growth rate. Summing (5.15), (5.17) and (5.20), the total rate of biomass production by the model tree is

$$G = G_f + G_r + G_p$$
$$= Y_G(P - R_M)$$
$$= \left[\frac{z(W_f + W_r + W_p)}{L_c} + \beta_1 W_f\right] \frac{dH}{dt} + \frac{W_f}{T_f} + \frac{W_r}{T_r} \tag{5.25}$$

Hence, the height growth rate of the model tree is

$$\frac{dH}{dt} = \left[\frac{L_c}{z \sum W_i + \beta_1 W_f L_c}\right] \left[Y_G(P - R_M) - \frac{W_f}{T_f} - \frac{W_r}{T_r}\right] \tag{5.26}$$

where $\sum W_i = W_f + W_r + W_p$. Fundamentally, height growth of a tree is achieved through the elongation of one or more terminal shoots. For our model tree, height growth is achieved through the elongation of active pipes, which occurs simultaneously with the production of active-pipe, leaf, and fine-root biomass. As such, height growth is just one aspect of the coordinated overall growth of the model tree as constrained by the pipe model and crown allometry.

The overall rate of production is driven by the rates of photosynthesis and respiration. Equations (3.4) and (3.9) from Chap. 3 provide these metabolic rates in terms of the components of biomass:

$$P = \sigma_f W_f$$

and

$$R_M = r_{M_f} W_f + r_{M_r} W_r + r_{M_p} W_p$$

where σ_f is the leaf specific rate of photosynthesis and r_{M_i} is the specific rate of maintenance respiration of the ith component of biomass. An effect of hydraulic limitation on P can also be modeled (see (3.40)).

The constraints on structure determine how the total production of the model tree is allocated among the leaf, fine-root, and active-pipe biomass. Recall that λ_f, λ_r, and λ_p are the allocation fractions of the total production. These fractions obtain by substituting (5.26) into (5.15), (5.17) and (5.20) and dividing by $Y_G(P - R_M)$, i.e.,

$$\lambda_f = \frac{\left[\dfrac{zW_f}{z\sum W_i + \beta_1 W_f L_c}\right]\left[Y_G(P - R_M) - \dfrac{W_f}{T_f} - \dfrac{W_r}{T_r}\right] + \dfrac{W_f}{T_f}}{Y_G(P - R_M)} \tag{5.27}$$

$$\lambda_r = \frac{\left[\dfrac{zW_r}{z\sum W_i + \beta_1 W_f L_c}\right]\left[Y_G(P - R_M) - \dfrac{W_f}{T_f} - \dfrac{W_r}{T_r}\right] + \dfrac{W_r}{T_r}}{Y_G(P - R_M)} \tag{5.28}$$

$$\lambda_p = \frac{\left[\dfrac{zW_p + \beta_1 W_f L_c}{z\sum W_i + \beta_1 W_f L_c}\right]\left[Y_G(P - R_M) - \dfrac{W_f}{T_f} - \dfrac{W_r}{T_r}\right]}{Y_G(P - R_M)} \tag{5.29}$$

Note that $\lambda_f + \lambda_r + \lambda_p = 1$, as required for conservation of mass.

5.2.3 *Net Growth Rates*

The rates of loss of leaf, fine-root, and active-pipe biomass depend on the rate of crown rise [see (5.16), (5.18), and (5.21)]. The aggregate rate of loss of active biomass (kg DW yr^{-1}) by the model tree from senescence or deactivation is

$$\sum S_i = \left[\frac{z\sum W_i}{L_c} - \beta_2 W_f\right]\frac{dH_c}{dt} + \frac{W_f}{T_f} + \frac{W_r}{T_r}, \qquad \text{for } i = \text{f, p, r} \tag{5.30}$$

Note: Stem wood that converts from active to inactive is counted as a loss from the pool of live biomass even though it remains embedded within the stem.

The net growth rate of the aggregate live biomass (kg DW yr^{-1}) obtains by subtracting (5.30) from (5.25), i.e.,

$$\begin{aligned}\frac{\mathrm{d}W}{\mathrm{d}t} &= \sum G_i - \sum S_i \\ &= \left(\frac{z\sum W_i}{L_c} + \beta_1 W_f\right)\frac{\mathrm{d}H}{\mathrm{d}t} - \left(\frac{z\sum W_i}{L_c} - \beta_2 W_f\right)\frac{\mathrm{d}H_c}{\mathrm{d}t}, \qquad \text{for } i = \mathrm{f, p, r}\end{aligned} \tag{5.31}$$

In contrast to the rate of production, we cannot express the rates of loss and net growth of live biomass solely in terms of the traits of the model tree, because the rate of crown-rise, $\mathrm{d}H_c/\mathrm{d}t$, is influenced by external factors.

A crown-rise model was introduced in Sect. 2.4. A key assumption of the model is that the maximum crown length of an average tree is constrained by the tree's height and the average distance to the adjacent neighbours, X (m), i.e., $\max\{H - H_c\} = \alpha_X + \beta_X X$. Crown length is shorter than the maximum if the crown-base height of a model tree, H_c, is higher than $H - \alpha_X - \beta_X X$, in which case the crown will lengthen from height growth while the crown base stays where it is (see Eq. (2.69)). Converting this verbal description to an instantaneous rate of crown rise,

$$\frac{\mathrm{d}H_c}{\mathrm{d}t} = \begin{cases} 0, & \text{if } H_c > H - \alpha_X - \beta_X X \\ \dfrac{\mathrm{d}H}{\mathrm{d}t} - \beta_X \dfrac{\mathrm{d}X}{\mathrm{d}t}, & \text{if otherwise} \end{cases} \tag{5.32}$$

Three situations where crown-base height may be higher than $H - \alpha_X - \beta_X X$ are: (i) before closure, when $H_c = 0$ and $H - \alpha_X - \beta_X X < 0$, (ii) following pruning, which suddenly lifts the crown-base height, and (iii) following thinning, which suddenly increases the spacing. Given these possibilities, knowledge of the average spacing does not eliminate the need for crown-base height as a state variable.

Our present task is to complete the dynamic model of a single tree, taking implicit account of the spatial limitations exerted by adjacent trees. One way is to modify the crown-rise model for this purpose by substituting potential or actual crown width for spacing. Thus, let X_c be the width of the potential ground coverage area of the model tree's crown before closure and the actual ground coverage area after closure. Then

$$\frac{\mathrm{d}H_c}{\mathrm{d}t} = \begin{cases} 0, & \text{if } H_c > H - \alpha_X - \beta_X X_c \\ \dfrac{\mathrm{d}H}{\mathrm{d}t} - \beta_X \dfrac{\mathrm{d}X_c}{\mathrm{d}t}, & \text{if otherwise} \end{cases} \tag{5.33}$$

We can use a control parameter, u, to control the rate of change of X_c. Most simply,

$$\frac{dX_c}{dt} = u \tag{5.34}$$

where, for example, $-0.02 \leq u \leq 0.02$. A value of $u > 0$ causes gradual lengthening of the model tree's crown, which is indicative of emergence. A negative value of u, by contrast, will cause the crown length of the model tree to gradually shrink, which is indicative of gradual encroachment and suppression of the model tree by adjacent neighbours. For the status quo, $u = 0$, which causes crown length to remain constant, with the rate of crown rise equalling the rate of height growth, after closure.

5.3 Cross-Sectional Growth

Foresters and most forest ecologists define the basal area of a tree, B (m^2), to be the cross-sectional area at breast height, bh (m). Breast height is usually specified to be either 1.3, 1.37, or 1.4 m above ground, depending on regional convention. The stem diameter at breast height, abbreviated as dbh (m), equals $2\sqrt{B/\pi}$ and is arguably the most important forest inventory variable. Indeed, one might argue that a dynamic model of a tree is incomplete if it does not provide a predicted time-course of dbh. However, we need not limit our focus to the growth of the most popular cross-sectional area and diameter. Instead, we can formulate the dynamic equations that provide the growth rate of cross-sectional area at any fixed height h on the model tree, where $0 \leq h \leq H$. First, however, we consider the aforementioned minor complication concerning juvenile and mature wood.

The wood that grows on a main stem between the tip and the base of the crown of many tree species is variously called *juvenile wood*, *crown wood*, or *core wood*. Among other anatomical differences, juvenile wood is distinguished by being less dense than the so-called *mature wood* that forms beneath the crown on the main stem in the same year (Zobel and Sprague 1998). Some investigators have also recognised a zone of transition wood between the juvenile wood and the mature wood (e.g., Clark III and Saucier 1989). Naturally, over time, the region of juvenile wood formation rises with the crown, typically resulting in a central core of juvenile wood that runs from the tip of the crown to the base of the stem. Beneath the crown, the central core is surrounded by the denser mature wood, the amount increasing from the base of the crown downward toward the base of the stem. The average density of wood follows a similar basipetal trend (e.g., Repola 2006).

The trends seen in real stems turn out to be remarkably simple to achieve in a model tree that is based on pipe model theory. We previously characterised an active pipe as a continuous strand of wood that extends through the area-preserving network of branches, stem, and coarse roots to connect leaf biomass to fine-root biomass. When a new active pipe is added to the model tree, we assume that the portion that runs through the central stem above the crown base consists of juvenile wood and the portion below consists of denser mature wood. The observed pattern of a central juvenile core—surrounded by mature wood below the crown—emerges as the crown of the model tree rises and more new active pipes are added.

Our formulation of this process is constrained by the fundamental premise of the pipe model: active stem biomass per unit length at height h is proportional to the total leaf biomass above that height (see Sect. 4.3). Thus, let a_s and a_m, respectively, be the cross-sectional areas of the within-crown and below-crown segments of a single active pipe, and let ρ_s and ρ_m be the corresponding bulk densities. The product of the cross-sectional area and bulk density provides the biomass per unit length. Since we are dealing with two segments of the same pipe, we require that the biomasses per unit length are the same, i.e., $a_\mathrm{s}\rho_\mathrm{s} = a_\mathrm{m}\rho_\mathrm{m}$, so

$$a_\mathrm{m} = a_\mathrm{s}\,\frac{\rho_\mathrm{s}}{\rho_\mathrm{m}} \tag{5.35}$$

Consequently, in any given year, cross-sectional growth of the stem below the crown of the model tree differs from the growth at the crown base by the factor $\rho_\mathrm{s}/\rho_\mathrm{m}$. Notably, the carbon balance of the model tree is not affected by any of this owing to the constraint that biomass per unit length of juvenile wood equals that of mature wood.

Let us now formulate a general model of cross-sectional growth. Over the course of a tree's life span, from germination to senescence, a fixed height h may occur first above, then within, and finally below the crown. Accordingly, we provide a cross-sectional area growth rate for each of these cases in our formulation. A key assumption of the model, to this point, is that $A_\mathrm{c} = \beta_A L_\mathrm{c}^z = \beta_A(H - H_\mathrm{c})^z$. We now broaden this assumption so that $A(h)$—the aggregate active-pipe area at height h on the central stem within the crown—scales with the zth power of crown length above height h i.e.,

$$A(h) = \begin{cases} 0, & \text{if } H \le h \\ \beta_A(H - h)^z, & \text{if } H_\mathrm{c} \le h < H \end{cases} \tag{5.36}$$

The second case of this model does not necessarily follow from pipe model theory, but it accords with our analysis of average-pipe length in the main stem above the crown base. We can now calculate z from the form factor of the main stem above the crown base, ϕ_s, which is defined as the ratio of the stem volume,

$$\beta_A \int_{H_\mathrm{c}}^{H} (H - h)^z \mathrm{d}h = \frac{\beta_A(H - H_\mathrm{c})^{z+1}}{z+1} = \frac{A_\mathrm{c}L_\mathrm{c}}{z+1} \tag{5.37}$$

to the cylindrical volume, $A_\mathrm{c}L_\mathrm{c}$. Consequently,

$$\phi_\mathrm{s} = \frac{1}{z+1}, \qquad z > 1 \tag{5.38}$$

and, conversely,

$$z = \frac{1}{\phi_s} - 1, \qquad 0 < \phi_s < 1 \tag{5.39}$$

The time derivative of (5.36) provides the rate of increase in cross-sectional area:

$$G_{A(h)} = \left[\frac{zA(h)}{H-h}\right]\frac{dH}{dt}, \qquad \text{if } H_c \leq h < H \tag{5.40}$$

Substituting the original relation, $A(h) = \beta_A(H-h)^z$, into (5.40) provides a better formulation for our purposes:

$$G_{A(h)} = \left[z\beta_A\,(H-h)^{z-1}\right]\frac{dH}{dt}, \qquad \text{if } H_c \leq h < H \tag{5.41}$$

An important distinction between these two formulations is that (5.40) requires a starting value of $A(h) > 0$, whereas (5.41) works with a starting value of $A(h) = 0$. Inasmuch as 0 is the value of $A(h)$ while $h \geq H$, it is also the starting value of $A(h)$ when H rises above h and, simultaneously, the growth of $A(h)$ begins.

The growth rate of $A(h)$ beneath the crown obtains by adjusting the growth rate, G_{A_c}, at $h = H_c$ with (5.35) to account for greater wood density. Consolidating the model,

$$G_{A(h)} = \begin{cases} 0, & \text{if } H \leq h \\ z\beta_A(H-h)^{z-1}\dfrac{dH}{dt}, & \text{if } H_c \leq h < H \\ \dfrac{\rho_s z\beta_A}{\rho_m}(H-H_c)^{z-1}\dfrac{dH}{dt}, & \text{if } 0 \leq h < H_c \end{cases} \tag{5.42}$$

Basal area growth rate, dB/dt, equals $G_{A(h)}$ for $h = \text{bh}$. However, because our formulation provides cross-sectional growth at any height on the model-tree stem, we can model the development of the stem profile by keeping track of the growth at a specific set of fixed heights. Ordinarily, we would track cross-sectional growth at heights h_i $(i = 1, 2, \ldots, m)$, where h_i is the height of the tree's ith annual node.

Support for this model is provided by empirical fits to loblolly pine (*Pinus taeda* L.) data (Valentine et al. 2012), and by its general agreement with extensive observations of stem form and taper (e.g., Gray 1956).

5.4 Summary of the Model

The active biomass of the model tree comprises: W_f, leaf biomass; W_p, active woody biomass, and W_r, fine-root biomass. Photosynthesis and respiration fuel the growth and maintenance of the biomass. Relationships among the components of biomass and the rates by which carbon is allocated to the growth of new biomass are provided by the pipe model, supplemented by crown allometry that relates leaf biomass to crown length. Crown length, in turn, is regulated by the average width of the potential crown coverage area, which, initially, is based on the average distance to adjacent trees.

As it turned out, the variables that define the state of the active biomass of the dynamic model tree at any time t are the tip height, H, the crown-base height, H_c, and the average width of the potential crown coverage area, X_c. All of the components of active biomass can be calculated from these three variables, i.e., W_r and W_p, respectively, can be expressed in terms of W_f with (5.5) and (5.7) and W_f, can be expressed in terms of the zth power of crown length with (5.4) and (5.10), i.e., $W_f = \eta_s \beta_A L_c^z$. Moreover, since we have not included any external forcing other than the spatial constraint in the present model, the rates of photosynthesis and respiration, and the carbon allocation fractions also can be calculated from just H and H_c. Consequently height growth rate can be expressed as a function of just tree height and crown length or, alternatively, tree height and crown-base height.

Before crown rise begins, the central stem of the model tree comprises active juvenile wood; after the onset of crown rise, the stem below the crown ordinarily comprises active and inactive juvenile wood, and active mature wood. After further crown rise inactive mature wood is added to the mix. The model tracks the inactive woody biomass of the main stem, W_q, and the model tree's basal area, B, which is the sum of active and inactive cross-sectional areas at breast height. The model also provides the cross-sectional areas of a sequence of heights from the stem base to the tip, which together provide the shape and size of the stem, and how it changes over time. The inactive biomass and the cross-sectional areas below the crown base depend on the growth history of the model tree. Thus, given the diameter of the crown coverage, X_c, and its rate of change, the solution of the model comprises time-courses of tip height, crown-base height, inactive stem volume, basal area, as well as time-courses of cross-sectional areas at the annual height nodes.

5.5 R Script

The solution of the model comprises time-courses of height, crown height, inactive stem volume, and basal area, as well as time-courses of cross-sectional areas at the annual height nodes.

5.5.1 Setup

Let us solve the model with the R package deSolve for period (m) of 80 years:

```
>     library(deSolve)
>     m <- 80
```

The allometric exponent, z, controls the relative rate of active pipe elongation versus stem expansion at the crown base due to the production of new active pipes:

```
>   z <- 1.8
```

Parameter values for branches, stem, and coarse roots ρ_i, ϕ_i, η_i, (i = b, s, t) are mostly taken from Mäkelä (1997) for Scots pine. First, ρ_i is wood density (kg DW (m^{-3})):

```
>  rho_b <- 400  # branches
>  rho_m <- 500  # mature stem wood
>  rho_s <- 400  # juvenile stem wood
>  rho_t <- 400  # coarse roots
```

Next, ϕ_i is the form factor:

```
>  phi_s <- 1 / (1 + z)
>  phi_b <- .75 - phi_s
>  phi_t <- 0.25
```

Finally, η_i is the pipe-model ratio (kg leaf m^{-2}):

```
>  eta_b <-  400
>  eta_s <-  460
>  eta_t <- 1200
```

We use these values to calculate β_{1_i} and β_{2_i}, and also β_1 and β_2:

```
>  beta_1_b <- rho_b * phi_b / eta_b
>  beta_1_s <- rho_s * phi_s / eta_s
>  beta_1_t <- rho_t * phi_t / eta_t

>  beta_2_b <- -rho_b * phi_b / eta_b
>  beta_2_s <-  rho_s * (1 - phi_s) / eta_s
>  beta_2_t <-  0

>  beta_1 <- beta_1_b + beta_1_s + beta_1_t
>  beta_2 <- beta_2_b + beta_2_s + beta_2_t
```

Other parameters include leaf and fine-root life spans, T_f and T_r (yr); and the ratio of fine root to leaf mass, α_r:

```
>      T_f <- 3.33
>      T_r <- 1.25
>  alpha_r <- 0.25
```

The physiological parameters include: σ_f (kg C (kg leaf)$^{-1}$ yr^{-1}), the specific rate of photosynthesis; a_σ (m^{-1}), the reduction in σ_f per unit crown length owing to hydraulic limitation; Y_G (kg DW (kg C)$^{-1}$), the yield of dry matter per unit carbon assimilated; and r_j (j = f, r, w) (kg C (kg DW)$^{-1}$ yr^{-1}), the specific rates of maintenance respiration:

```
>   sigma_f <- 4
>   a_sigma_Hc <- 0.01
>   a_sigma_Lc <- 0.02
>   Y_G <- 1.54
>   r_f <- .25
>   r_r <- .25
>   r_w <- 0.025
```

Both the time of initiation and the rate of crown rise depend, in part, on $H_c(0)$ (m), the starting value of the crown base height; $X_c(0)$ (m), the starting value of the potential width of the crown coverage area; and α_x (m) and β_x, the parameters that convert X_c into maximum crown length:

```
>     Hc_0 <- 0
>     Xc_0 <- 2.5
> alpha_x <- 1.45
>   beta_x <- 1.65
```

We also assign starting values to tree height, $H(0)$ (m), and the cross-sectional area of sapwood at the crown base, $A_c(0)$ (m^2), for the calculation of the parameter, β_A:

```
>      H_0 <- 0.1
>     Ac_0 <- 0.000003
>   beta_A <- Ac_0 / (H_0  - Hc_0)^z
```

A control parameter, u (m), determines the rate of crown rise relative to the rate of height growth after closure, e.g., $-0.02 \le u \le 0.02$, with negative values associated with suppression of the model tree (crown shortening), and positive values associated with emergence (crown lengthening). The other possibility, $u = 0$, provides a constant crown length after the start of crown rise.

```
>   u <- 0.0
```

The dynamic model is formulated in a function called `model_one`. This function contains the differential equations that we solve to provide the time-courses of the model tree's tip height, H, crown-base height, H_c, basal area, B, and inactive wood biomass, W_q (kg DW). The function also contains differential equations that provide the time-courses of cross-sectional area at the annual height nodes. Because the nodes are not known at the outset, our strategy is to first solve the model, in what we call `run 1`, for the time-courses of H, X_c, H_c, B, and W_q. The time-course of tip height, $H(t)$, provides the annual height nodes, so once these nodes are known, we solve the model again, in `run 2`, to obtain the cross-sectional areas at the annual nodes. It is possible to do everything in one run, but the code to do that is more complicated.

```
>   model_one <- function( t, y, parms ) {with( as.list( c( y ) ), {
+
+ # preliminary calculations
+
+    L_c <- H - H_c                                        # crown length
```

```
+     L_p <- beta_1 * H + beta_2 * H_c               # active pipe length
+     W_f <- eta_s * beta_A * L_c^z                  # leaf biomass
+     W_r <- alpha_r * W_f                           # fine root biomass
+     W_p <- L_p * W_f                               # active wood biomass
+     W_s <- (beta_1_s * H + beta_2_s * H_c) * W_f  # stem wood biomass
+       W <- W_f + W_r + W_p                         # total live biomass
+       P <- sigma_f * (1 - a_sigma_Hc * H_c         # photosynthesis with
+                       - a_sigma_Lc * L_c ) * W_f  #  limitation
+     R_M <- r_f * W_f + r_r * W_r + r_w * W_p       # rate of maintenance
                                                     #  respiration
+       G <- Y_G * (P - R_M)                         # net growth rate
+      Hs <- alpha_x + beta_x * Xc_0 + Hc_0          # tree height at closure
+       Q <- H^20 / (Hs^20 + H^20 )                  # numerical switch to
                                                     #  turn on crown rise
+ rho_ms <- rho_s / (rho_s + Q * (rho_m - rho_s))   # gradual transition
+                                                    #  from juvenile to
+                                                    #  mature wood at the
+                                                    #  crown base
+
+ # height growth rate
+       dH <- max(0, (L_c / (z * W + beta_1 * W_f * L_c))
+             * (G - (W_f / T_f) - (W_r / T_r)) )
+
+ # rate of change of crown coverage
+       dXc <- u
+
+ # rate of crown rise
+       dHc <- Q * max(0, dH - beta_x * dXc)
+
+     if( run == 1 ){
+
+ # basal area growth rate
+        dB <- 0
+          if(1.3 < H) {
+           if(1.3 < H_c) {
+             dB <- rho_ms * beta_A * z * (L_c)^(z - 1) * dH
+           } else {
+            dB <- beta_A * z * (H - 1.3)^(z - 1) * dH
+           }
+          }
+
+ # rate of increase in deactivated stemwood biomass
+        dWq <- (z * (W_s / L_c) - beta_2_s * W_f) * dHc
+
+        return( list( c( dH, dXc, dHc, dB, dWq ) ))
+
+     } # end of if( run == 1 )
+
+ # Active cross-sectional area growth rates at h[i]
+     G_Ah <- rep(0, m+2)
+     j <- floor(t+2)
+     G_Ac <- rho_ms * beta_A * z * (L_c)^(z - 1) * dH
+     for ( i in 1:j ) {
+       if( h[i] < H ) {
+         if( h[i] < H_c ) {
+            G_Ah[i] <- G_Ac
```

```
+          } else {
+             G_Ah[i] <- beta_A * z * (H - h[i])^(z - 1) * dH
+          }
+       }
+    }
+   return( list( c( dH, dXc, dHc, G_Ah ) ))
+   }) }
>
> # end model_one
```

5.5.2 Solution

For run 1, we create a vector called yini that contains the starting values for the state and response variables:

```
> # Initial values of state
>
>  yini_a <- c(    H = H_0,           # tree height (m)
+               X_c = Xc_0,           # width of crown coverage (m)
+               H_c = Hc_0 )          # crown base height (m)
>  yini_b <- c(   B = 0,              # basal area (@ 1.4m) (m^2)
+               W_q = 0 )             # inactive stem mass (kg}
>    yini <- c( yini_a, yini_b )
```

The deSolve package contains a function called ode that integrates the differential equations in model_one from time $t = 0$ to $t = m$ using the starting values in the vector yini. The resultant time-course vectors are arranged as columns in a matrix called out.

```
>    run <- 1
>    times <- seq(0, m, by = 1)
>    out <- ode( func = model_one, y = yini, parms = null,
          times = times )
```

Here are the last 5 rows of the solution (at times 96–100), rounded to 3 decimal places:

```
> round(out[(m-3):(m+1),],3)

     time      H X_c    H_c     B     W_q
[1,]   76 34.104 2.5 28.506 0.033 192.182
[2,]   77 34.263 2.5 28.665 0.033 194.341
[3,]   78 34.421 2.5 28.823 0.033 196.484
[4,]   79 34.576 2.5 28.978 0.033 198.609
[5,]   80 34.730 2.5 29.132 0.033 200.719
```

For convenience, we extract and name the time-course vectors:

```
>         t <- out[1:(m+1),1]
>       H_t <- out[1:(m+1),2]
>      Xc_t <- out[1:(m+1),3]
```

```
>      Hc_t <- out[1:(m+1),4]
>       B_t <- out[1:(m+1),5]
>      Wq_t <- out[1:(m+1),6]
```

5.5.3 The Stem Profile

Because we now have the annual height nodes, it is a simple matter to generate the stem profile year by year or decade by decade. The time-course vector `H_t` contains $m+1$ annual heights starting with the initial tip height, `H_0`, at time $t=0$. Missing from this vector is the height, 0 m, of the first node, which is located at the ground line. Thus, we create a vector, h, of length $m+2$ that comprises the heights of the annual nodes including the first node:

```
>    h <- c(0, H_t)
```

We also create a second vector, Ah, that contains the $m+2$ starting values of the cross-sectional areas at the nodal heights, i.e., `Ah[i]` is the cross-sectional area at height `h[i]`. We assume that all of these areas are zero, except at the ground line where the area is calculated with (5.10):

```
>         Ah <- rep(0, m+2)
>      Ah[1] <- beta_A * H_0^z   # at the ground line
>      names(Ah) = paste('Ah',1:(m+2),sep = '')
```

In the solution, the cross-sectional area `Ah[i]` remains zero until the tip height of the model tree, H, reaches the nodal height `h[i]`. After that, the area increases in accordance with (5.36).

The vector of starting values for the second run is:

```
>      yiniAh <- c(yini_a, Ah)
```

The solution is arranged in a matrix called `out2`:

```
>       run <- 2
>      out2 <- ode( func = model_one, y = yiniAh, parms = null,
          times = times )
```

We now graph the half-decadal stem profiles, i.e., the profiles at $t = 0, 5, 10, \ldots, m$. At time $t = j$ $(0 \leq j \leq m)$, the tip height of the model tree equals the height of annual node $j+2$. The cross-sectional areas at nodes 1 through $j+2$, respectively, are found in row $j+1$, columns 5 through $j+6$ of the matrix `out2`.

```
>      pdf(file="model_one_radius.pdf",width=3,height=9)
>   # graph the radial profile at t = m
>
>       r <- 100 * sqrt(out2[m+1,5:(m+6)] / pi)   # radii (cm)
>       plot( r, h, type = "l", xlab = "Radius (cm)",
           ylab = "Height (m)" )
```

```
>   # graph the radial profiles at t = 0, 5, 10,...
>
>       j <- 0
>       while( j < m ){
+          lines( 100 * sqrt(out2[(j+1),5:(j+6)] / pi), h[1:(j+2)] )
+          j <- j + 5
+       }
>       dev.off()
```

Graph the corresponding stem cross-sectional area by height:

```
>      S.label = expression(paste('Sectional area (dm'^{2},')',
              sep =''))
>     # postscript(file="model_one_xsection.eps",width=3,height=9)
>             pdf(file="model_one_xsection.pdf",width=3,height=9)
>       plot( out2[m+1,5:(m+6)] * 100, h, type = "l",
+                                 xlab = S.label,  ylab = ' ' )
>                              #  xlab = S.label,
                                  ylab = 'Height (m)' )
>       j <- 0
>       while( j < m ){
+          lines( out2[(j+1),5:(j+6)] * 100, h[1:(j+2)] )
+          j <- j + 5
+       }
>       dev.off()
```

5.5.4 Response Variables and Graphs

The time-course vectors of response variables are calculated from the time-courses of the state variables, for example,

```
>       Lc_t <- H_t - Hc_t
>       Lp_t <- beta_1 * H_t + beta_2 * Hc_t
>       Ac_t <- beta_A * Lc_t^z
>      dbh_t <- 200 * sqrt( B_t / pi)    # dia at breast height (cm)
>      dcb_t <- 200 * sqrt(Ac_t / pi)    # dia at crown base (cm)
>       Wf_t <- eta_s * Ac_t
>       Wr_t <- alpha_r * Wf_t
>       Wb_t <- (beta_1_b * H_t + beta_2_b * Hc_t) * Wf_t
>       Ws_t <- (beta_1_s * H_t + beta_2_s * Hc_t) * Wf_t
>       Wt_t <- (beta_1_t * H_t + beta_2_t * Hc_t) * Wf_t
>       Wp_t <- Wt_t + Ws_t + Wb_t
>       Ww_t <- Wp_t + Wq_t
```

Next we calculate the time-course vectors of the components of the allocation fractions, $G_i(t)/G(t)$:

```
>        P_t <- sigma_f * (1 - a_sigma_Hc * Hc_t - a_sigma_Lc) * Wf_t
>       RM_t <- r_f * Wf_t + r_r * Wr_t + r_w * Wp_t
>        G_t <- Y_G * (P_t - RM_t)
>        num <- G_t - (Wf_t / T_f) - (Wr_t / T_r)
>        dem <- z * (Wf_t + Wr_t + Wp_t) + beta_1 * Wf_t * Lc_t
```

```
>    lamf_t <- (z * Wf_t * (num / dem) + (Wf_t / T_f)) / G_t
>    lamr_t <- (z * Wr_t * (num / dem) + (Wr_t / T_r)) / G_t
> # lamp_t <- (z * Wp_t + beta_1 * Wf_t * Lc1_t) * (num / dem) / G_t
>    lamp_t <- 1 - lamf_t - lamr_t
```

Graph the time-courses of the components of tree size (dbh, dcb, H, H_c, L_c):

```
>       pdf(file="model_one_size.pdf",width=4.8,height=4.8)
>       maxy <- ceiling( max( H_t[m+1],dbh_t[m+1] ) * 12 ) / 10
>       plot(  t, dbh_t, type = "l", lty = 1, xlab = "Time (yr)",
+        ylab = "Size",    ylim = c(0,maxy), xlim = c(0,m+10) )
>       lines( t, H_t,    type = "l", lty = 2 )
>       lines( t, dcb_t, type = "l", lty = 3 )
>       lines( t, Hc_t,   type = "l", lty = 4 )
>       lines( t, Lc_t,   type = "l", lty = 5 )
>       text(m+6,    H_t[m], bquote(H))
>       text(m+6, dbh_t[m], bquote(dbh))
>       text(m+6, dcb_t[m], bquote(dcb))
>       text(m+6,  Hc_t[m]+1, bquote(H[c]))
>       text(m+6,  Lc_t[m], bquote(L[c]))
>    dev.off()
```

R code for other graphs is similar.

5.5.5 *Results*

Any particular solution of the model depends on the values of the parameters and the starting values of the state variables. Nonetheless, we can draw some preliminary conclusions about the model's behavior from a single run, starting with the time-courses that correspond to measures of size (Fig. 5.1). The model-tree's

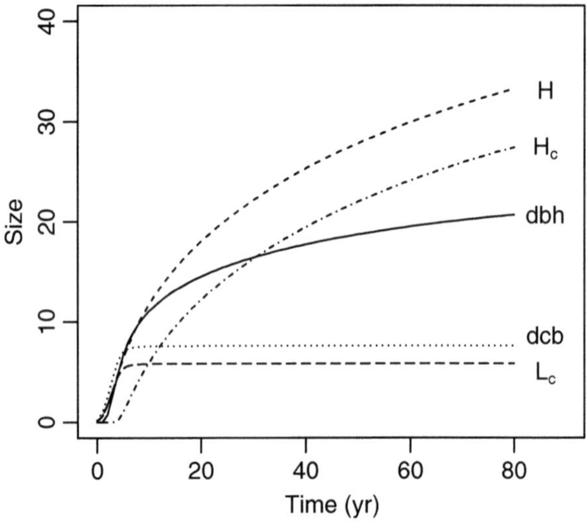

Fig. 5.1 Time-courses for the model tree: tip height, H (m); crown-base height, H_c (m); diameter at breast height, dbh (cm); diameter at crown base, dcb (cm); and crown length, L_c (m)

height growth initially accelerates from its initial value of 0.1 m yr^{-1} and then slows to provide a sigmoid time-course in agreement with both theory and general observation. Crown-base height, H_c, is initially 0 m, and then tracks tip height when $H - \beta_x X_c(0)$ becomes positive. The maximum diameter of crown coverage, X_c (m), is a constant because we specified the value of the control parameter, u, to be 0. Consequently, crown length equals the height of the model tree before crown rise commences, then crown length stays constant afterwards. Diameter at the crown base, dcb, increases with the $(z/2)$th power of crown length until crown rise commences, and then remains constant because crown length is constant. Diameter at breast height, dbh, equals 0 until the tip height of the model tree increases to breast height, 1.3 m, then dbh scales with the $(z/2)$th power of the length of crown above breast height. After the base of the crown rises above breast height, the dbh time-course shows a slower rate of increase, partly because all further increases in dbh is due to the growth of mature wood, which has higher density and, therefore, less volume per unit mass, than the juvenile wood that grows above the crown base.

The radial and cross-sectional profiles of the central stem of the model tree are depicted in Fig. 5.2 at times $t = 0, 5, 10, \ldots, 80$ years. The radial profile is, of course, just half of the diameter profile. The central core of juvenile wood that runs from the tip to the base of the central stem is apparent in both the radial and cross-sectional area profiles. The cross-sectional profile beneath the crown base tends to become increasingly linear with respect to stem length over time, thereby approximating a frustum of a quadratic paraboloid. In this regard, stem form development resulting from our interpretation of the pipe model, which allows for both juvenile and mature wood, is consistent with both the theory and the extensive observations on stem form reported by Gray (1956).

Tree mass (kg DW) is of interest in and of itself, and it is also the measure by which trees figure into the global carbon cycle, under the usual assumption that the mass of carbon in a tree equals 50% of the dry weight of the tree.

The total wood mass, W_w, is the sum of the masses of inactive wood, W_q, and active wood, W_p, the latter comprising the masses of the active wood in branches, central stem, and coarse roots. All of the inactive wood of the model tree resides within the central stem. The total mass of wood in the central stem is the sum of the inactive wood mass and the central stem's active wood mass, W_s. The time-courses shown in Fig. 5.3 indicate that all the wood mass of the model tree is active wood mass from time $t = 0$ until the onset of crown rise. After that, the mass of inactive wood increases to first exceed the mass of active wood in the stem and then exceed the mass of all the active wood in the model tree. Eventually, the central stem contains the lion's share of the wood mass, and most of that is inactive wood mass.

The rate of production of each component of live biomass of the model tree—leaf, fine root, or active wood—depends on the fraction of the net assimilated carbon allocated to that component. The time-courses of these fractions—λ_f, leaf; λ_r, fine root; λ_p, active wood—undergo rapid change in the early stages of growth, and then settle down to rather small rates of change after the onset of crown rise (Fig. 5.4).

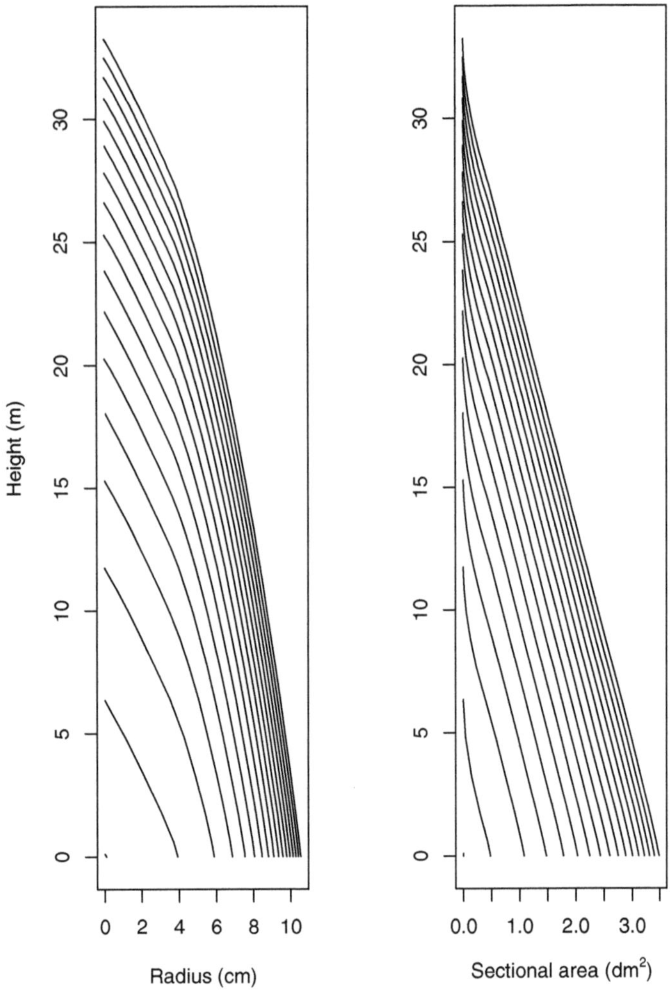

Fig. 5.2 Radial profiles (left) and cross-sectional area profiles (right) of the model tree at times $t = 0, 5, 10, \ldots, 80$ years

5.5.6 Sensitivity

It is always a good idea to examine the degree to which the solution of a model changes when the parameter values and the initial values of the state variables are altered. The graphs in Figs. 5.1, 5.2, 5.3, 5.4 provide results for $z = 1.8$ and $u = 0$. For sake of example, we now systematically vary the values of these two parameters to demonstrate the sensitivity of the model solution to changes in values of z and u. The results are provided in Figs. 5.5 and 5.6.

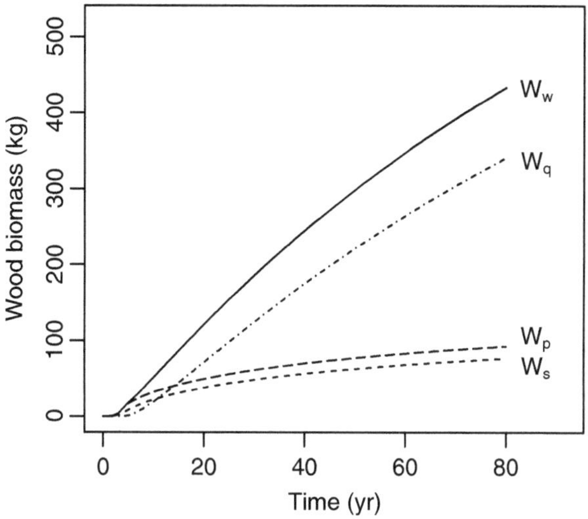

Fig. 5.3 Time-courses for the model tree: total wood mass, W_{w}; total inactive wood mass, W_{q}; total active pipe mass, W_{p}, in branches, stem, plus coarse roots; and total active active wood mass in the stem, W_{s}

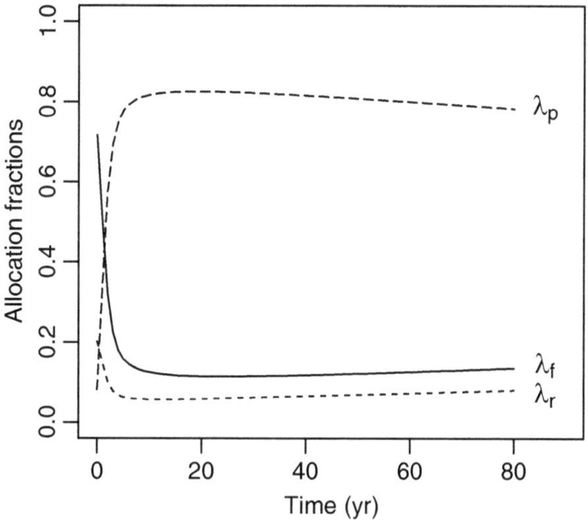

Fig. 5.4 Time-courses for the model tree: fraction of production allocated to active wood, λ_{p}; leaves, λ_{f}; and fine roots, λ_{r}

The parameter u controls the maximum width of ground coverage by the model tree's crown. A negative value, i.e., $u = -0.02$, causes the coverage area to shrink over time and, coincidentally, it causes the crown length to shorten, which, in turn, causes the stem cross-sectional area at the crown base (active-pipe area), A_{c}, to shrink. And, since foliage biomass, W_{f}, is proportional to A_{c}, productivity also suffers. Conversely, a positive value, i.e., $u = 0.02$ causes opposite effects. Interestingly, a larger fraction of the annual production is allocated to wood growth in the model tree undergoing suppression ($u = -0.02$), compared to the emergent tree ($u = 0.02$). However, the overall amount of annual wood production is larger in emergent tree, which accounts for its greater yield of wood biomass.

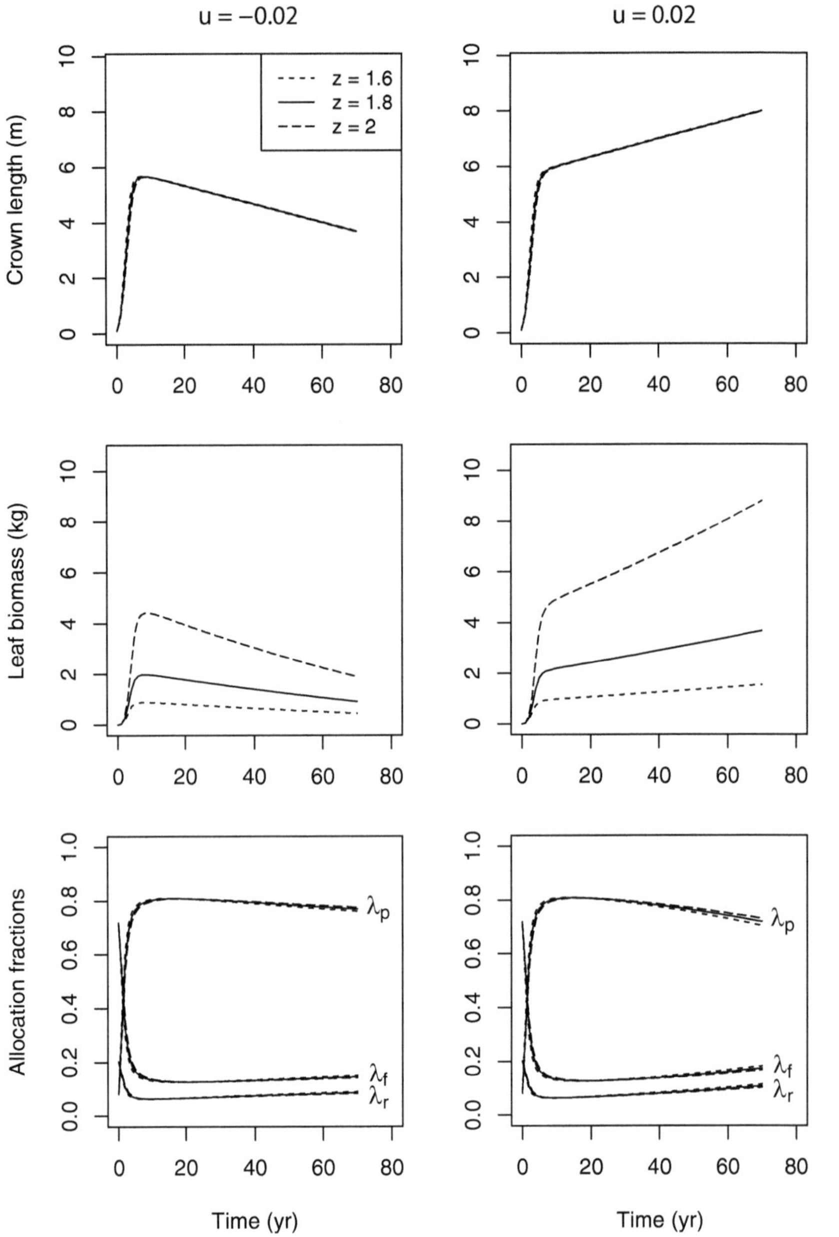

Fig. 5.5 Predicted time-course of crown length, L_c (m); leaf biomass, W_f (kg DW); and allocation fractions for production of new leaves, λ_f; fine roots, λ_r; and active wood, λ_p with $u = -0.02$ (left column) or $u = 0.02$ (right column). Results for $z = 1.6$ (short dash), $z = 1.8$ (solid) and $z = 2$ (long dash) in each of the six panels

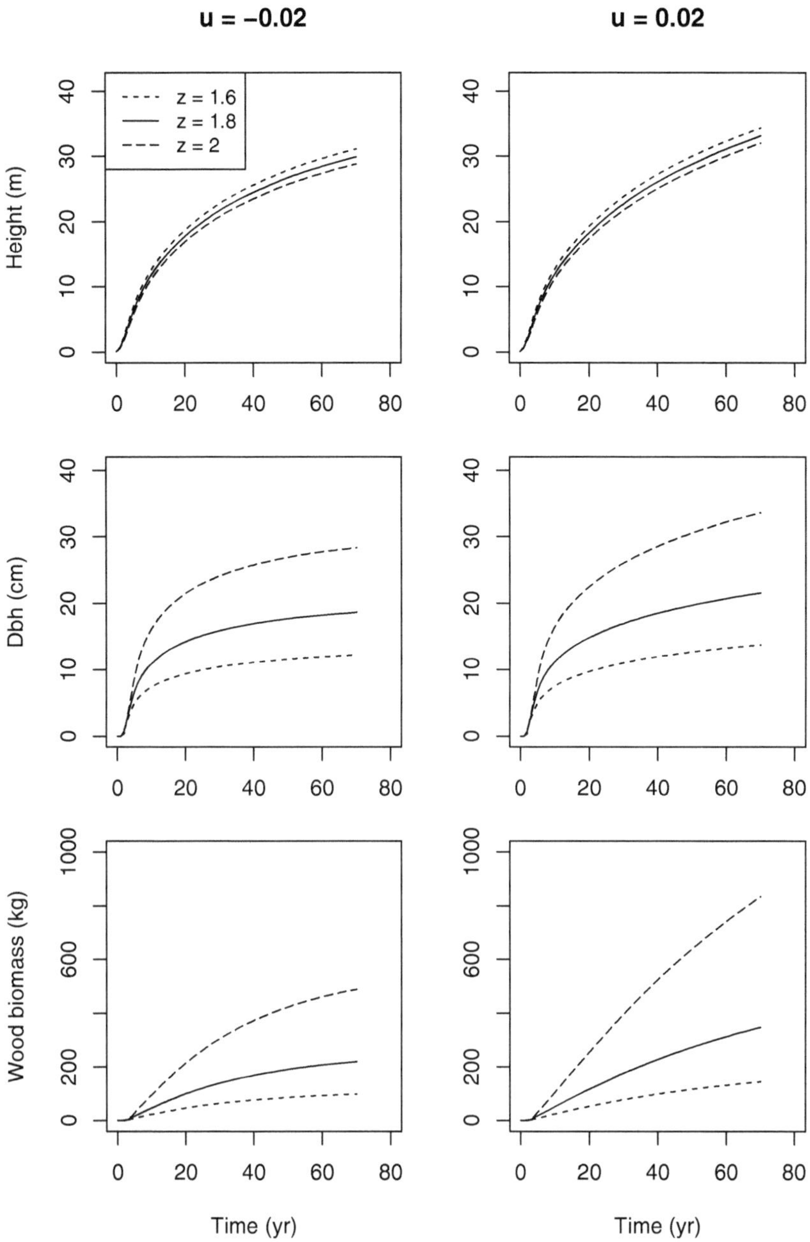

Fig. 5.6 Predicted time-course of tree tip height, H (m); diameter at breast height, dbh (cm); and wood biomass, W_w (kg DW), calculated with $u = -0.02$ (left column) or $u = 0.02$ (right column). Results for $z = 1.6$ (short dash), $z = 1.8$ (solid) and $z = 2$ (long dash) in each of the six panels

The parameter z is the exponent of the relation between active pipe area and crown length, i.e., $A_c \propto L_c^z$. As such, z serves to regulate the tradeoff between primary growth—the extension of roots and shoots—and secondary growth—the expansion of branches, stems, and coarse roots. Increasing the value of z causes large increases in leaf biomass, dbh, and wood biomass. By contrast, crown length, which is dependent only on space in the model, is unaffected by the value of z. Tree height shows a modest increase with increasing z, and the fraction of production allocated to wood shows a very small decrease.

5.6 Redux and Reuse

The model of tree dynamics presented in this chapter—apart from the crown-rise mechanism—is largely a synthesis from pipe-model and carbon-balance theories. Key ingredients of the tree-dynamics model were extracted from previously published models, most notably a model of crown allometry (Mäkelä and Sievänen 1992), the tree- and stand-level models of Mäkelä (1986, 1997)—the latter known as CROBAS—and the height-growth and stem-growth models of Valentine and Mäkelä (2005) and Valentine et al. (2012, 2013)—collectively called the Bridging model. Our approach to modelling stem form and taper in this chapter—differentiating between juvenile and mature wood in a manner consistent with the specific pipe length of pipe-model theory—is, so far as we know, original.

The core equations developed in this chapter are used, in the next chapter, in a model of stand dynamics based on a mean tree. The stand-dynamics model incorporates: (1) a more sophisticated model of photosynthesis, (2) the effects of competition for light and space among model trees, and (3) tree mortality from self-thinning calculated with a rule that we derive from the crown-rise model. The stand-dynamics model, in turn, is part of a more sophisticated model described in Chap. 7 that optimises carbon and nitrogen balances to maximise height growth rate. Empirical estimation of the parameters of the core equations is discussed in Chap. 8.

5.7 Exercises

5.1 Retrieve the R script, `crise_mod_runs.R`, from the electronic supplementary material of this chapter, and place it in a convenient folder (work directory). Solve the model in R with the command: `source("crise_mod_runs.R")`. Assess the effect of the control parameter, u, by changing its value.

5.2 Suppose that a model stand is thinned in some year t, doubling the spacing. What is the effect on N_t? How would you alter the script to implement a simulated thinning?

Chapter 6
Competition

The term competition is used to describe individual tree or stand-level reactions to situations where the trees must share limited resources. Resource limitation at the stand level constrains total growth, productivity, and the maximum biomass that a site can support. In this chapter we focus on the reactions of individuals to resource limitation. We consider the distribution of resources among individuals in crowded stands, plastic reactions of trees to competition, and the combined effects of environment and plasticity on resource acquisition. We review empirical models for the effects of resource limitation on competition-induced mortality (self-thinning) in stands, and we introduce physiological approaches to tree × tree interactions on the differentiation of trees in spatially explicit stands.

6.1 Introduction

This chapter focuses on the reactions of individual trees to resource limitation. For individual trees, growth is usually a prerequisite for survival; resource limitation leads to crowding, which may cause suppression and eventually mortality of some individuals. In addition to light, water, and nutrients, physical space—room to expand—is an important resource that limits individual growth under competition.

Foresters have considered competition at the stand level in terms of density-yield relationships and self-thinning trajectories. Empirical models of individual tree growth usually employ a competition index to scale growth rate of each tree, based on the size of, and distance to, each of the tree's adjacent neighbours.

Eco-physiological approaches to competition view tree growth as a dynamic process of resource acquisition and allocation in a local environment, as affected by neighbouring trees (Fig. 6.1). As the tree and its neighbours grow in size and mass, the local environment providing the resources is in continuous change. The

A. Mäkelä, H. T. Valentine, *Models of Tree and Stand Dynamics*,
https://doi.org/10.1007/978-3-030-35761-0_6

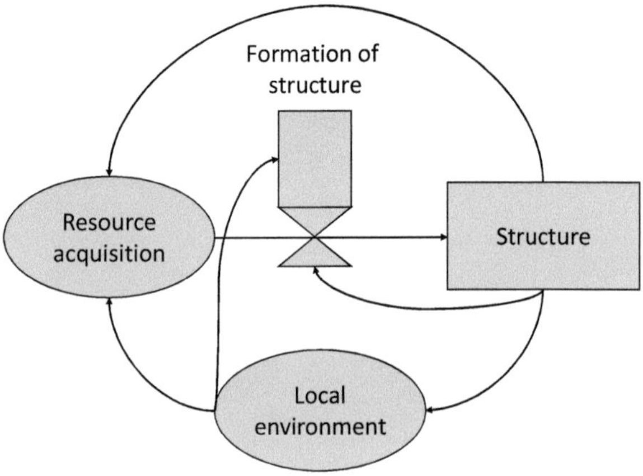

Fig. 6.1 Schematic presentation of the interaction of structure and function in tree growth. Tree structure develops as a result of eco-physiological functioning that controls resource acquisition and allocation of resources to the formation of structure. The local environment is influenced by the structure of neighbouring trees relative to the subject tree

change is determined by how fast the resources are acquired by the competing trees and how they are allocated to form new structures.

Unless the initial size distribution of a stand is extremely even, competition usually leads to an initial increase of variability in the sizes of trees, as the slightly bigger and/or better positioned individuals gain in growth relative to their less fortunate neighbours. This is soon followed by a decrease in variability as mortality occurs among the most suppressed individuals. How fast the differentiation takes place depends on the degree and nature of tree interactions. If all individuals affect the amount of resources available to their neighbours in the same way, e.g., relative to their size, the competition is called *symmetrical* or two-sided. By contrast, *asymmetrical* or one-sided competition refers to a situation where large trees reduce the resources of small trees more than proportionally to their size. For example, tall trees shade short trees, but not vice versa.

Competition constitutes a major selective force in the evolution of plant populations, causing a strong selective pressure on growth strategies that enhance survival under resource limitation. An individual plant may move through different competitive positions (from open-grown to suppressed) during its lifetime, so survival requires the ability to acclimate to different local environments. Consequently, trees have evolved wide plasticity of structural and functional traits, which allow them to forage resources efficiently in different competitive positions.

The vast majority of research into competition in tree stands has focused on above-ground competition. Above-ground processes obviously lend themselves to observation more readily than those below-ground, but as we will see below, above-ground competition also seems to be functionally more challenging than root competition. Above-ground competition will also be our main focus in this chapter. We consider the distribution of resources among individuals in crowded stands, plastic reactions of trees to competition, and the combined effects of environment and plasticity on resource acquisition. We review empirical models for the effects of resource limitation on self-thinning of stands, and we introduce physiological approaches to tree × tree interactions on the differentiation of trees in spatially explicit stands.

6.2 Setting the Scene: Effects of Competition on Growth and Mortality

Models including competition effects usually start off from a situation where the tree population consists of individuals of variable size, such that their growth rates and resource acquisition rates differ and may cause a further increase in the variability of size with time. How the variability was created is less frequently studied, although some models of succession with seedling recruitment also address this question (Bugmann 2001; Strigul et al. 2008). Competition between species or between genetically variable individuals of the same species essentially depends on the character and magnitude of their differences in resource acquisition efficiency. In that case differentiation of individuals in the population is largely due to their genetic properties. Competition between genetically identical (or similar, as it may be in practice) individuals, on the other hand, is molded by the interactions of the individuals through changes in their mutual environments that affect their future growth and chances of survival. This sets up a highly dynamic scene, but requires that some external driver will create the initial size distribution that leads to competitive interactions.

An example of differentiation during stand development is provided by a loblolly pine spacing experiment in Virginia and North Carolina (Fig. 6.2). Measured time-courses for loblolly pine data derive from four plots from one experimental

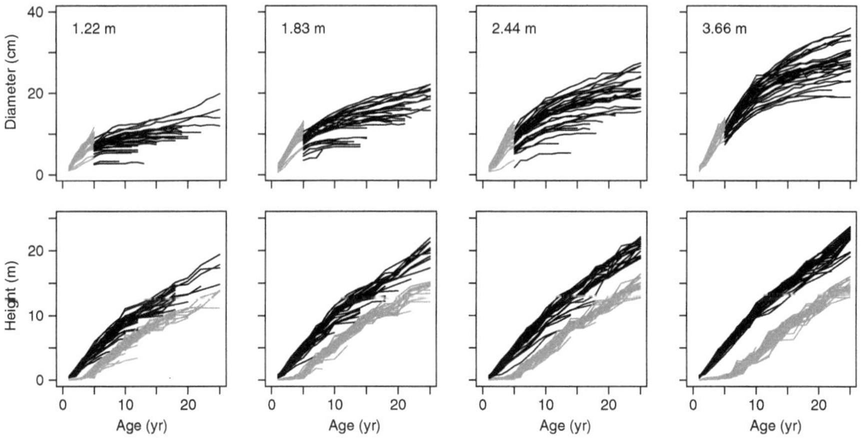

Fig. 6.2 Time courses from loblolly pine spacing trials. Top row: basal diameter (cm), gray; dbh (cm), black. Bottom row: crown-base height (m), gray; total height (m), black. The initial spacing (m) is provided. (From Valentine et al. 2012)

location, each plot with an initial spacing differing from the others. All the trees represented in the figure derived from a common seed source and were planted at the same time. Yet the trees within each plot show remarkable divergence in the sizes of their respective diameters with the passage of time, with less among-tree variation evident in the time courses of total height and crown height (Valentine et al. 2012).

Here we focus on within-species competition. We begin with aspects of resource acquisition, acclimation and mortality under competition that we consider important for understanding the different modelling approaches that are reviewed in the subsequent sections.

6.2.1 Resource Acquisition

A tree's primary organs of resource acquisition are its leaves and fine roots. The stem system, however, is no less essential because its branches serve to separate the leaves, facilitating the absorption of light energy to drive photosynthesis. Similarly, the system of coarse roots separates and distributes the fine roots within the so-called root plate. The fine roots—often in association with mycorrhizal fungi—absorb water and dissolved nutrients, especially nitrogen, that are likewise essential for photosynthetic processes, transport mechanisms, and growth.

Given well placed buds from which to do so, trees generally will extend shoots and grow new leaves where the light intensity meets some threshold of sufficiency (e.g., Snow 1931; Sprugel 2002).

Below ground, the conditions for fine-root growth and ramification seem quite broad. As Pregitzer (2002) noted, "[fine roots] seem to be everywhere in the soil, penetrating the rotting leaf litter, wedged between stones and proliferating in worm casts. Millions of lateral branches serve as the carbon depot for mycorrhizal hyphae that ramify away for the root tip to forage widely for essential nutrients." Nonetheless, the prevailing view is that a tree will concentrate its fine roots in patches of soil where water and nutrients are most available. Of course any adjacent neighbouring trees are also doing these same things at the same time, so we rationalize that, together, these trees are competing for the available light, water, and nutrients.

The primary effect of competition on trees is a reduction in their rates of resource acquisition. At the stand level, crowding causes the mean resources available to each tree to diminish. At the level of individuals, those with a better competitive status get a larger proportion of the total resources than their smaller neighbours. This causes the growth rates of trees to slow due to competition.

Light is the focal resource in above-ground competition (see Sect. 6.4). Because light comes from above and is shaded downward, the light capture per unit leaf area is larger in taller trees. Even small deviations of efficiency are detrimental if they lead to reduced height growth and hence increased suppression. This suggests that above-ground competition of plant populations is one-sided. Relative biomass growth rates increase with increasing plant biomass, and the size distribution of plants competing for light becomes more variable over time (Weiner 1990; Sorrensen-Cothern et al. 1993; Kokkila et al. 2006).

Below-ground resources show less of a directional pattern. Although water and nutrients from above-ground litter also penetrate the soil from the surface downward, there is important lateral and upward movement of water and dissolved nutrients in the direction of decreasing water potential, with a tendency to even out the moisture profile in the soil.

On the other hand, there is a lot of small-scale variability in moisture and nutrients created, e.g., by mounds on the soil surface, piles of woody debris, or stoniness. Nutrient-rich pockets in the soil may result from clusters of decomposing organic material, such as dead animals or plant parts, or uneven distribution of weathering minerals. This variability differs from the variability of radiation in that it is largely random and not affected by tree-to-tree interactions. However, the patchiness of the soil environment is probably an important factor for seedling growth and survival, giving a head start to plants from those seeds that landed on a moister and richer spot than the others.

When trees grow, their root systems expand and thus overcome the small-scale variability of soil properties. The spatial distribution of fine root growth seems less restricted than that of leaves which needs to be attached to mechanically stable branching structures. Trees have also been found to form root connections that further contribute to equalising their root-specific uptake rates of water and nutrients (Caldwell et al. 1998; Warren et al. 2008).

All this means that although crowding reduces the overall access of plants to nutrients and water, the spatial distribution of below-ground resources can generally be regarded as fairly even, implying that below-ground competition is symmetrical. There is empirical evidence that plant populations with no above-ground competition do not exacerbate initial size differences (Weiner 1990; Weiner et al. 1997). This suggests that in these experiments the availability of soil resources was uniform to all fine roots. Rajaniemi (2003) found, however, that a patchy distribution of soil nutrients led to increasing size differences in an experiment on three herbaceous plants, implying asymmetrical competition.

To what extent soil resources are uniformly available to all trees remains unresolved. However, while patchiness may be the prevailing feature in stands of seedlings, causing initial size differences to be created, this is more a question of a random process than that of tree interactions through the soil medium. When trees grow larger, their roots systems also grow, and the initial differences are evened out in the process. The main effects of below-ground competition, therefore, seem to occur through an overall limitation of resources rather than suppression of individuals.

6.2.2 Acclimations

Natural selection has provided trees with numerous acclimations to situations where competition or resource limitation tends to reduce their metabolic rates. An important acclimation to the prevailing light level is the increase of specific leaf area along with increasing shade downward in the canopy (e.g. Poorter et al. 1995; Ishii et al. 2007). Morphologically, this creates a transition from thicker sun leaves in high light to thinner shade leaves in lower light. In effect, compared to sun leaves, shade leaves stretch their masses over larger areas. The payoff of this adaptive strategy is a tendency toward equilibration of the photon absorption per unit investment of leaf mass (e.g., Freschet et al. 2013). Of course there is a limit to this strategy, because if a leaf is stretched too thin, it becomes structurally unstable.

Other responses to spatial variation in incident light intensity include evolved branching patterns that determine how leaves are arranged within a crown (Horn 1971). Shaded branches tend to stretch out and display their leaves horizontally, while branches and leaves in full light assume a more erect position (Fig. 6.3). Trees also acclimate to their current competitive rank. The spatial density of leaves—or the coefficient of proportionality in Eq. (4.40)—tends to be lower in suppressed trees than in dominants (Mäkelä and Vanninen 1998), as it pays to reduce self-shading within the crown that is already heavily shaded by neighbours. Sprugel (2002) provided evidence that trees in dominant positions invest a greater proportion of resources to produce leaves higher in the crown—where most of the light is—whereas suppressed trees continue to produce leaves at low light levels, since that is their only option. Sprugel concluded that strategy provides the maximum carbon fixed per unit of leaf biomass or nitrogen invested.

In order to maintain a good light position, trees will generally have to grow taller. Some, mainly broadleaved species, are also extremely efficient in lateral growth to occupy gaps left by falling neighbours (Dieler and Pretzsch 2013). In either case, unlike their root systems, the crowns of neighbouring trees only rarely intermingle. This means that at some point of crown extension, only upward growth is possible

Fig. 6.3 Effect of light availability on crown shapes. Left: A suppressed Norway spruce tree in the understory. Right: Top of a young Scots pine tree growing in full light. The crown shapes were traced from photographs

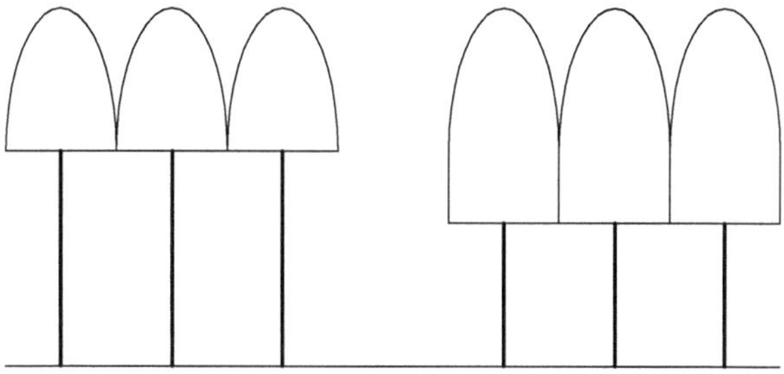

Fig. 6.4 Crown rise patterns in shade-intolerant (left) and shade-tolerant (right) species

unless mortality occurs to release space for the surviving trees. The competition for light thus becomes competition for space, where growing taller and placing new leaves ever higher up seems to be the best route to survival.

When crowns of neighbouring trees abut each other to form a continuous stand canopy, close approaches of crowns are prevalent especially among larger trees. This is usually attributed to abrasion that occurs during wind sway. The radial expansion of a crown generally ceases where it abuts an adjacent crown. The shape and volume of a tree's crown is thus constrained by the crowns of abutting neighbours, and vice versa.

Ordinarily, in this situation, the crown base of a tree is also the base of the abutment zone. As trees grow in height, their abutment zones move upward commensurately. However, some shade-tolerant species might be more realistically described in terms of an upper crown that reaches down to the abutment zone, and a lower crown that is no longer expanding radially but nevertheless maintains its leaves and vital functions (Fig. 6.4). Non-uniform spacing leads to asymmetrical crowns.

Below ground, a fine root absorbs water and nutrients through its surface. Thus, foraging for nutrient rich patches may be best done with fine roots with large specific root length, i.e., long thin roots that provide a large foraging range and absorptive surface for a given investment in root mass (e.g., de Kroon et al. 2012). Other below-ground acclimations to resource limitation include growing roots where soil moisture and nutrient concentrations are higher (Vogt et al. 1995; Hodge 2004). However, unlike crowns, root systems of trees do intermesh.

6.2.3 Suppression and Self-Thinning

Because one-sided competition widens the size distribution of individuals and increases the differences in their competitive status, there comes a point when the most suppressed individuals cannot grow and hence survive any longer. The functional and structural acclimations may prolong but not stop this process. For example, the acclimation to maintain height growth helps smaller trees to keep their light-use efficiency in par with their larger neighbours. However, this occurs at the cost of the supporting structures that need to be built to remain close to the top of the canopy, and leads to a reduction in crown ratio. The relative cost of height growth increases with decreasing crown ratio, because it implies a higher proportion of growth to be allocated to wood instead of leaves (Chap. 5). Trees with smaller crown ratios also have a smaller proportion of leaf mass in the total biomass (Sect. 4.3), implying a reduced relative growth rate of above-ground biomass regardless of the fact that the light capture by leaves is maintained.

A tree undergoes suppression when its rate of crown rise is in synchrony with its adjacent neighbours, but its height growth rate is slower than that of its neighbours. Consequently the live crown length of the suppressed tree shrinks over time, which eventually leads to the situation where the slow-growing tree's topmost branches—possibly its only branches—abut the bottom branches of the faster growing neighbours. When the slow growing tree's live crown shrinks to zero length, the tree is dead from suppression. At the stand level, suppression mortality is called *self-thinning*.

6.2.4 Implications for Modelling Competition

Because competition is about tree interactions in a confined space, a model accounting for competition must somehow be able to describe these interactions. The preceding chapters of this book have dealt with either stand level models or models of individual trees applied to the mean tree of the stand. *Mean-tree models* describe competition from the point of view of how resource limitation affects mean-tree growth and survival. In order to explicitly describe competition and the consequent differentiation between trees, we need to extend our analysis to models that depict trees of different sizes, possibly at different locations and showing different local densities in the stand.

While physiological forest models often operate on mean tree growth derived from stand-level material fluxes, most empirical models are individual-based, tree-level, or individual-tree models that describe the growth of multiple trees interacting with each other in the same area. Individual-tree models are further separated into *spatial* and *non-spatial* classes. Spatial models take account of the locations and sizes of neighbouring model trees to gauge the competitive influences on any particular model tree. Non-spatial models represent trees of different sizes as diameter or height distributions with the assumption that all trees of a size class in a stand or patch exert a similar, average influence on the trees of another size class. This is referred to as the *mean field assumption*.

Individual-based models explicitly account for the impacts of neighbours on growth, either simply on growth rates or also on structural acclimations such as slenderness, crown ratio or crown shape. In process-based models these impacts are mechanistic, environment-mediated effects on material fluxes, whereas in empirical models they are often defined through competition indices that adjust the growth of trees according to their relative size and position in the canopy. In addition to the impacts on growth, tree-level models also include a description of tree survival which is related to the competitive position of the tree.

In the following sections, we review approaches to competitive interactions in models with different degrees of complexity. We move from mean-tree models to tree-level models and cover models that consider different aspects of competition, from survival through resource limitation to structural acclimations. Finally, we review some more advanced approaches, mainly including structural acclimations in spatial models.

6.3 Simple Stand-Level Approaches to Competition

When resources are limiting at the stand level, no further investment to leaf or fine-root mass or area will increase the resource acquisition of the whole stand. Simple mass-balance models demonstrate that total leaf mass (Eqs. (3.17) and (3.18)) and fine-root mass (Eqs. (3.27) and (3.30)) eventually reach dynamic equilibria where net growth vanishes. However, some individuals will continue to grow while others are bound to give way. The simplest models of competition are defined at stand level, and they are concerned with the growth of the mean tree in a situation where all stand-level resources are tied up, in other words, mean tree growth requires that the stand is undergoing self-thinning.

In forestry, this problem has mainly been considered empirically in a geometrical framework focusing on crowding in space. Classical models in this category include the Yoda self-thinning rule and the Reineke model. The geometrical-empirical approach considers competing trees as three-dimensional objects packed in a space restricted in two dimensions. Growth can therefore be maintained as long as it is solely directed upwards. The various approaches differ in their assumptions on how individual tree shapes respond to crowding. The Yoda approach effectively postulates that trees of different sizes retain their geometric similarity, while the Reineke model allows for changing of shape as trees grow taller.

6.3.1 *The Yoda Rule*

Let us consider a stand of plants where the average plant biomass is $\bar{W}$ (kg) and the live plant density—the number of live plants per unit ground area—is $\mathcal{N}$ (m^{-2}). The self-thinning rule of Yoda et al. (1963) defines a maximum for $\bar{W}$ that scales with $\mathcal{N}$ raised to the $-3/2$ power. When average plant biomass increases toward the maximum, $\bar{W}_{\max}$, self-thinning ensues, so that the stand density is reduced and, simultaneously, the maximum average plant biomass is increased. Thus, the self-thinning rule posits a reciprocal relationship between maximum average plant biomass and the number of plants per unit ground area raised to the 3/2 power, i.e.,

$$\bar{W} \leq \bar{W}_{\max}$$

where

$$\bar{W}_{max} = a\mathbb{N}^{-3/2} = \frac{a}{\mathbb{N}^{3/2}} \tag{6.1}$$

Yoda et al. (1963) expressed the view that this quantitative relation is so universally found in pure stands of various plant species that it may be termed the three-halves power law of self-thinning. Although some usage of this label persists, more commonly (6.1) is called *the self-thinning rule* or, more explicitly, *the Yoda rule*. White (1981) indicated that (6.1) should be considered a boundary, i.e., "the maximum permissible combination of density and mean plant biomass (aerial parts only considered) which it is empirically possible to achieve in plant populations."

To evoke a literal interpretation of the self-thinning boundary, we note that the reciprocal of stand density is equivalent to average ground area per plant, A, and, therefore, $\mathbb{N}^{-3/2} = (A^{-1})^{-3/2} = A^{3/2}$. Moreover, the square root of the average ground area per plant is equivalent to the average spacing, X, so $A^{3/2} = X^3$. Ergo, at the boundary line, average plant mass is proportional to the cube of the average spacing among the plants, i.e.,

$$\bar{W}_{max} = aX^3 \tag{6.2}$$

It is now easy to discern that the constant, a, has units of kg DW m^{-3}.

A graphic display of the boundary line is usually presented with logarithmic axes, so that W_{max} plots against $\mathbb{N}$ with a slope of $-3/2$. White (1981) indicated that a self-thinning trajectory of an actual stand, i.e., a graph of $\bar{W}$ versus $\mathbb{N}$ over a sequence of time, may closely approach the boundary, but should not cross the boundary. Weller (1990) suggested that an actual trajectory should be called a *dynamic thinning line* to distinguish it from the theoretical boundary (Fig. 6.5a).

In the last half century, fascination about the rule has generated numerous analyses, interpretations, and exposés (e.g., Zeide 1987; Norberg 1988; Reynolds and Ford 2005), not to mention debate about its accuracy and usefulness (e.g., Osawa and Sugita 1989; Lonsdale 1990; Weller 1990, 1991). Part of the rule's allure has been attributed to its tying production ecology on the left-hand side of the equation to population ecology on the right. Many investigators of self-thinning in stands of trees have redefined $\bar{W}$ as average stem biomass or stem volume.

Westoby (1981) suggested an alternative formulation of the Yoda rule that obtains by multiplying both sides of (6.1) by $\mathbb{N}$. This results in a rule that relates total plant biomass per unit ground area, $\mathbb{W} = \bar{W}\mathbb{N}$ (kg m^{-2}), to the number of plants per unit ground area, $\mathbb{N}$, i.e., $\mathbb{W} = a\mathbb{N}^{-1/2}$.

Above we noted, as have others before us (e.g., Crawley 2007), that growth drives self-thinning, so it seems prudent to re-write Westoby's version of the Yoda rule thus:

$$\mathbb{N} \leq \mathbb{N}_{max}$$

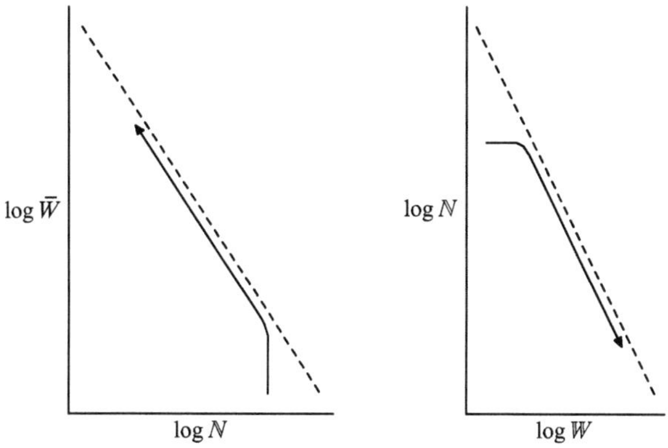

Fig. 6.5 Depiction of a boundary line (dash) and a dynamic thinning line (solid) for the Yoda rule. Left: Yoda's formulation. Right: Westoby's formulation

where

$$N_{\max} = \left(\frac{a}{W}\right)^2 \tag{6.3}$$

which indicates a straightforward reciprocal relationship between the maximum number of plants per unit ground area and the square of the total biomass per unit ground area (Fig. 6.5b).

To use the rule to model a decrease in stand density from an increase in tree biomass, we would need additional information, e.g., a carbon-balance model, to provide the amount of increase in tree biomass. Then, for example, we might assume that little or no self-thinning occurs from time t to $t + \Delta t$ if $N_t < (a/W_{t+\Delta t})^2$; otherwise, active self-thinning is occuring. Hence,

$$N_{t+\Delta t} = \begin{cases} N_t, & \text{if } N_t \le \left(\dfrac{a}{W_{t+\Delta t}}\right)^2 \\ N_t \left(\dfrac{W_t}{W_{t+\Delta t}}\right)^2, & \text{if otherwise} \end{cases} \tag{6.4}$$

This particular formulation provides an abrupt transition from no self-thinning to active self-thinning that tracks the boundary. The model could be tweaked to provide a gradual transition. However, the Yoda rule is just a predicted stand-level outcome of lower-level dynamics. As was mentioned in the previous section, the predominant mechanism of self-thinning in forests involves the overtopping of the shrinking crowns of shorter trees by the expanding crowns of their taller adjacent neighbours. Timing of the onset and the subsequent rate of self-thinning depend on the spatial arrangement of the trees and the differentiation of their crowns, which in turn depends on the degree to which rates of shoot growth vary among the neighbouring trees, which depends, in part, on the acquisition of resources below ground. It may be more profitable to model these lower-level dynamics to see what time-course of self-thinning emerges.

6.3.2 The Reineke Rule

Weller (1987) provided evidence from statistical analyses that the Yoda rule is seldom an accurate description of the dynamic thinning lines of stands of trees. One alternative is the Reineke rule, which preceded the Yoda rule by three decades (see Reineke 1933). The rule relates the stand density, N, to the quadratic mean diameter of the trees:

$$N \leq N_{\max}$$

where

$$N_{\max} = \frac{k}{{D_q}^{1.605}} \tag{6.5}$$

The quadratic mean diameter, D_q, is the diameter corresponding to the average basal area per tree; it is calculated from dbh's of, say, n trees thus:

$$D_q = \sqrt{\frac{\sum_i \mathrm{dbh}_i^2}{n}} \tag{6.6}$$

The Reineke rule was developed by an American forester, L.H. Reineke, to define an index of stand density. The Reineke rule was always intended to be a boundary line. However, subsequent work has revealed that the exponent often deviates from the published value of 1.605 (e.g. Zeide 1985).

6.3.3 Summary: Geometrical-Empirical Approach

The geometrical-empirical approach to mean-tree mortality at the self-thinning line has been very widely applied in both empirical and process-based models. The same approaches have also been modified for individual-based models. The models have been found to be remarkably consistent with empirical data, yielding stable self-thinning lines for different species (Pretzsch et al. 2014).

The theoretical basis of the Yoda model derivation was, at least implicitly, that trees are self-similar objects (see Chap. 4). If this were the case, then the exponent of D_q of the Reineke rule should be 2 rather than 1.605. An exponent less than 2 implies that as the basal diameter of trees increases, the maximum stem density becomes progressively larger than that expected from the assumption of self-similarity. This is consistent with the assumption that crowding is more directly related to crown diameter than basal diameter, because the crown ratio of trees in stands usually decreases with increasing basal diameter. This observation was utilised by Laasasenaho and Koivuniemi (1990) who improved the precision of the rule for Norway spruce (*Picea abies* L. Karst.) plantations by substituting quadratic mean diameter at the crown base for D_q.

6.4 Models with Competition for Light

The shading of light by neighbouring trees is the main agent of competitive interactions in a vast majority of process-oriented forest growth models. A lot of physiological and environmental research has been carried out on light interception in canopies of different spatial configurations (Norman and Welles 1983; Grace et al. 1987; Oker-Blom et al. 1989; Wang and Jarvis 1990; Medlyn 2004), but the representation of shading in stand growth models has usually been non-spatial and based on some version of the Lambert-Beer law applied to crowns at variable heights. This situation corresponds to the assumption of *horizontally homogenous* canopies. These are canopies where a vertical leaf distribution of each tree (size class) is defined, but the leaf area at any given height is assumed to be evenly distributed over the whole stand area.

Non-spatial shading models are used in growth models that consist of a list of trees or tree size classes affected by the shading cast by their neighbours. Here we review a group of models that include shading by neighbours as a key factor influencing the growth rate of the subject tree, either directly or through photosynthesis. Because competition for light is asymmetrical, this leads to differentiation of the trees in different size classes. A mortality rule is then applied to first determine the overall mortality, then select the trees that are lost to mortality, with probabilities increasing with decreasing growth rate.

6.4.1 Competition for Light in Gap Models

Gap models, first introduced by the JABOWA model (Botkin et al. 1972), are a class of forest models intended for describing the process of succession. JABOWA was the first forest model to explicitly include competition for resources between individuals (of multiple species) as part of its growth description. To portray the succession process, gap models have sub-models for different tree and population processes such as growth, mortality (gap creation), seed dispersal and seedling recruitment. All these processes depend on environmental drivers through phenomenological equations. The now well established gap model paradigm defines a framework for describing these successional processes on the basis of the following four axioms (Bugmann 2001):

1. Forests are represented as a collection of small patches with variable successional stage and age.
2. Patches do not interact with other patches.

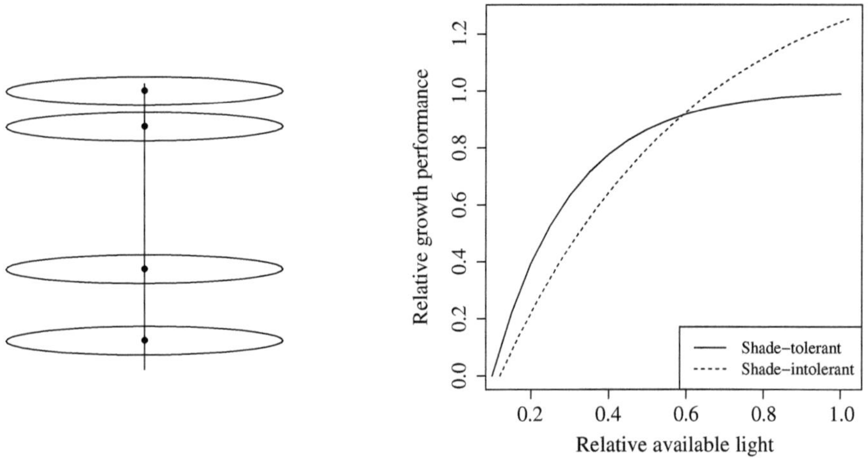

Fig. 6.6 Effect of shading by neighbours in gap models. Left panel: Leaf area of each tree in a gap is placed as a uniform, thin disk at the top of the tree. Right panel: The growth of shade tolerants and shade intolerants reacts differently to the relative availability of light. Re-drawn from Bugmann (2001)

3. All patches are homogeneous in size and resource level.
4. The leaves of each tree are concentrated at the tree top as an infinitesimally thin disk.

The two latter axioms imply total one-sided competition for light: the taller trees shade the shorter trees. The growth rate of each tree on a patch is determined as a phenomenological function of the tree's shade-tolerance class—tolerant or intolerant—and of the amount of light absorbed by its leaf disk, as calculated by the Lambert-Beer model (Fig. 6.6).

The gap model approach, first presented in the JABOWA model, has subsequently been adopted by many successors. However, in later developments of gap models, especially after the 1990s, one or more of the axioms has been relaxed and the equations for the key processes have been developed, generally towards more process-based methods (Bugmann 2001; Fischer et al. 2016). The concept of gap models still covers models of forest succession that include the four key sub-models mentioned above.

The shading function remains an important, though not the only, determinant of species competitive status in a mixed stand undergoing succession. Other factors include impacts of temperature and soil resources on growth, mortality and recruitment. The relative impacts on species vary from one environment to another. However, in applications to single species stands, competition for light essentially determines the relative growth rates and the consequent development of size distributions of individuals (Alam et al. 2010).

In the original gap models, competition for light affects the growth rate of stem diameter, and all other tree variables are allometric functions of diameter.

The models do not therefore consider structural acclimations. However, some more recent gap models separate diameter and height growth to allow stem slenderness to vary with stand density (Fischer et al. 2016; Lindner et al. 1997).

The original gap models also describe density-dependent mortality as a function of stem diameter, the smaller diameters having a lower probability of survival (Bugmann 2001).

6.4.2 Models with Photosynthesis

Instead of concentrating all leaf area at one vertical location, tree crowns can be considered as vertical distributions of leaves in a horizontally homogeneous canopy. This allows us to account for the overlapping of crowns as they shade each other. We demonstrate this here using an example of two size classes with heights H_1 and H_2 and crown base heights H_{c_1} and H_{c_2}. We can then apply the Lambert-Beer model separately to the parts of the canopy occupied by either one or both of the size classes. Assume, first, that $H_1 > H_2$ and $H_{c_1} > H_{c_2}$. The amount of light absorbed by tree class 1 above the top of tree class 2 is (see (4.51))

$$I_{\text{abs},1} = I_0 \left(1 - e^{-k_i L_{11}}\right) \tag{6.7}$$

where k_i is the effective light extinction coefficient of the canopy of tree class i, I_0 is the incident radiation above the canopy, and L_{11} is the leaf area of size class 1 located above the top of size class 2. It is important to notice that the effective extinction coefficients of the different size classes may be different because of differences in the clustering of leaves between different competitive positions. The radiation remaining to enter the canopy at height H_2 is $I_0 e^{-k_1 L_{11}}$. This is further absorbed by both tree classes forming the canopy down to height H_{c_1}:

$$I_{\text{abs},2} = I_0 e^{-k_1 L_{11}} \left(1 - e^{-k_1 L_{12} - k_2 L_{21}}\right) \tag{6.8}$$

where L_{12} and L_{21} are the leaf areas of size class 1 and 2, respectively, between heights H_2 and H_{c_1}. It is usually assumed that the total absorbed light is distributed between the trees or groups of trees in proportion to their leaf area, implying that

$$I_{\text{abs},2\text{i}} = I_{\text{abs},2} \frac{L_{i2}}{L_{12} + L_{21}} \tag{6.9}$$

where $I_{\text{abs},2\text{i}}$ is the light absorbed by tree class i between heights H_2 and H_{c_1} and L_{i2} is the corresponding leaf area.

The bottom layer of size class 2 absorbs a part, $I_{\text{abs},3}$, of the light remaining below H_{c_1}:

$$I_{\text{abs},3} = I_0 e^{-k_1 (L_{11} + L_{12}) - k_2 L_{21}} \left(1 - e^{-k_2 L_{22}}\right) \tag{6.10}$$

The same logic can be applied to any number of partially overlapping canopy layers, allowing us to consider the distribution of absorbed photosynthetic light among tree size classes with specified vertical distribution, while technically assuming that the horizontal distribution is homogeneus. One of the first studies considering the competition between tree size classes in a stand using photosynthetic light as a basis was that by Mäkelä and Hari (1986). Härkönen et al. (2010) used the approach to a mixed stand where the effective extinction coefficients were evaluated according to Eq. (4.52) (Duursma and Mäkelä 2007).

Forrester et al. (2014) developed summary equations of mutual shading for stands of considerable heterogeneity on the basis of a detailed 3-dimensional light interception and photosynthesis model, Maestra (Grace et al. 1987; Wang and Jarvis 1990; Medlyn 2004). The summary model is similar to Eqs. (6.7)–(6.10): the photosynthetically active radiation (PAR) absorbed by each layer is calculated starting with the highest layer, then reducing the available PAR for each layer by the absorbed PAR of all higher layers. The fraction of the available radiation absorbed by subgroup i at layer j is

$$f_{ij} = \lambda_{v_i} \lambda_{h_j} \left(1 - e^{-\sum_i k_{\mathrm{H}i} L_i}\right) \tag{6.11}$$

where $k_{\mathrm{H}i}$ is the homogenous canopy extinction coefficient of subcanopy i, L_i is its respective leaf area index, and λ_{v_i} and λ_{h_j} are factors correcting for vertical and horizontal heterogeneity, respectively. Forrester et al. (2014) provide estimated values of $k_{\mathrm{H}i}$ for a number of species and approximate equations for the correction factors as functions of $k_{\mathrm{H}i}$, L_i, and relative canopy heights. They show fair correspondence between their approach, the Maestra model and measurements in four forest ecosystems located in different vegetation zones. Forrester et al. (2017) applied this method to a multilayer and multi-species version of the 3-PG model (Landsberg and Waring 1997). The availability of light affected photosynthesis and hence growth rate of the trees representing different canopy layers. Mortality was calculated using the Yoda rule for the whole stand but with species-specific parameters.

6.4.3 *Concluding Remarks on Competition in Process-Based Models*

The models described above are typical examples of including competition in process-based models. While including impacts of tree interactions on growth rate, they often ignore any acclimative effects that are manifested as a feedback from resource limitation to tree structure. For example, we saw that, in standard gap models, tree height is a function of tree diameter, leading to a pre-determined height-to-diameter ratio that does not depend on competitive status. In many photosynthesis-based models the shape of trees is not even considered, as size is represented only by diameter (Landsberg and Waring 1997). Since acclimations generally act so as to counteract some adverse effects on growth, ignoring them will likely overestimate the effects of shading on growth, unless the models are parameterised using empirical data (Forrester et al. 2017). In addition, such models will not be able to realistically represent the variability of tree structure with competitive position. Whether or not this is significant depends on the purpose of the model. However, in forestry and particularly when making predictions about timber yield, one of the key impacts of competition is the effect of competitive status on crown ratio and, consequently, on the height-to-diameter ratio. How this can be incorporated in process-based modelling is discussed in the following section.

6.5 Models with Structural Plasticity

In our process-based approach to modelling (Chap. 5), the crown-length dynamics and any resultant cross-sectional growth at breast height are key determinants of growth. Through parallel analysis of structure and function, they are defined as the integrated results of the rates of photosynthesis, respiration, carbon allocation, organ turnover, and branch loss as influenced by the environment and the competitive forces wrought by neighboring trees. Crown ratio and the related height-to-diameter ratio are therefore inherent in the makeup of the model.

The focus of this section is to explore the interaction of crown rise—and hence crown ratio—and stand density, controlled by mortality. We begin by presenting a general rule based on crown length, the idea of which is related to the Yoda and Reineke models, in the sense that it is based on the space requirements of trees in a crowded stand. However, here we use a more direct biological argument related to the space required by crowns, and allow for variable crown ratios instead of self-similar trees as in the original Yoda approach. We first apply the rule to a mean-tree model, then move on to an individual tree model where the sizes of trees are specified but not their locations. The same model is applicable regardless of whether the individual-tree model classifies as spatial or non-spatial.

6.5.1 A Crown-Length Rule

A self-thinning rule based on crown length is inferred by the crown-rise model described in Sect. 2.4. The crown-rise model is based, in part, on the assumption that the crown-base height, H_c (m), is constrained by the average spacing, X (m), when the stand is closed and crown rise has begun, i.e.,

$$H_c = H - \alpha_x - \beta_x X \tag{6.12}$$

where H (m) is average tree height, and α_x (m) and β_x are parameters. The average crown length, $L_c \equiv H - H_c$, is

$$L_c = \alpha_x + \beta_x X \tag{6.13}$$

and the spacing, expressed in terms of crown length, is

$$X = \frac{L_c - \alpha_x}{\beta_x} \tag{6.14}$$

Because stand density (trees m^{-2}) is $\mathcal{N} \equiv 1/X^2$, we can transform (6.14) into a self-thinning boundary line for a closed stand undergoing crown rise, that is,

$$\mathcal{N} \le \mathcal{N}_{max}$$

where

$$\mathcal{N}_{max} = \left(\frac{\beta_x}{L_c - \alpha_x}\right)^2, \qquad L_c > \alpha_x \tag{6.15}$$

Self-thinning, under this rule, is forced by an increase in average crown length, which is achieved through height growth.

As was mentioned, self-thinning and the resultant change in spacing ordinarily occur after stand closure and the differentiation of tree crowns, when faster growing trees with elongating crowns overtop or severely shade slower growing trees with shrinking crowns. Because the larger trees survive the self-thinning, the average rate of crown rise is less than the rate of height growth, which results in an increase in average crown length. Two questions that must be addressed are: (a) when does a stand differentiate sufficiently for self-thinning to commence, and (b) how much slower is the rate of crown rise compared to the rate of height growth during self-thinning?

We addressed both of these questions in Sect. 2.4.3. For (a), let $X(0)$ and $H_c(0)$, be the respective starting values of average spacing and crown-base height. The average tree height at time $t \ge 0$ is $H(t)$. Under the assumptions of the crown-rise model, the stand is open at time t if

$$H_c(0) > H(t) - \beta_x X(0) \tag{6.16}$$

Closure of the stand occurs when tree height increases to the point where

$$H_c(0) = H(t) - \beta_x X(0) \tag{6.17}$$

and the onset of crown-rise occurs when the tree height increases further to the point where

$$H_c(0) = H(t) - \alpha_x - \beta_x X(0) \tag{6.18}$$

Thus let $H_x = H_c(0) + \beta_x X(0)$ denote the average tree height when the model stand first closes. Following a silvicultural thinning at time $t = T$, substitute $H_x = H_c(T)+\beta_x X(T)$. We assume that self-thinning commences when the average height of a crown base, H_c, rises to approach $k_x H_x$, where k_x is a parameter.

To address question (b), we choose to work with spacing rather than stand density because the math is simpler. The time derivative of (6.14) is:

$$\begin{aligned}\frac{dX}{dt} &= \frac{1}{\beta_x}\frac{dL_c}{dt}\\ &= \frac{1}{\beta_x}\left(\frac{dH}{dt} - \frac{dH_c}{dt}\right)\end{aligned} \tag{6.19}$$

Following the approach described in Section 2.4.3, we assume that crown rise is a fraction of average height growth rate, i.e., $dH_c/dt = (1-u)dH/dt$ where $0 < u \leq 1$. Substituting into (6.19) provides the rate of increase in spacing that results from the self-thinning of the model stand, i.e.,

$$\frac{dX}{dt} = \begin{cases} 0, & \text{if } H_c \leq k_x H_x \\ \dfrac{u}{\beta_x}\dfrac{dH}{dt}, & \text{if otherwise} \end{cases} \tag{6.20}$$

The solution of the model is identical to (2.75). Utilisation of a switching function, (2.76), from Section 2.4.3 smooths the transition from the first to the second case:

$$\frac{dX}{dt} = S_x \frac{u}{\beta_x}\frac{dH}{dt} \tag{6.21}$$

Note that this model predicts that the average spacing among trees increases linearly with average tree height in an actively self-thinning stand.

6.5.2 A Mean-Tree Model with Crown Rise and Self-Thinning

Below, we present a mean-tree model that integrates the effects of shading on photosynthesis, the effects of crowding on crown rise, and the combined effects of shading and crowding on tree growth and mortality.

We assume that (a) a model stand has density $N(t)$ (m^{-2}) at time t, and (b) the dynamics of each model tree within the model stand is described by the tree-level model that we formulated in Chap. 5. The model stand, in effect, comprises identical model trees that grow in size and decrease in number over time. We also assume that the stand density is constant until after the onset of crown rise. The rate of mortality from self-thinning is driven by the rate of height growth in accordance with the crown-length rule.

To begin, we assume that an average model tree is summarised by three state variables, viz., tree height, H (m); height of crown-base, H_c (m); and the average between-tree distance, i.e., spacing, X (m). Tree basal area, B (m^2) and inactive stem biomass, W_q (kg DW) are response variables. A starting value is needed for each of the state variables. We use a light-use-efficiency approach combined with the Lambert-Beer equation for shading to model the rate of photosynthesis.

Calculation of the rates of change for the state variables requires the following preliminary calculations:

$L_c = H - H_c$	Crown length (m)
$W_f = \eta_s \beta_A L_c^z$	Leaf biomass (kg DW)
$W_r = \alpha_r W_f$	Fine-root biomass (kg DW)
$W_s = (\beta_1 H + \beta_2 H_c) W_f$	Active wood biomass (kg DW)
$\mathbb{N} = 1/X^2$	Stand density (m^{-2})
$\mathbb{L} = a_{SL} W_f / X^2$	Leaf area index
$P = \mathbb{P}_0(1 - \exp(-k_p \mathbb{L}))/\mathbb{N}$	Tree photosynthesis rate (kg C yr^{-1})
$R_M = r_{M_f} W_f + r_{M_r} W_r + r_{M_p} W_p$	Tree maint resp rate (kg C yr^{-1})
$G = Y_G(P - R_M)$	Tree growth rate (kg DW yr^{-1})
$H_r = \alpha_x + \beta_x X(0) + H_c(0)$	Ave height at crown-rise
$S_r = H^{n_r}/(H^{n_r} + H_r^{n_r})$	Switch on crown rise
$H_x = \beta_x X(0) + H_c(0)$	Ave height at closure
$S_x = H_c^{n_x}/[H_c^{n_x} + (k_x H_x)^{n_x}]$	Switch on self-thinning

The rates of change in the state variables:

$$\frac{dH}{dt} = \frac{L_c}{z(W_f + W_r + W_p) + \beta_1 W_f L_c} \left(G - \frac{W_f}{T_f} - \frac{W_r}{T_r} \right) \tag{6.22}$$

$$\frac{dX}{dt} = S_x \frac{u}{\beta_x} \frac{dH}{dt} \tag{6.23}$$

$$\frac{dH_c}{dt} = \max \left\{ 0, S_r \left(\frac{dH}{dt} - \beta_x \frac{dX}{dt} \right) \right\} \tag{6.24}$$

$$\frac{dW_q}{dt} = \left(\frac{z W_s}{L_c} - \beta_{2_s} W_f \right) \frac{dH_c}{dt} \tag{6.25}$$

$$\frac{dB}{dt} = \begin{cases} 0, & \text{if } H \leq 1.3 \\ z\beta_A(H-1.3)^{z-1}\dfrac{dH}{dt}, & \text{if } H_c \leq 1.3 < H \\ \dfrac{\rho_s}{\rho_m} z\beta_A(H-H_c)^{z-1}\dfrac{dH}{dt}, & \text{if } 1.3 < H_c \end{cases} \tag{6.26}$$

Calculation of stand-level biomass components is straightforward, for example, $\mathbb{W}_i(t) = \mathbb{N}(t)W_i(t)$ for $t = t_0, t_1, t_2, \ldots, t_m$. Likewise, basal area of the model stand is $\mathbb{B}(t) = \mathbb{N}(t)B(t)$.

We solved the mean-tree model with five different initial spacings, viz., $X(0) =$ 1.22, 1.86, 2.44, 3.05, 3.66 m. For all five runs we used the same starting values for both tree height, $H(0) = 0.1$ m, and for crown-base height, $H_c(0) = 0$. In lieu of specifying a value for β_A, we set the cross-sectional area at the crown base at $A_c(0) = 0.000003\,\mathrm{m}^2$ and calculated $\beta_A = A_c(0)/H_c^z(0)$. The values assigned to the fundamental parameters are provided in Table 6.1. We calculated the values of the composite parameters, i.e., β_1, β_2, and β_{2_s}, with the formulas provided in Chap. 5. Results are depicted in Figs. 6.7, 6.8, 6.9, 6.10, and 6.11.

6.5.3 A Tree-Level Model with Crown Rise and Self-Thinning

A tree-level model—based on the theory reviewed in this book—can be formulated through minor changes to the mean-tree model. Typically one would define an initial model-tree count for a defined area of ground and calculate the tree density, $\mathbb{N}$ (trees m^{-2}). Each model tree needs a starting height (m) and a specific rate of photosynthesis. Varying the values the specific photosynthetic rates among the model trees provides the means for differentiation of height growth and consequent self-thinning. A positive correlation between starting height and specific photosynthetic rate is reasonable in this regard. Another reasonable assumption, and one upon which the crown-rise model was originally based, is that the crown-base height is the same for all the model trees. Thus, a crown-base height, $\bar{H}_c$ (m), common to all model trees is calculated with the crown-rise model, $\bar{H}_c = \max\{0, \bar{H} - \alpha_x - \beta_x X\}$, from the mean of the model-tree heights and the average spacing, $X = \sqrt{1/\mathbb{N}}$ (m). A more elaborate crown-rise model provides for individualized crown-base heights

$$H_{c_i} = \max\{0, \bar{H} - \alpha_x - \beta_x X + \gamma_x(H_i - \bar{H})\} \tag{6.27}$$

Individual crown-base heights may occur above or below the stand average (Valentine et al. 2013). The expectation, in concert with the analysis of Sprugel (2002), is that the shorter trees will have the lower crown-base heights.

A projection period ordinarily is divided into m discrete time steps, e.g., years. For each time step, 1 to m, we require the requisite preliminary calculations and solution of (6.22), (6.25), and (6.26) for each of the model trees in turn (we suggest

Table 6.1 Parameter values used in the mean-tree model solutions

Symbol	Value	Unit	Definition
$\mathbb{P}_0$	3.0	kg C (m^2 ground)$^{-1}$ yr^{-1}	Max photosynthesis
a_{SL}	3.0	m^2 (kg DW)$^{-1}$	Specific leaf area
k_p	0.6	–	Light extinction
r_{M_f}	0.25	kg C (kg DW)$^{-1}$ yr^{-1}	Leaf respiration
r_{M_r}	0.25	kg C (kg DW)$^{-1}$ yr^{-1}	Fine-root resp
r_{M_p}	0.025	kg C (kg DW)$^{-1}$ yr^{-1}	Live wood resp
Y_G	1.54	kg DW (kg C)$^{-1}$	Growth efficiency
z	2.0	–	Exponent
α_r	0.25	–	Fine root:leaf
ϕ_s	$1/(1+z)$	–	Upper stem form factor
ϕ_b	$0.75-\phi_s$	–	Branch form factor
ϕ_t	0.25	–	Root form factor
ρ_b	400	kg DW m^{-3}	Branch wood density
ρ_m	500	kg DW m^{-3}	Mature wood density
ρ_s	400	kg DW m^{-3}	Juvenile density
ρ_t	400	kg DW m^{-3}	Root wood density
η_b	400	kg DW m^{-2}	Branch pipe-model ratio
η_s	460	kg DW m^{-2}	Stem pipe-model ratio
η_t	1200	kg DW m^{-2}	Root pipe-model ratio
α_x	1.45	m	Crown-rise parameter
β_x	1.65	–	Crown-rise parameter
k_x	2	–	Switch parameter
n_x	4	–	Switch exponent
n_r	4	–	Switch exponent
u	0.25	–	Control parameter

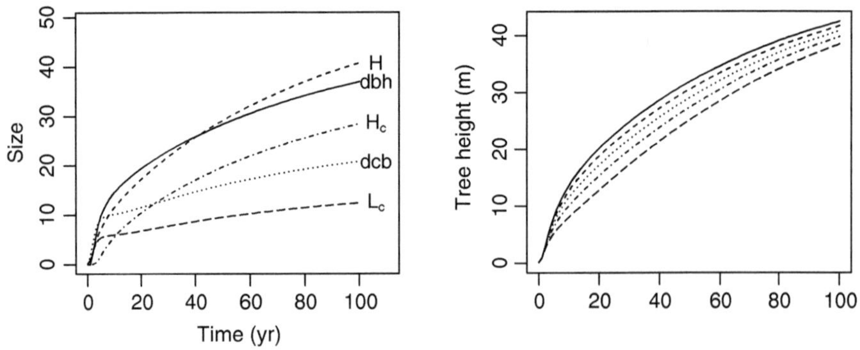

Fig. 6.7 Left: Time-courses of average model-tree dimensions resulting from a starting spacing of 2.44 m. Right: Time-courses of average model-tree height resulting from each of the five starting spacings: 1.22 m, long dash; 1.86 m, dash dot, 2.44 m, dot, 3.05 m, short dash, 3.66 m, solid

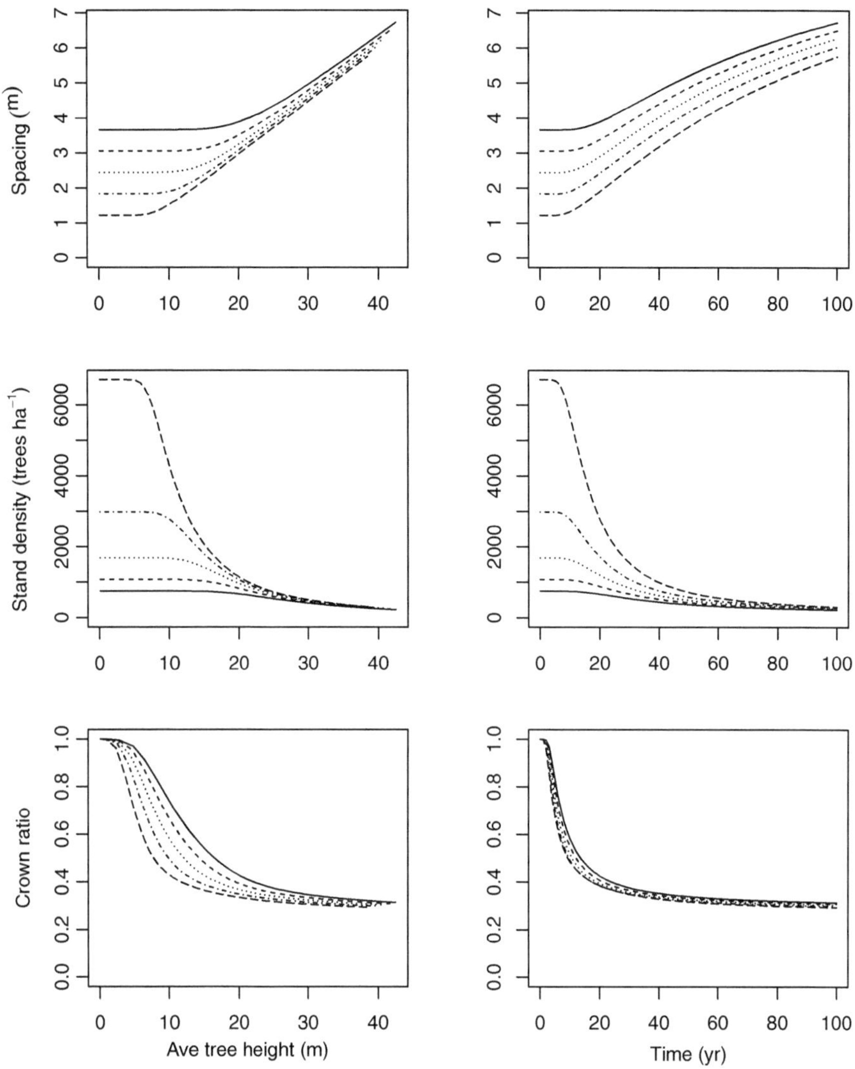

Fig. 6.8 Time courses of average spacing, stand density, and crown ratio versus average model-tree height (left) and time (right). Initial spacing: 1.22 m, long dash; 1.86 m, dash dot, 2.44 m, dot, 3.05 m, short dash, 3.66 m, solid

Euler's method), and then the recalculation of the individual tree heights. Any model trees whose crown lengths have shrunk to ≤ 0 or to some prescribed minimum are presumed dead from self-thinning, thus reducing N and increasing the average spacing, X.

A review of empirical non spatial models of stand growth, including those that provide for multiple species, is provided by Weiskittel et al. (2011).

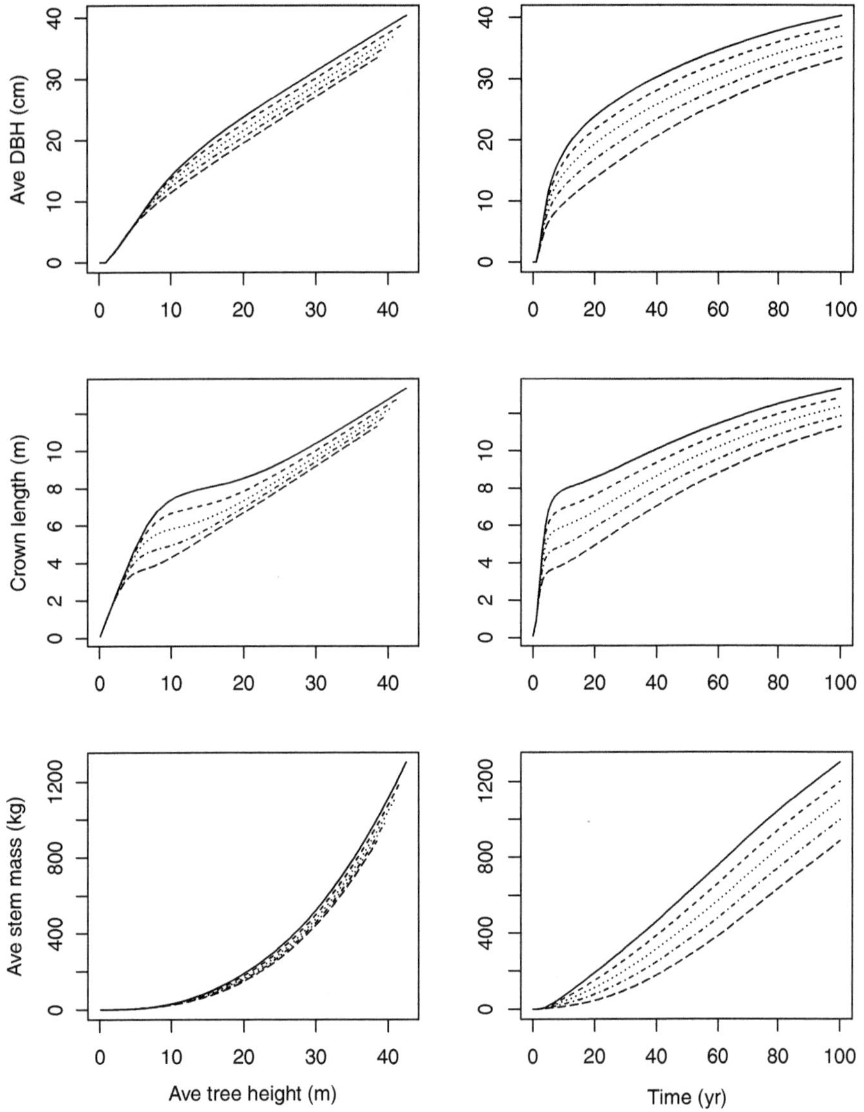

Fig. 6.9 Time courses of average dbh, crown length, and stem mass versus average model-tree height (left) and time (right). Initial spacing: 1.22 m, long dash; 1.86 m, dash dot, 2.44 m, dot, 3.05 m, short dash, 3.66 m, solid

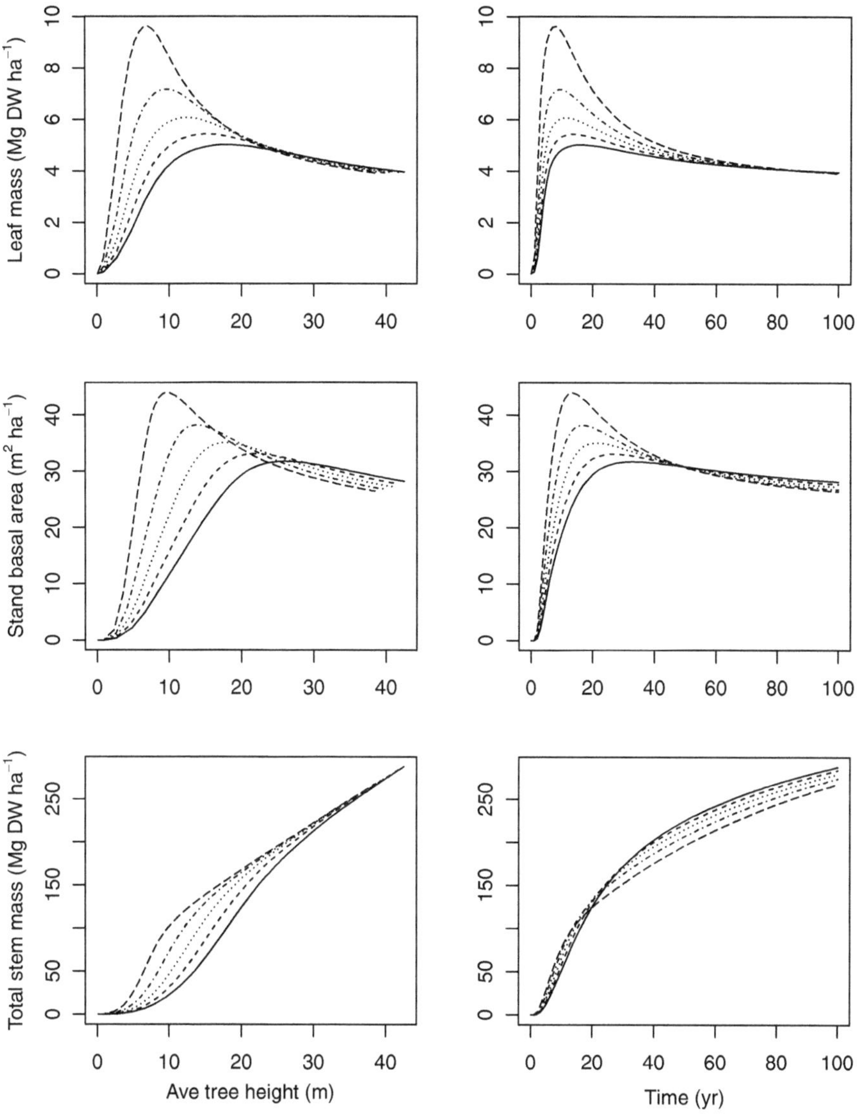

Fig. 6.10 Time courses of average leaf mass, stand basal area, and total stem mass of the model stand versus average model-tree height (left) and time (right). Initial spacing: 1.22 m, long dash; 1.86 m, dash dot, 2.44 m, dot, 3.05 m, short dash, 3.66 m, solid

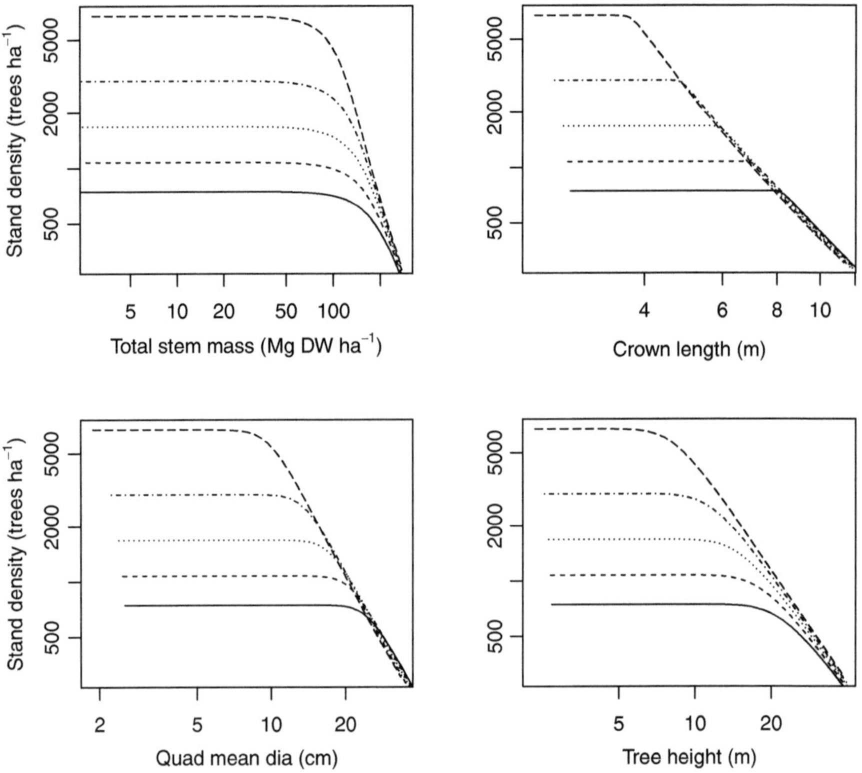

Fig. 6.11 Development of stand density and stand and tree characteristics in self-thinning canopies described with a mean-tree approach. Initial spacing: 1.22 m, long dash; 1.86 m, dash dot, 2.44 m, dot, 3.05 m, short dash, 3.66 m, solid

6.5.4 Summary: Crown Rise and Space

The above models represent simple but efficient approaches that combine the effects of tree interactions on growth rate and structural acclimations. They make use of the Yoda and Reineke type of models, however, acknowledging that not whole trees but rather their crowns are the objects that compete for space in the canopy of a forest stand. That these assumptions already provided a logical description of growth patterns in an even-aged stand for the mean tree (Figs. 6.8, 6.9, 6.10, and 6.11), suggests that the feedback between crown rise and stand density is a significant factor in stand dynamics.

Tree interactions in more complex stands, such as stands comprising multiple species with differing shade tolerances, with each species possibly represented by trees with a wide assortment of sizes, are likely more versatile than just a feedback between crown rise and self-thinning. We will briefly review some spatial approaches to both homogenous and complex stands.

6.6 Spatial Approaches

Spatial models use the locations and sizes of the model trees to simulate crown dynamics. The model stand may be surrounded with a buffer containing additional model trees to circumvent unwanted edge effects on the model trees of interest. An elegant mechanistic scheme for bounding model-tree crowns in three dimensions was devised by Mitchell (1975). In recent years, researchers codified morphological plasticity (lean and sweep) in stems, which provides more realistic crown dynamics, especially pertaining to broad-leaved species (e.g., Purves et al. 2008). García (2014) contends that allowing for or simply recognising plasticity in non-spatial models largely obviates the need for spatial models for most purposes, though spatial models are useful for very detailed studies of tree and crown dynamics. Spatial analyses also come in useful if the model is extended to other processes in addition to growth, such as regeneration or habitat requirements for forest fauna.

6.6.1 Spatial Crown-Rise Model

The crown-rise model also generalizes to three-dimensional crowns (Valentine et al. 2000). Suppose, for example, that X_{ij} is the distance between adjacent model trees i and j. Let H_i and H_j denote the respective tree heights. Further, let r_{ij} be the distance from model tree i toward model tree j where the crown-length frame of model tree i abuts that of model tree j. This point occurs at distance $X_{ij} - r_{ij}$ from model tree j, therefore

$$H_i - \frac{r_{ij}}{\tan\phi} = H_j - \frac{(X_{ij} - r_{ij})}{\tan\phi} \tag{6.28}$$

so that

$$r_{ij} = \frac{X_{ij} + (H_i - H_j)\tan\phi}{2} \tag{6.29}$$

Model tree i would be overtopped by model tree j if r_{ij} were less than 0. Likewise, model tree j would be overtopped by model tree i if r_{ij} were greater than X_{ij}. However, as depicted in Fig. 6.12, the crown-length frame of the model tree i abuts the crown-length frame of model tree j at a distance $0 < r_{ij} < Xij$ from model tree i.

In the two-dimensional depiction in Fig. 6.12, the crown-base heights of each of three model trees differ from one side to the other. In reality, crown-base height may vary in several directions around the central stem. One way to proceed is to pixelate the ground area of the model stand and use trigonometry to determine which model tree's crown, if any, covers each pixel (see García 2014).

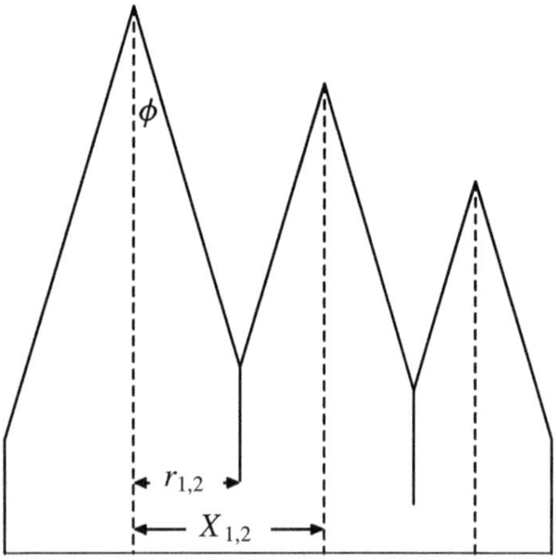

Fig. 6.12 Asymmetrical crown-length frames develop in spatial models

6.6.2 *Perfect Aggregation*

A vast majority of forest growth models were intended for projections of managed stands that usually follow even-aged management of a single-species tree crop. The spatial pattern of a managed stand is usually quite uniform, while competitive interactions occur between trees of different crown height. Most models describing managed stands are therefore individual-tree, non-spatial models. There is theoretical evidence that in such cases a spatial approach adds little to the dynamics of the stand.

Kokkila et al. (2006) considered the aggregation of a spatial model of even-aged pine stand growth (ACROBAS). The tree model in ACROBAS is essentially the model presented in Chap. 5. They applied the concept of *perfect aggregation* to analyse if the model could be aggregated such that a stand-level analysis could produce essentially the same dynamics as the spatial tree-level model. Two simplifications of the model were analysed, a mean tree model (MTM) and a distribution-based model (DBM) where size differences were allowed. Both simplifications were non-spatial, i.e., the spatial configuration of the stand was considered random. The results demonstrated that when the stand configuration in the spatial model was initialised as random (Poisson-distribution), there were hardly any differences between the models. However, minor differences were caused by tree mortality which was not totally random as smaller trees were more likely to die than big ones. Secondly, as the stand configuration tended toward regularity as a result of selective mortality, the total growth in the spatial model was somewhat faster than that in the non-spatial models. Two variables, mean height and stand density, aggregated almost perfectly.

Perfect aggregation means that the same functional relationship, possibly with transformed parameter values, applies both at the micro (individual) scale and the macro (stand) scale. Kokkila et al. (2006) attributed the existence of the almost perfect aggregation to stand-level resource limitation. This was demonstrated through a simple model with carrying capacity. While carrying capacity is fixed, the mean size and number of individuals may vary over a wide range of combinations within the same site, leading, for example, to dense populations with slender trees or sparse populations with thick trees. This provides an example of perfect aggregation as presented by Iwasa et al. (1987). They showed that a two-variable system characterised by the mean size (x) and number (N) of individuals could be perfectly aggregated to a one-variable system with $y = Nx$ if the growth and mortality of individuals were constrained by some form of carrying capacity.

Let us consider a stand consisting of N individuals in an area A. We assume that the trees have a range of mutual interaction, $R \leq A$. Let us denote by n_i the number of trees that fall in the range of interaction of tree i, by $\bar{x}_i$ the mean size of trees interacting with tree i, and by x_i the size of tree i. Assume that the effect of neighbours on tree i is essentially a density effect within the range of interaction, i.e., it is a function of $n_i\bar{x}_i/R$. Further, assume that the relative growth rate of tree size is restricted by carrying capacity K. The following model is an example satisfying this requirement:

$$\frac{\mathrm{d}x_i}{\mathrm{d}t} = rx_i\left(1 - \frac{n_i\bar{x}_i}{RK}\right), \qquad i = 1, 2, \ldots, N \tag{6.30}$$

Let us further assume that (1) the spatial distribution of the trees is random, and (2) tree size and tree location are independent of each other. The perfect aggregation result by Iwasa et al. (1987) implies that for each patch, there is an aggregated model with $y_i = n_i\bar{x}_i/R$ (as long as the patches are not sparser than a certain minimum density).

We shall first analyse the mean value of y_i in a stand where the above assumptions 1 and 2 hold. We denote $\bar{x}_i = \bar{x} + \epsilon_i$, where $\bar{x}$ is stand mean and ϵ_i is the deviation of $\bar{x}_i$ from the stand mean, and ϵ_i has zero mean. This implies the following mean value of y_i:

$$\begin{aligned} y_i &= \frac{1}{NR}\sum_{i=1}^{N} n_i\,(\bar{x} + \epsilon_i) \\ &= \frac{\bar{x}}{NR}\sum_{i=1}^{N} n_i + \frac{1}{NR}\sum_{i=1}^{N} n_i\epsilon_i \end{aligned} \tag{6.31}$$

Because n_i is randomly distributed, its expected mean value over the stand is NR/A, so the first sum on the right-hand-side reduces to $\bar{x}N/A$. Secondly, because tree number and tree size were assumed independent, the expectation value of the second sum on the right-hand-side is zero. The mean value of y is hence

$$\bar{y} = \bar{x}\frac{N}{A} \tag{6.32}$$

We can also express y_i using δ_i, the deviation of its mean: $y_i = \bar{y} + \delta_i$. δ_i is defined relative to the mean of y_i in patch i and it has zero mean. This allows us to write the growth equation of tree i as

$$\frac{dx_i}{dt} = rx_i\left(1 - \frac{\bar{y} + \delta_i}{K}\right), \qquad i = 1, 2, \ldots, N \tag{6.33}$$

The mean growth rate of trees in the stand is

$$\begin{aligned}\frac{1}{N}\sum_{i=1}^{N}\frac{dx_i}{dt} &= \frac{1}{N}\sum_{i=1}^{N} rx_i\left(1 - \frac{\bar{y} + \delta_i}{K}\right) \\ &= \frac{r}{N}\left(1 - \frac{\bar{y}}{K}\right)\sum_{i=1}^{N} x_i - \frac{r}{KN}\sum_{i=1}^{N} x_i\delta_i \end{aligned}\tag{6.34}$$

The tree sizes are random and independent of location, such that x_i and δ_i are independent, and the last sum on the right-hand side vanishes. This yields for the mean growth rate of trees in the stand:

$$\frac{1}{N}\sum_{i=1}^{N}\frac{dx_i}{dt} = r\left(1 - \frac{\bar{y}}{K}\right)\bar{x} \tag{6.35}$$

Utilising (6.32), this can easily be seen to be

$$\frac{d\bar{x}}{dt} = r\bar{x}\left(1 - \frac{N\bar{x}_i}{AK}\right) \tag{6.36}$$

which is the individual-tree growth equation applied to the mean tree in the stand. The system therefore has a perfect aggregation. Alternatively, the aggregated equation can be written in terms of the patch mean of the aggregate variable y,

$$\frac{d\bar{y}}{dt} = r\bar{y}\left(1 - \frac{\bar{y}}{K}\right) \tag{6.37}$$

This result suggests that carrying capacity is a very powerful constraint for growth models. If a stand model, however complex otherwise, incorporates carrying capacity that limits growth and tree interactions due to density effects, then the above results hold promise for finding a nearly perfect aggregation for spatial models for competition.

6.6.3 *Perfect Plasticity Approximation*

The study by Kokkila et al. (2006) corroborates the intuitively reasonable claim that as long as stands are fairly homogeneous in their spatial structure, an explicit geometry of the stand will add little to stand dynamics. Indeed, the general rationale for using spatial models for the growth of managed even-aged forests is to account for irregular, clustered spatial structures created by small-scale variation in soil properties or by disturbances, such as gap formation due to falling of large trees, or mortality of seedlings through browsing by moose. Such irregularities are suggested to cause not only reduction in growth, but also a reduction in the economic value of timber, as trees round gap edges invest a lot of growth into large branches while others in clusters suffer from strong suppression.

The strongest arguments for spatial models have been presented by ecologists studying complex unmanaged stands. For example, Crawley (2007) reported of a model simplification exercise where a spatially explicit, detailed model of tree growth was tested against equally detailed measurements of tree and environment variables, including seed production and dispersal as well as timber production (Pacala et al. 1996). The model was simplified by reducing the number of variables and parameters, yet trying to retain as much explanatory power as possible. The conclusion was that as long as the model's spatially explicit character was retained, a very simple dynamic model with only one resource, namely light, was sufficient. Removing the spatial information and moving to a mean-field model, however, dramatically decreased the explanatory power of the model.

Other modellers have come to an opposite conclusion. More recent mechanistically and physiologically oriented developments within the gap model paradigm have retained the original non-spatial representation and yet report results of applications complying very well with data of fairly complex stands (Lasch et al. 2005; Fischer et al. 2016). García (2017) presents a theory-based argument that complex "systems have [emergent] properties at a macroscopic scale (e.g., at the stand level) that are hard to explain simply from microscopic properties (e.g., tree level)". It is therefore difficult to derive stand-level behaviour from a large number of complex details at the lower level without encountering chaotic behaviour that has lost the essential information about the macrolevel system. García (2017) goes on to propose and test a model defined at a coarse scale (stand-level means of cohorts) through aggregation.

Strigul et al. (2008) presented a new approach to dealing with complex stands, the Perfect Plasticity Approximation (PPA). Under PPA the spatial configuration of stands becomes redundant, because trees are always assumed to locate their leaves in the gaps in the forest regardless of the location of their stems (Fig. 6.13). This is thought to be accomplished through leaning stems and with branches growing into gaps left by dead trees. This effectively results in a horizontally homogeneous canopy, such that a non-spatial model including size classes provides an adequate description of stand growth. Strigul et al. (2008) demonstrated the approach by comparing stand development in a spatial model, SORTIE (Pacala et al. 1993), with a version of the same model that assumed plastic crowns, and further, with a fully non-spatial model based on PPA.

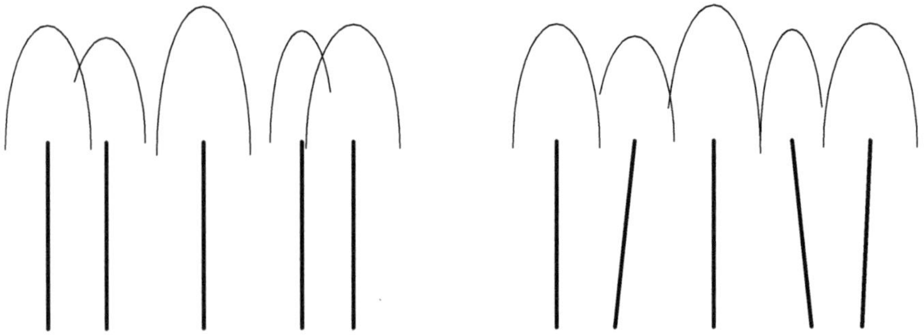

Fig. 6.13 The Perfect Plasticity Approximation assumes that tree crowns are located to fill all gaps, regardless of the location of their stems. Stem locations in the left panel are central to the crowns, whereas in the right panel PPA has been applied to locate crowns so as to fill all canopy gaps

The PPA model applies to stands with multiple species, and it describes their growth, mortality and seedling recruitment. In its simplest form, PPA uses the following assumptions (Strigul et al. 2008; Purves et al. 2008; Bohlman and Pacala 2012):

1. Tree crowns are flat disks located at the top of the stem; disk radius is an allometric function of diameter at breast height.
2. Tree height is an allometric function of tree diameter.
3. A stand consists of cohorts born each year and undergoing growth and mortality.
4. Trees are divided into canopy trees and understory trees.
5. All canopy trees have all their crowns exposed to light.
6. Understory trees are those whose crowns are not exposed to full light.

The model coincides with a continuous version of the Leslie matrix model of the development of size distributions (Strigul et al. 2008). The growth and mortality rates can be defined appropriately, but in a very simple version of the model the absolute growth rates and relative mortality rates were constant, defined separately for the canopy and understory trees (Purves et al. 2008). Despite the remarkably simple structure, the model was able to reproduce the development of basal area and species distributions in large-scale applications to temperate forests in Canada (Purves et al. 2008) as well as to tropical forests in Panama (Bohlman and Pacala 2012). Both studies were based on extensive data collection for estimating the parameter values.

García (2014) tested the applicability of PPA to even-aged, single-species stands with crowns defined as 3-dimensional geometric shapes that were also assumed to shade each other. He demonstrated that with these more realistic crown shapes, PPA was a less accurate description of whole-stand dynamics and tree-to-tree interactions. Nevertheless, tree size was by far the most important indicator of relative productivity, the spatial position only accounting for about 10% of resource capture.

The PPA approach suggests that an adequate description of forest population dynamics and succession is possible without considering spatial effects. This is a clear advantage for large-scale predictions and analyses of forest succession, especially if impacts of climate on the process rates can be incorporated. However, it is likely better suited for temperate and tropical forests than their boreal and Mediterranean counterparts, where stands are often sparse due to resource limitations other than light, and where growth rates may be too slow to maintain perfect plasticity for prolonged periods after disturbance. As the model should always be chosen for a particular purpose, there still seems to be room for analysing the effects of irregularity of structure also with spatial models of forest growth.

6.7 Exercises

6.1 It was noted in Sect. 6.3 that the Yoda model, (6.1), implicitly assumes self-similarity of tree form. Does the Yoda rule follow from this self-similarity assumption? Consider this by assuming self-similarity of trees in a stand and consider trees of variable size at the self-thinning line. Do you need additional assumptions?

6.2 In a European boreal forest the PAR radiation reaching the top of the canopy during the growing season varies approximately in the range 5500–7000 mol m^{-2} yr^{-1}. Leaf mass in a closed-canopy pine forest is 4000–6000 kg ha^{-1}. Specific leaf area (all-sided) is about 14 m^2 kg^{-1} and the extinction coefficient for all-sided leaf area is $k = 0.25$. Light-use efficiency of photosynthesis is assumed to be $\beta = 0.00027$ kg C mol^{-1}.

a. Calculate the annual photosynthesis of the whole stand using Eq. (6.7) and converting the absorbed PAR radiation to photosynthates using the light use efficiency given above.
b. Consider the stand as consisting of two size classes with appropriate heights, crown heights and proportions of total leaf mass. How is the total photosynthesis divided by the size classes, following the theory of Sect. 6.4.2?
c. As an acclimation to shading, the specific leaf area has been found to approximately double from full to zero light. How would this affect the distribution of annual photosynthesis between the size classes you considered above?

6.3 Consider the crown rise model presented in Eqs. (6.16)–(6.18). What is the significance of parameters α_{x} and β_x? How do you think the process would change if $\beta_x > 1$ and / or $\alpha_{\mathrm{x}} < 0$?

6.4 What quantities could variables x and y represent in Eqs. (6.30)–(6.37)?

Chapter 7
Tree Structure Revisited: Eco-Evolutionary Models

In this chapter we introduce the eco-evolutionary approach to modelling and review its applications for determining carbon allocation to tree structure in different environments. We begin with an introduction to optimisation ideas and a discussion of how they relate to the theory of evolution, and how such models should be used and interpreted. We then present some examples of models that utilise the eco-evolutionary approach to derive plant structure and carbon allocation. These include optimal crown structure, stem shape, different examples of the co-allocation of carbon and nitrogen, and a brief introduction to applications of game theory in the context of modelling tree structure.

7.1 Introduction

Since the early work on physiological modelling of forest stand growth (e.g., Ågren et al. 1980; McMurtrie and Wolf 1983), carbon allocation to different growth components has been recognised as a key problem (Landsberg 1986; Cannell and Dewar 1994; Le Roux et al. 2001; Poorter et al. 2011; Franklin et al. 2012; Merganičová et al. 2019). Although our ability to quantify and model processes related to carbon acquisition of trees has increased considerably (Pretzsch et al. 2008; Fontes et al. 2010), the question of carbon allocation persists. Franklin et al. (2012) argued that allocation is not a process as such, but rather a consequence of several parallel processes. No matter how well we understand the individual processes, this does not seem to help us with the problem of allocation, unless we use some additional information or principle that tells us how the different processes or aspects are to be put together.

Electronic Supplementary Material The online version of this chapter (https://doi.org/10.1007/978-3-030-35761-0_7) contains supplementary material, which is available to authorized users. The videos can be accessed by scanning the related images with the SN More Media App.

A. Mäkelä, H. T. Valentine, *Models of Tree and Stand Dynamics*,
https://doi.org/10.1007/978-3-030-35761-0_7

In the preceding chapters of this book, we have assumed that allocation is constrained by some structural rules, including the pipe model, fine-root mass to leaf mass ratio, and prescribed crown allometry. These assumptions allowed us to solve for the allocation of carbon to the different functional parts of the tree, (5.27)–(5.29), leading to reasonable growth predictions from the carbon balance. So far, we justified the structural rules employed in Chap. 5 mainly by empirical evidence and theoretical allometric arguments, reviewed in Chap. 4. Here, we consider the proposition that such regularities follow from natural selection.

That tree structure should have been shaped by evolution is beyond doubt: natural selection has favoured those forms that balance different aspects of plant structure and functioning in the most efficient way from the point of view of resource capture, survival and successful reproduction (e.g., Niklas and Kerchner 1984; Pigliucci 2005). Structures promoting survival under resource limitation have likely been subject to strong selective pressures, giving rise to constrained growth strategies. Because resource availability may vary between individuals of the same species, or even during the lifetime of one individual, efficient strategies will also have to show plasticity under environmental shifts.

The Darwinian theory of evolution is regarded as fundamental to our understanding of biology and ecology, but its influences on growth modelling have not been straightforward. Several modellers have used heuristic decision-rule approaches to growth allocation on the basis of evolutionary arguments. For example, the priority principle of competing sinks postulates that different plant functions and organs have different growth demands that are satisfied from the carbon pool following an order of priorities (e.g., Waring and Schlesinger 1985; Bossel 1996). Another example is the assumption that allocation of carbon is driven by the objective of balancing the carbon and nitrogen contents in the plant, such that allocation to roots is increased when the plant is nitrogen limited (Reynolds and Chen 1996; Landsberg and Waring 1997; Running and Coughlan 1988; Running and Gower 1991).

Darwinian theory is explicitly employed in so-called *eco-evolutionary models* that look for the fittest forms by means of mathematical optimisation (Mäkelä et al. 2002; Dewar et al. 2009; Franklin et al. 2012). This assumes that we can formulate the basic functioning of the organism using a (set of) mechanistic model(s) that leaves some parameters unresolved, typically those concerning some linkages or trade-offs in the functioning of the organisms. The optimisation problem is solved by maximising a function, which is assumed to represent the fitness of the organism, to evaluate the unknown parameters for a range of environmental constraints. The resulting optimal plant behaviour is regarded as an emergent property that can be used to constrain the linkages between components in the underlying mechanistic model.

In this chapter, we review the eco-evolutionary approach to carbon allocation. We begin with a more formal introduction to the optimisation ideas, including a discussion of how they relate to the theory of evolution and how such models should be used and interpreted. We then present some examples of models that utilise the eco-evolutionary approach to derive plant structure and carbon allocation.

7.2 Rationale for Optimisation

As noted by Mäkelä et al. (2002), an optimisation model of tree function is formulated in terms of: (1) adaptive traits that vary within a feasible range; (2) environmental driving variables; (3) a (dynamic) model of tree processes, dependent on the adaptive traits and the environment; and (4) an objective function, dependent on tree processes. The optimisation entails finding a feasible adaptive trait that maximises the objective function when tree processes adhere to the postulated model under the specified environmental conditions.

The two ingredients of the optimisation model that are not part of, say, a mechanistic model, are the objective function and the adaptive traits. The objective function serves as a proxy of Darwinian fitness – usually some key trait or rate that *we assume* relates to the likelihood of a tree surviving to reproductive maturity. For example, an objective function might be defined as the rate of carbon gain or height growth, because the faster growing trees tend to be the ones that survive suppression-induced mortality. An adaptive trait represents the plasticity of the tree's response to environmental factors. In a mechanistic model, the trait might be defined as a parameter with a fixed value, but in the optimisation model, the value of the trait is optimised in order to maximise the objective function.

Optimisation models are not intended to prove that a particular plastic trait is an important adaptation or that a particular expression of the trait is optimal. Models do not prove anything. Instead, given the domain of possible solutions and their functional significance, an optimisation model predicts what the trait should be like if it were optimal. This method leads to testable hypotheses about the optimality of the trait in question under the hypothesized constraints. Moreover, the model provides a quantitative tool for assessing the role of constraints versus selection in relation to particular traits. If the model can be shown to represent reality fairly, it can be used to make predictions, just like any other model that can be tested against data. The method also can be used for analysing what would happen, at least in the short term, if selective pressures were modified as a result of environmental change.

The optimisation method is particularly useful where the balancing of parallel processes is not understood mechanistically, or where a mechanistic explanation would require a model with unwieldy complexity. The physiology of photosynthesis, nutrient acquisition, water uptake, and transpiration, for example, are each reasonably well understood in isolation. optimisation may help find rules that govern the balancing of the different processes in a way that is evolutionarily stable (Dewar et al. 2009). It is when this balancing is carried out by means of structural adaptation or acclimation that the optimisation problem will lead to rules concerning carbon allocation.

7.3 Crown Structure

7.3.1 The Evolutionary Significance of Crown Architecture for Carbon Allocation

Crowns are plastic. Shoots appear to grow into light. If an opening is created, branches of surrounding trees will soon fill it. Also, the main direction of light appears to have an impact on crown shape. Trees in low-latitude savannas, where light comes directly from above, have wide horizontal crowns, whereas boreal conifers, where the direction of sun rays is more from the side, tend to have very narrow and tall column-like crowns. A classic text describing the plastic adaptation of trees to different light environments is the book by Horn (1971).

These observations have led to the proposition that crown structure should form so as to allow for maximum light capture under any prevailing conditions. In a pioneering study to formally test this hypothesis, Oker-Blom and Kellomäki (1982) utilised a light interception model of the type described in Sect. 4.6.1 to analyse the effect of crown shape on the amount of light absorbed by open-grown trees during the growing season at different latitudes. Crown shape was defined as the ratio of crown length (L_c) to crown radius (R) in conical crowns in which leaves were evenly distributed. However, contrary to expectation, no optimal crown shape could be derived from these premises. The efficiency of light absorption increased both towards longer columnar crowns and towards flatter disc-like crowns at all latitudes, with minimum light absorption found at length to radius ratios between $L_c/R = 1$ and $L_c/R = 4$. Some indication of possible north-south differences in optimum crown form was nevertheless suggested by the result that the L_c/R ratio corresponding to the minimum light absorption increased with decreasing latitude, such that tall columnar crowns were relatively more beneficial for light absorption in higher than in lower latitudes.

The result by Oker-Blom and Kellomäki (1982) suggests that crown shape is not solely determined on the basis of maximisation of light absorption. Other factors, such as mechanical strength under wind force and snow loads, also have been discussed as possible traits playing a role in natural selection (Bruning 1976). From the point of view of carbon economy and allocation, the benefits from light absorption have to be weighed against the related investment in the construction of the supporting branching network (see Fig. 6.1).

Ford (1992) presented a theoretical framework for analysing the interactions of structure and function in tree crowns, summarised in the following four axioms:

Axiom I	Structural features determine radiation interception and, hence, photosynthate produced.
Axiom II	Leaf growth is constrained in both its quantity and spatial distribution by branch amount and morphology.
Axiom III	The growth of support tissue, i.e., branches, required by leaves and the maintenance of support tissue through respiration influence the amount of leaves produced.
Axiom IV	Production of large branches decreases trunk production.

The first axiom is related to the potential benefits of crown structure to the carbon economy of trees, while Axioms II–IV state the constraints and costs incurred when leaves are constructed and displayed to receive solar radiation. Balancing the costs and benefits evokes a trade-off between photosynthetic production and the branching network needed to support the leaves, which suggests that an optimal structure could be found.

In order to find the crown structure that maximises the net carbon gain, we need to quantify the effect of leaf distribution within the crown on: (a) photosynthetic production, and (b) the related carbon requirement of the woody support tissue. This can be done at different levels of detail. Mäkelä and Sievänen (1992) considered the problem in terms of tree-level biomass and structure variables, whereas Sterck and Schieving (2007) utilised a detailed 3-dimensional crown model for the optimisation. Both studies employed similar assumptions about light absorption and both constrained crown structure using the pipe model. These analyses are reviewed in the following sections.

When trees are competing with their neighbours, the crown form of best fit may not be the one that maximises their carbon gain at each moment of time. To avoid being overtopped by neighbours, trees may have to allocate carbon to height growth more than would seem necessary if the neighbours remained at their current height. Accounting for competition will pose additional constraints on crown form (Iwasa et al. 1985; Lindh 2016). These constraints will be discussed in Sect. 7.6.

7.3.2 Crown Allometry

As seen in Chap. 4, the pipe model provides a relationship between leaf mass and the cross-sectional area of the sapwood in trees, but it does not inform us about the length-to-diameter ratios in branches and stems. On the other hand, the fractal model of tree structure (see (4.40), Sect. 4.5.3) predicts an allometric relationship between total leaf mass and crown dimensions, which is also corroborated by empirical evidence and was utilised in the growth model of Chap. 5 to derive the rate of height growth in pipe-model trees. The study by Mäkelä and Sievänen (1992) offers a possible evolutionary explanation for the observed crown allometry.

Because the ratio of leaf mass to sapwood area is fixed in pipe-model trees, the allocation of growth between leaves and wood is essentially determined by the length growth of the stem and branches. The problem considered by Mäkelä and Sievänen (1992) was to find the optimal growth of crown length in an open-grown tree with constant crown shape (L_c/R ratio) which followed the pipe model structure and was constrained by the carbon balance equations presented in Sect. 3.2. The open-grown assumption was made for simplicity to avoid modelling crown rise, and the crown shape was assumed constant to reduce the dimensions of the problem.

A trade-off arises, because any investment of carbon to expand a crown has an impact on the degree of self-shading within the crown. Self-shading is minimised in sparse crowns where the relative investment into woody structures is large. Shorter branches, on the other hand, require less allocation to woody growth but lead to dense crowns that may be inefficient in light capture due to strong self-shading (Fig. 7.1). In addition to the pipe-model structure and the carbon balance, the key equations of the derivation include the allocation of carbon to leaf mass and the dependence of photosynthesis on crown size and total leaf mass. Mäkelä and Sievänen (1992) used an approximate form of the leaf-mass allocation derived from the pipe model under carbon balance by Mäkelä (1986):

$$\lambda_f(t) = \frac{1 - b - \beta \hat{H}(t)}{(1 + \alpha_r) + \rho_p \hat{H}(t)/\eta_p} \tag{7.1}$$

where $\hat{H}(t)$ is mean pipe length in the crown, b is the proportion of leaves available for reuse due to the senescence of the fine roots attached to those leaves, β is the proportion of leaves shed per unit length of sapwood pipes available for reuse, α_r is the ratio of fine-root mass to leaf mass, ρ_p is sapwood density and η_p is the ratio of leaf mass to active pipe (sapwood) area. The equation describes the decline of allocation to leaf mass with increasing mean pipe length.

The photosynthetic production of the crown was assumed to depend on the leaf mass (proportional to leaf area) density on the crown envelope as follows:

$$P(t) = \frac{\sigma_C S_A}{k_G} \left[1 - e^{k_G W_f / S_A}\right] \tag{7.2}$$

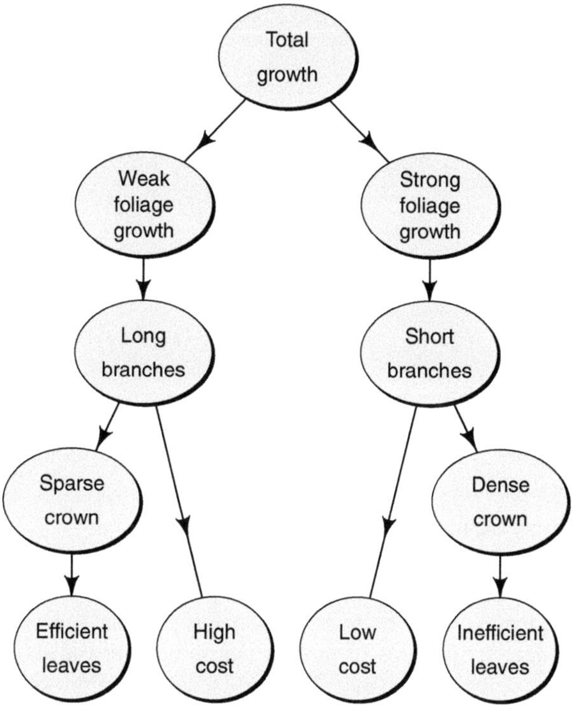

Fig. 7.1 Schematic presentation of the tradeoffs in carbon allocation between leaves and branches

where σ_C is annual rate of photosynthesis per unit unshaded leaf mass, S_A is crown surface area, W_f is leaf mass and k_G is a generalised extinction coefficient that depends on the L_c/R ratio and the mean angle of solar radiation. This equation is the limiting case of photosynthesis using the Lambert-Beer model with extinction coefficient k_{eff}, as in (4.52), when the stand-level leaf area index approaches zero, as it would in the case of open-grown trees.

The objective of the optimisation was to find a height growth strategy, $u(t)$, that maximises the expected life-time net production:

$$J = \int_0^T e^{-rt} G(t)\, \mathrm{d}t \tag{7.3}$$

where $G(t)$ is net production at time t, T is the latest time of interest in the problem and r is probability of survival per unit time. The constraints to the problem were the dynamic equations for leaf mass and height:

$$\frac{\mathrm{d}W_f(t)}{\mathrm{d}t} = \lambda_f(t)G(t) - s_f W_f \tag{7.4a}$$

$$\frac{\mathrm{d}\hat{H}(t)}{\mathrm{d}t} = u \tag{7.4b}$$

where $G(t)$ is as in (3.7),

$$G(t) = Y_G(P(t) - R_M(t))$$

λ_f and $P(t)$ are as in (7.1) and (7.2), R_M is as in (3.9),

$$R_M(t) = r_{M_f} W_f(t) + r_{M_r} W_r(t) + r_{M_p} W_p(t)$$

fine root biomass is as in (4.13):

$$W_r(t) = \alpha_r W_f(t)$$

and woody biomass is calculated on the basis of the pipe model:

$$W_p(t) = \rho_p \hat{H}(t) W_f(t) / \eta_p.$$

where the notation is as in (7.1).

The above equations define a dynamic optimisation problem, which can be solved using optimal control theory (Bar-Yam 2012). Mäkelä and Sievänen (1992) used numerical methods to find the optimal time course for crown height. The optimal height growth strategy was to first grow at an increasing rate, then gradually slow down after a peak at an early age. The height growth rate was accompanied by a similar growth rate of leaf mass, such that an exponential relationship was found between tree (crown) height and total leaf mass (Fig. 7.2).

Both the optimal peak height growth and its time of occurrence were sensitive to values of parameters characterising the environment. Increasing the annual foliage-specific photosynthesis rate or decreasing the ratio of fine-root to leaf mass both

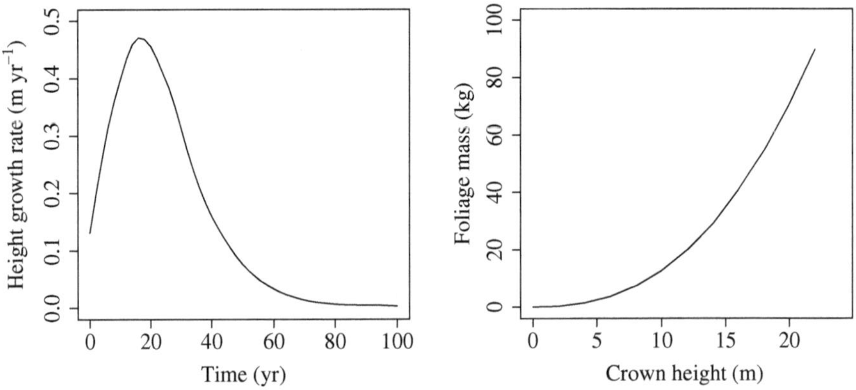

Fig. 7.2 Left panel: Optimal growth rate of crown height. Right panel: Functional relationship between crown height and leaf biomass in the optimal trajectory. Parameter values correspond to Scots pine in Finland. (Redrawn from Mäkelä and Sievänen 1992)

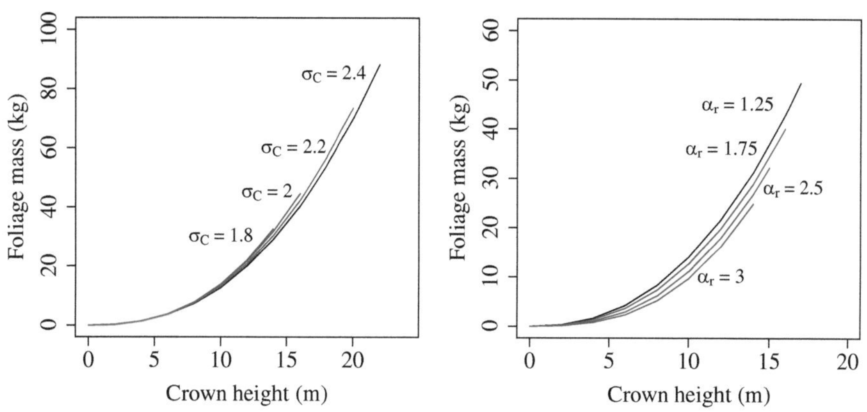

Fig. 7.3 Sensitivity of the allometric relationship between crown height and total leaf mass along the optimal trajectory to parameter values. Left panel: Sensitivity to the unshaded leaf-specific photosynthetic rate σ_C (kg C (kg DW)$^{-1}$ yr^{-1}). Right panel: Sensitivity to the fine-root to leaf mass ratio α_r. (Redrawn from Mäkelä and Sievänen 1992)

increased height growth rate and reduced the age at which the growth rate peaked. However, the relationship between crown height and total leaf mass was insensitive to these parameter changes (Fig. 7.3). This suggested that a close approximation of the optimal height growth should be derivable from the constraint that crown height and total leaf mass satisfy the relationship found in the optimum solution.

Mäkelä and Sievänen (1992) assumed that on the optimal trajectory, total leaf biomass could be approximated with a power function of crown height:

$$W_f(t) = \beta_W \hat{H}^z(t) \tag{7.5}$$

where β_W and z are parameters. Height growth could then be calculated by differentiating the equation with respect to time and solving for $d\hat{H}(t)/dt = u$:

$$u = \frac{1}{\beta_W \hat{H}^{z-1}} \frac{dW_f}{dt} \tag{7.6}$$

This way a *feedback rule* was defined where height growth depends on the state of the tree at any moment of time (after substituting (7.4a) for dW_f/dt), rather than on the particular parameter values applied. It is therefore more practical for applications than the open strategy; moreover, it appears to be more informative biologically. The feedback strategy produced a height growth pattern that was only marginally suboptimal compared with the open strategy. The optimum value for the allometric exponent z depended somewhat on the parameter values but was centered around $z = 2.5$ (Mäkelä and Sievänen 1992).

The feedback rule represents an allometric relationship between total leaf mass and crown length (or crown surface area), equivalent in form to the result derived from the fractal model, that is (4.40). Similar to the fractal model, the optimisation assumed area-preserving branching and constant crown shape, but in contrast, it required no specification of the type of branching structure. Furthermore, the fractal model assumed that leaf-mass distribution was volume filling, in other words, evenly distributed within the crown volume. In large crowns, this would mean that the inner leaves would become increasingly shaded by the outer layers of the crown, which would lead to diminishing returns from any investment into leaf biomass in terms of new carbon photosynthesized. On the other hand, studies ignoring the cost of supporting structures have found that in order to display the maximum amount of leaf area to the incoming light rays, leaves should be evenly distributed on the crown surface (Honda and Fisher 1978; Borchert and Slade 1981). In line with empirical observation (Zeide 1998; Duursma et al. 2010), the optimum result falls between these two extremes. This reflects the existence of a trade-off between the cost of investment into woody structures and the benefit of displaying leaves to light, as envisaged by Ford (1992).

It is interesting that although the optimum problem was simplified to open-grown trees only, the resulting allometric structure of crowns seems to gain empirical support in trees growing in canopies as well (Zeide 1998; Duursma et al. 2010). Canopy trees have much smaller crown ratios (crown length over tree height), but the crowns themselves seem to be constructed much in the same way regardless of canopy position and crown ratio (Ilomäki et al. 2003; Kantola and Mäkelä 2006), the most notable difference being a somewhat lower leaf density in the more suppressed trees (Mäkelä and Vanninen 1998). This has also been found to be optimal in pipe-model trees when the structure has been considered more explicitly as above, as we will see in the following section.

7.3.3 *Optimal Crown Shape and Leaf Density*

Sterck and Schieving (2007) studied optimal crown shapes in pipe-model crowns using a 3-dimensional plant model consisting of basic growth units, metamers, and their spatial distribution. A metamer is the most basic building block of a model crown, consisting of a woody internode, node, and one or more leaves and axillary meristems.

Following the pipe model structure, Sterck and Schieving (2007) assumed that each leaf in the crown was attached to a unit pipe running from the base of the leaf to the base of the crown. When a leaf dies, the corresponding unit pipe also dies but remains in the crown structure as a disused pipe. The metamers are therefore bundles of live and disused pipes, with or without leaves attached to them (Fig. 7.4). Branches without any living leaves were assumed to be shed by the tree. New metamers were allowed to grow where meristems were displayed, in other words, at the tip of existing live metamers. In addition, it was assumed that meristems re-grow where branches have fallen off.

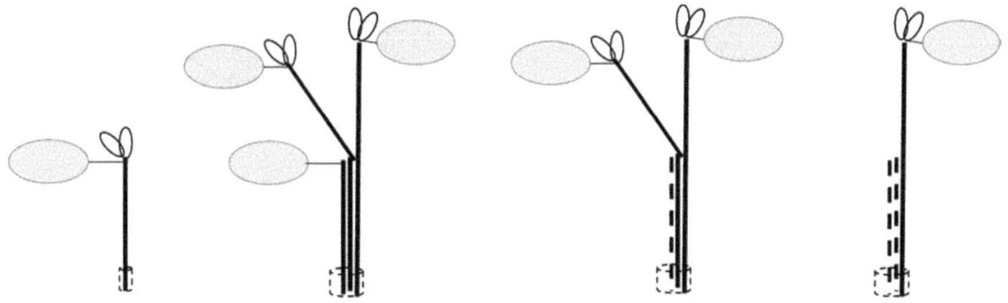

Fig. 7.4 A metamer consists of a leaf, a segment with one pipe element connected to the leaf, and an axillary and apical meristem. The meristems may produce new metamers that develop a unit pipe from the leaf to the stem base. A new apical metamer extends the parent segment, and a new axillary metamer creates a branching. When a leaf is shed, the attached pipe dies (dotted line). The parts of the pipe are also dropped which are in segments that do not contain any living pipes. (Redrawn from Sterck and Schieving 2007)

The 3-dimensional structure of the crown was created assuming rules about branching and orientation of the metamers, while the length of metamers as well as leaf length and width were assumed constant. The carbon balance of the tree consisted of the carbon gain by leaf photosynthesis, which was assumed either constant (no shading) or dependent on the light environment of the leaf, maintenance respiration of leaves and live pipes, growth respiration and loss of meristems due to senescence. Shading was considered with respect to light rays reaching the crown in three dimensions.

Under these basic constraints on tree structure and function, Sterck and Schieving (2007) explored the possible rules that could influence the development of crown size and shape. The alternative basic growth rules were: (1) random allocation of metamers, and (2) allocation of metamers to meristem locations that maximised the net carbon gain of the growing unit pipe system over its lifetime. The influence of light environment on the optimal allocation pattern was also studied. In practice, the allocation was carried out one meristem at a time: once total growth had been evaluated from the carbon balance, new metamers with their leaves and attached unit pipes were created according to the rule until all the photosynthates had been used up, and a new growth cycle started.

The random allocation of metamers produced irregularly shaped crowns that were not able to reach a stable equilibrium but sooner or later started to decline in growth and died (Fig. 7.5). This was because in this case, the mean pipe length was allowed to increase without control and therefore at some point it exceeded the level under which net growth remains positive. The factors of pipe length that reduce future growth potential include increased demand of maintenance respiration relative to production, and increased demand of growth allocation to wood instead of leaves (Mäkelä 1986).

The optimised allocation of metamers resulted in crowns that were regular in shape (Fig. 7.5). In the case with no shading, the crowns became more or less spherical and the leaves were rather evenly and densely distributed, resembling the

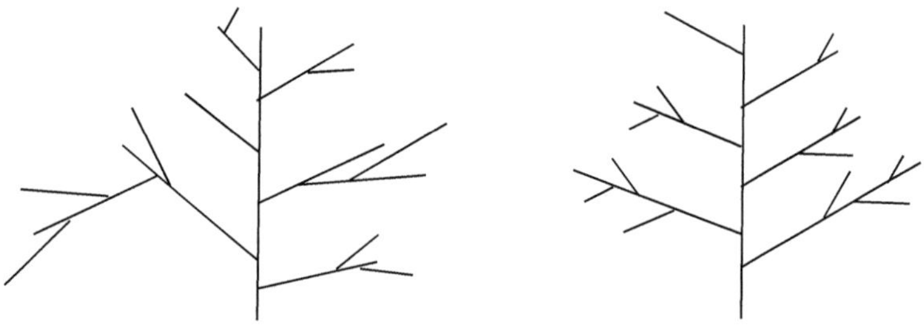

Fig. 7.5 An illustrative presentation of the tree shapes created using random (left) or optimal (right) rules of metamer allocation in crown extension. (Drawn on the basis of Sterck and Schieving 2007)

volume-filling crowns of the fractal model. When self-shading was included the tree distributed its leaves relatively more on the outer surface of the crown, to keep leaves more exposed to photosynthetic light.

Different assumptions about surrounding canopy environment had implications on crown shape. An open-grown tree in a constant light environment grew a rather wide, close-to-spherical crown, whereas a tree exposed to a vertical light gradient, on the other hand, developed a more oblong crown where a relatively large proportion of the leaves were allocated to the top of the crown. A tree growing under the main canopy (constant light environment with low light) grew a sparse and somewhat irregular crown with leaves mainly on the outer surface.

We note, in passing, that the assumptions underpinning our core model differ somewhat from those used in the Sterck-Schieving study. We assume, for the core model, that a pipe does not die when its leaf matter dies, unless the pipe occurs at the crown base during crown rise. Otherwise, the pipe extends as new leaf matter emerges. In a more elaborate version of our core model, we allow for a fraction of the pipes to die throughout the crown concurrent with development (see Valentine et al. 2012).

7.3.4 Crown Structure: Summary

The two studies reported above defined tree crowns at different levels of detail and used different methods of simulation and optimisation, but they both shared the same basic assumptions on carbon balance and pipe model structure. In both studies, these assumptions led to two main conclusions. Firstly, height growth—or more generally, extension growth—is limited but a steady state can be reached if carbon allocation between extension growth and leaf growth is balanced. Secondly, this balanced growth requires an allometric relationship between crown size and total leaf mass.

In the tree-level study by Mäkelä and Sievänen (1992), the assumption on self-shading by the crown was essential for deriving the allometric relationship between crown size and total leaf mass, because it established the only link between branch length and leaf mass. Had this constraint been ignored, it would have been optimal for the tree to place all leaves in one point and still produce an amount of photosynthates that was proportional to total leaf mass. The 3D model by Sterck and Schieving (2007), on the other hand, made an *a priori* assumption about metamer structure, including metamer length and leaf area. This allowed for the optimisation of crown shape even in the absence of self-shading. Because each individual metamer had a prescribed structure and thus a prescribed carbon gain in the absence of shading, the metamers had to be arranged in such a way that the carbon consumption was minimised. This was achieved by near-symmetrical crowns where branches bifurcate as often as possible, creating the shortest mean pipe length.

The result of crown symmetry was still valid in the case when self-shading was assumed, if the light environment was also assumed symmetrical. The 3D model allowed for the analysis of non-symmetrical environments as well, suggesting that the foliage distribution should be denser where light is stronger. In the study by Sterck and Schieving (2007), the optimum crown density was also clearly related to the level of shading, whereas the results by Mäkelä and Sievänen (1992) indicated almost no effect (Fig. 7.3). This could be related to the assumption of fixed metamer size. Empirical evidence also supports the conclusion that increasing shade tends to make crowns sparser, suggesting that some constraints in metamer structure may provide a limit to the plasticity of crowns.

Interestingly, neither of the models reviewed here provided any clues as to how crown ratio would vary among different environments. More generally, attempts to derive optimal foliage distribution when crown ratio is allowed to vary have been less successful unless competition between trees has been considered by means of evolutionary games (Iwasa et al. 1985; Lindh 2016). We return to these aspects in Sect. 7.6.

7.4 Stem Form

The benefit of optimal investment of carbon substrates for the construction of stems and coarse roots has been acknowledged for more than a century. Newnham (1965) analysed forms of tree stems in British Columbia in light of mechanical theories advanced by K. Metzger in 1894 and by Gray (1956). Newnham noted that, "the main function of the tree, according to Metzger, is to expand its crown and root systems as much as possible and to produce seed, and because strengthening of the stem diverts building material from these organs of the tree, the stem must have the highest possible bending strength with the most economical expenditure of material." Metzger assumed that branch-free portions of stems are solidly fixed in the ground and that they manifested uniform resistance to bending from wind forces, which led to his conclusion that stems should tend to form in the shape of cubic paraboloids. Gray (1956), on the other hand, argued that soil is a weak medium that allows the base of the stem and buttress roots to move, so quadratic paraboloidal form not only meets the needed strength requirements, but also saves 20% in building materials, compared to cubic paraboloidal form. As it turned out, the results of Newnham's analysis supported the more economical model of Gray, which, nonetheless, supported the principal of optimality as espoused by Metzger. The model that we developed in Sect. 5.3 in concert with pipe model theory also predicts stems that approximate quadratic parabolic form.

7.5 Co-allocation of Carbon and Nitrogen

7.5.1 *Functional Balance*

A functional balance refers to a situation where two or more resource acquisition and/or transport processes contributing to plant growth are balanced, so as to match the demand of the related substance in plant functions. The concept was first introduced to explain empirical observations of root-shoot ratios in non-woody plants (White 1935; Brouwer 1962; Davidson 1969) and it was based on the assumptions that: (a) carbon and nitrogen are used in a constant ratio for dry matter growth, and (b) the assimilation of each element is in balance with its utilisation. If these assumptions are correct, then the shoot and root biomasses are related as follows:

$$\sigma_{\mathrm{N}} W_{\mathrm{r}} = [\mathrm{N}]\sigma_{\mathrm{C}} W_{\mathrm{f}} \tag{7.7}$$

where [N] is nitrogen to carbon ratio of plant biomass, σ_{N} and σ_{C} are the net specific assimilation rates of nitrogen and carbon, respectively, and W_{r} and W_{f} are total fine-root and leaf masses.

It is conceivable that such a balance is actually optimal for a plant growing under limited resources. Plants investing more resources than needed for satisfying the demand, into either leaves or roots, would be wasteful as the excess could not be utilised due to the strict proportional need of the two substances. Wasting resources would lead to slower growth and therefore poorer fitness within the population. The functional-balance assumption thus allows us to infer the root-shoot ratio of plants from the nitrogen to carbon ratio of plant tissue and the tissue-specific net assimilation rates of nitrogen and carbon.

For practical applications of the simple idea expressed in (7.7), several complications arise. Firstly, the ratio of nitrogen to carbon in plant biomass is not strictly constant but also depends on the relative availabilities of the elements. Secondly, net nitrogen assimilation is affected by the resorption of nitrogen from senescent plant material. Thirdly, the net carbon assimilation rate is dependent on the ratio of maintenance respiration to photosynthesis, and, hence, tree structure. Finally, the dependence of gross root-specific nitrogen assimilation rate of soil nitrogen is still not very well understood.

7.5.2 Functional Balance During Exponential Growth

The above problems are minimised if we consider tree seedlings growing in a nutrient solution. In young seedlings, the loss of biomass through senescence can be ignored, and the biomass-specific net assimilation rates of both nitrogen and carbon can be taken to be approximately constant, leading to exponential growth. The uptake of nutrients and its relation to nitrogen availability can be monitored through nutrient addition into the solution.

This was the set-up applied by Torsten Ingestad in his pioneering nutrient experiments with seedlings of several deciduous tree species (Ingestad 1980; Ingestad et al. 1981; Ingestad and Ågren 1992). Nitrogen and other nutrients were added to the solution at different rates in different seedling lots. Furthermore, the nutrient addition rate was increased exponentially with time to allow for the growth of seedlings. This set-up led to different growth rates and different shoot-root ratios in the lots. However, the experiments also showed that the nitrogen concentration in the seedlings varied with the nitrogen input rate.

If the assumption of the strictly proportional demand of carbon and nitrogen in structural growth is relaxed, a new optimisation problem can be defined that solves for both tissue nitrogen concentration and shoot-root ratios simultaneously. However, additional assumptions are needed to constrain the additional trait to be optimised. Modelling studies have considered allowing the tissue nitrogen concentration to vary as a function of nitrogen supply (Mäkelä and Sievänen 1987), and including additional assumptions on the impacts of nitrogen on growth and production (Johnson and Thornley 1987; Hilbert 1990; Ågren and Franklin 2003).

Hilbert (1990) considered plants in the exponential growth phase, assuming that maintenance respiration was nil, so that total growth was proportional to carbon assimilation:

$$\frac{\mathrm{d}(W_\mathrm{f}+W_\mathrm{r})}{\mathrm{d}t} = Y_\mathrm{G}\sigma_\mathrm{C}W_\mathrm{f} \tag{7.8}$$

and that the rate of photosynthesis was a saturating function of nitrogen concentration in leaves:

$$\sigma_\mathrm{C} = \frac{P_\mathrm{max}([\mathrm{N}]-n_1)}{[\mathrm{N}]+n_2} \tag{7.9}$$

In the above, Y_G is growth efficiency, i.e., production of biomass per unit assimilation of carbon, P_max is maximum carbon assimilation rate, [N] is tissue nitrogen concentration (assumed the same for root and shoot), n_1 is concentration of non-photosynthetic nitrogen and n_2 is a saturation parameter.

The functional balance implies that for a given tissue nitrogen concentration, the shoot fraction of the plant, f_s, depends on the nitrogen and carbon assimilation rates:

$$f_s = \frac{\sigma_N}{\sigma_N + Y_G\sigma_C[N]} \tag{7.10}$$

Hilbert (1990) searched for the tissue nitrogen concentration that, under these conditions, maximised the relative growth rate, r, of the seedling,

$$r = \frac{1}{(W_f + W_r)}\frac{d(W_f + W_r)}{dt} \tag{7.11}$$

The *optimal* [N] depended on both the nitrogen and carbon assimilation rates:

$$[N]_{opt} = \sqrt{\frac{\sigma_N(n_1 + n_2)}{Y_G P_{max}}} + n_1 \tag{7.12}$$

Hilbert's approach is directly applicable to Ingestad's data (Ingestad 1980) and offers a potential explanation for the results. The ingenious aspect of Ingestad's experiment was that the nutrient addition to the growth solution in each nitrogen level increased exponentially with time, allowing for exponential growth of those seedling lots that were nitrogen limited. The results also showed that the plant relative growth rates and tissue nitrogen concentrations remained fairly constant after an initial acclimation period, which was not included in the analysis. We can therefore say that the seedlings were undergoing steady state exponential growth, with

$$W_f(t) = f_s W(0)e^{rt} \tag{7.13a}$$

$$W_r(t) = f_r W(0)e^{rt} \tag{7.13b}$$

$$W_N(t) = [N]W(0)e^{rt} \tag{7.13c}$$

where W is total seedling dry mass, W_N is seedling nitrogen content and t is time from the start of the experiment (Table 7.1).

To derive a comparison with the results by Hilbert (1990), the measurements can be further used to estimate $Y_G\sigma_C$ and σ_N. First, combining (7.8) and (7.11), we have

$$Y_G\sigma_C = \frac{r}{f_s} \tag{7.14}$$

This allows us to estimate $Y_G P_{max}$, n_1 and n_2 of (7.9) from Ingestad's data (Fig. 7.6).

Table 7.1 Measurements reported by Ingestad (1980). R_N: increase rate of nutrient addition to growth solution (% day^{-1}), where nutrient addition at time t (day) is $N(t) = N_0 e^{R_N t}$; [N]: nitrogen concentration in plant tissue (%), $W(t_0)$: seedling biomass (g) at start of experiment (time t_0), $W(t_f)$: seedling biomass (g) at end of experiment (time t_f), f_s: share of shoot biomass in total seedling biomass, $t_f - t_0$: duration of experiment (days), r: relative growth rate (day^{-1}) calculated with (7.13) from initial and final biomass and experiment duration. The "optimum" in the experiment refers to nutrient addition at which growth rate saturates, i.e., faster nutrient addition will no longer increase growth rate as shoot processes become limiting

R_N	[N]	$W(t_0)$	$W(t_f)$	f_s	$t_f - t_0$	r
5	1.54	0.42	2.31	0.65	35	0.049
7.5	1.94	0.18	2.52	0.69	35	0.075
10	2.21	0.10	3.63	0.70	35	0.103
15	2.85	0.78	7.41	0.78	14	0.161
"Optimum"	4.28	0.06	6.81	0.80	24	0.197

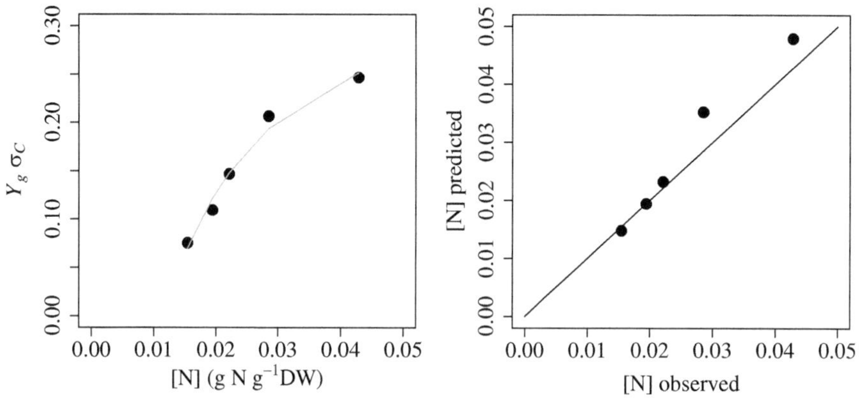

Fig. 7.6 Left panel: Foliage-specific net production rate (day^{-1}) as a function of tissue nitrogen concentration as estimated from Ingestad (1980) (points), with a best-fit non-linear regression line of (7.9). The fitted parameters were $Y_G P_{max} = 0.387$ day^{-1}, $n_1 = 0.0117\,\mathrm{g\,N\,(g\,DW)^{-1}}$, $n_2 = 0.00051\,\mathrm{g\,N\,(g\,DW)^{-1}}$. Right panel: Comparison of tissue nitrogen concentration from observations (Ingestad 1980) and predictions from the optimality model, (7.12), derived by Hilbert (1990)

Nitrogen uptake is the rate of change of nitrogen content, which by definition is equal to $\sigma_N W_r$,

$$\frac{dW_N}{dt} = r[N]W(0)e^{rt} = \sigma_N f_r W(0)e^{rt} \tag{7.15}$$

implying that

$$\sigma_N = \frac{r[N]}{f_r} \tag{7.16}$$

We now have all the components required for calculating the optimal nitrogen content derived by Hilbert (1990) (see (7.12)). Comparison of this with the measured [N] seems to corroborate the optimality theory of shoot and root partitioning during exponential growth (Fig. 7.6).

7.5.3 Optimal Canopy Density and Nitrogen Supply

Although the exponential growth phase is only a tiny fraction of a tree's life, the studies on shoot-root ratios during exponential growth demonstrate the significance of nutrient supply and assimilation to plant growth and structure. They also suggest that acclimations to other phases in the tree's life cycle could be studied under the paradigm of optimal co-allocation of carbon and nitrogen.

An important and relatively long-lasting phase can be defined by canopy closure. These situations can be analysed effectively by considering the C and N balances of the stand at steady state (Dewar 1996; Franklin 2007). Strictly speaking, the steady-state assumption does not apply to woody biomass, as stem elongation continues until stand senescence (Mäkelä and Valentine 2001). However, the elongation growth utilises a small fraction of NPP (Mäkelä 1986), so at appropriate timescales, the steady-state assumption can nonetheless serve as a realistic approximation for resource-limited stands.

The effect of N supply on steady-state canopy size has been the focus of several studies, often using eco-evolutionary arguments (McMurtrie 1991; Dewar 1996; Franklin and Ågren 2002; Franklin 2007). For example, Dewar (1996) derived a canopy nitrogen content that maximises NPP for a given leaf area index, and Franklin (2007) derived a leaf area index that maximises NPP for a given nitrogen content.

Essentially, these derivations are based on the observation that canopy photosynthesis tends to have a saturating response to an increase in foliar nitrogen content (as seen in Fig. 7.6), whereas maintenance respiration increases almost linearly with tissue N (Ryan 1995). Dewar (1996) presented a detailed physiologically based analysis that allowed him to quantify the optimum canopy N content under the assumption that canopies maximise NPP (Fig. 7.7).

7.5.4 Co-allocation of Carbon and Nitrogen in Closed Canopies

In the above canopy steady-state models, the optimum canopy N content was found as a trade-off between anabolic and catabolic processes, both dependent on the N content. However, they did not attach an explicit cost to the construction or maintenance of the root system that would be required to acquire the assumed N

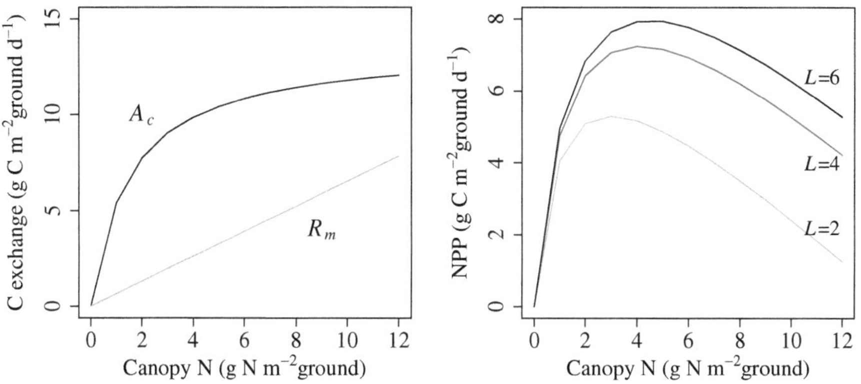

Fig. 7.7 Left panel: Daily rate of canopy photosynthesis (A_c) and maintenance respiration (R_m) as a function of canopy nitrogen content. Right panel: The difference between daily mean canopy photosynthesis and respiration as a function of canopy nitrogen content for different canopy leaf areas. (Redrawn from Dewar 1996)

content. By contrast, the exponential-growth studies focused on the relative carbon costs of N acquisition through the construction and maintenance of roots. However, they also assumed that the requirement for N can always be satisfied by suitable growth allocation to roots, regardless of the size of the plant. At the stand level, this assumption must be tempered by the reality of limited nitrogen availability. This requires expanding Hilbert's (1990) optimum functional balance approach to steady-state canopies.

Mäkelä et al. (2008b) explored the optimal co-allocation of C and N in a closed forest stand at steady state, where light or nitrogen may limit productivity. The approach was similar to Hilbert's, but adaptations were made to accommodate the steady state of large trees rather than exponential growth of seedlings. Firstly, as the growth rates of the components of live biomass are nil at steady state where production balances turnover, relative growth rates would make little sense and were replaced by actual production and turnover rates. Secondly, because steady-state canopies of trees may allocate a variable proportion of growth to wood, shoot growth was replaced by woody growth and leaf growth as separate components. The model derived by Mäkelä et al. (2008b) thus maximised net production by optimising the co-allocation of carbon and nitrogen to leaves, fine roots, and live wood, given the maximum uptake rate of nitrogen, and accounting for the costs of production and maintenance respiration. In the following, we review the main assumptions and results of this model.

Consider the aggregate biomasses of leaves, fine roots, and live wood in a stand of trees. Denote the respective dry matter densities (kg ha^{-1}) by $\mathbb{W}_{\mathrm{f}}$, $\mathbb{W}_{\mathrm{r}}$, and $\mathbb{W}_{\mathrm{w}}$, and their tissue nitrogen concentrations (kg N (kg component)$^{-1}$) by $[\mathrm{N}]_{\mathrm{f}}$, $[\mathrm{N}]_{\mathrm{r}}$, and $[\mathrm{N}]_{\mathrm{w}}$.

Let G be the rate of production of new dry matter (kg ha^{-1} yr^{-1}), and let λ_{f}, λ_{r}, and λ_{w}, respectively, be the proportional allocation of production to leaves, fine roots, and live wood, where $\lambda_f + \lambda_r + \lambda_w = 1$. The growth rates (kg ha^{-1} yr^{-1}) of the dry matter components are

$$\frac{dW_f}{dt} = \lambda_f G - \frac{W_f}{T_f} \tag{7.17}$$

$$\frac{dW_r}{dt} = \lambda_r G - \frac{W_r}{T_r} \tag{7.18}$$

$$\frac{dW_w}{dt} = \lambda_w G - \frac{W_w}{T_w} \tag{7.19}$$

where T_f, T_r, and T_w are the respective average longevities (yr) of tissue.

From a carbon balance perspective, the rate of dry matter production is

$$G = Y_G(P - R_M) \tag{7.20}$$

where P (kg ha^{-1} yr^{-1}) is the rate of photosynthetic production, R_M (kg ha^{-1} yr^{-1}) is the rate of maintenance respiration, and Y_G (kg component (kg C)$^{-1}$) is the conversion efficiency of carbon to dry matter in the growth process, including growth respiration. Maintenance respiration by component is proportional to the tissue nitrogen concentration, therefore

$$R_M = r_m[N]_f W_f + r_m[N]_r W_r + r_m[N]_w W_w \tag{7.21}$$

where r_m (kg C (kg N)$^{-1}$ yr^{-1}) is the nitrogen-specific respiration rate (Ryan 1991). Photosynthesis depends on the amount of leaf mass,

$$P = \sigma_f W_f \tag{7.22}$$

where σ_f (kg C (kg DW leaf)$^{-1}$ yr^{-1}) is the leaf-specific rate of photosynthesis.

Denote by W_N the areal density of nitrogen in the stand (kg N ha^{-1}). The N balance of the stand is

$$\frac{dW_N}{dt} = U - \frac{f_f[N]_f W_f}{T_f} - \frac{f_r[N]_r W_f}{T_r} - \frac{f_w[N]_w W_f}{T_w} \tag{7.23}$$

where U (kg N ha^{-1} yr^{-1}) is the nitrogen uptake rate, and f_f, f_r, and f_w are the fractions of tissue nitrogen that remain in senescent tissue. We assume that the nitrogen concentrations of fine roots and live wood are proportional to that of leaves, i.e., $[N]_r = n_r[N]_f$, and $[N]_w = n_w[N]_f$, where n_r and n_w are constant. Further, we denote by σ_r the fine-root-specific nitrogen uptake rate (kg N (kg fine root)$^{-1}$ yr^{-1}), such that

$$U = \sigma_r W_r \tag{7.24}$$

Both carbon assimilation and nitrogen uptake become less efficient with crowding, i.e., the specific rates σ_f and σ_r are reduced. As described in previous chapters, stand photosynthesis is usually represented by an exponential function that follows from the Lambert-Beer law (McMurtrie 1991). This function can be accurately approximated with a rectangular hyperbola, which provides a better alternative if an analytical solution of steady state is preferred. We therefore model σ_f as

$$\sigma_f = \frac{\sigma_{fM} K_f}{W_f + K_f} \tag{7.25}$$

where σ_{fM} is the light-saturated specific rate reached under unshaded conditions, and K_f (kg ha^{-1}) is the density of leaf dry matter that reduces the specific rate to 50% of the light-saturated rate. This leads to

$$P = \frac{\sigma_{fM} W_f K_f}{W_f + K_f} \tag{7.26}$$

which provides a saturating dependence of the rate of photosynthesis on leaf mass (and LAI, since LAI = $W_f \times$ SLA). Following Hilbert (1990), the light-saturated specific rate of photosynthesis is a function of foliar nitrogen concentration (as 7.9 but with modified notation):

$$\sigma_{fM} = \frac{\sigma_{fM0}[N]_p}{[N]_p + [N]_{ref}} \tag{7.27}$$

where σ_{fM0} (kg C (kg leaf DW)$^{-1}$ yr^{-1}) is the nitrogen-saturated specific rate of photosynthesis, $[N]_{ref}$ is the concentration of photosynthetically active nitrogen for which $\sigma_{fM} = \sigma_{fM0}/2$, and $[N]_p$ is the actual photosynthetically active nitrogen concentration in leaves, defined as

$$[N]_p = \max\{[N]_f - [N]_0; 0\} \tag{7.28}$$

where $[N]_0$ is the concentration of non-photosynthetic, or structural, nitrogen in the leaves.

Similarly, the specific rate of nitrogen uptake by fine roots, σ_r, is modeled by

$$\sigma_r = \frac{\sigma_{rM} K_f}{W_r + K_r} \tag{7.29}$$

where the maximum specific rate, σ_{rM} (kg N (kg fine root)$^{-1}$ yr^{-1}), depends on the availability of nitrogen in the soil. This leads to a nitrogen uptake rate by the stand that saturates as

$$U = \frac{\sigma_{rM} W_r K_r}{W_r + K_r} \tag{7.30}$$

where K_r is analogous to K_f of (7.26).

Up to here, the above model is similar to that of Hilbert (1990), the only difference being that the specific uptake rates of nitrogen and carbon both saturate with increasing active biomass at the canopy level as formulated in (7.25) and (7.29). As in Hilbert's model, (7.27), carbon uptake rate also saturates with respect to nitrogen concentration in the leaves.

Another difference is introduced by the inclusion of woody mass as a separate state variable and thus a separate sink of the assimilated carbon and nitrogen. This is incorporated, assuming that trees follow the pipe model, so the mass of live wood is related to foliar mass and mean pipe length, $\hat{H}_s$ (m), by

$$\mathbb{W}_w = \alpha_w \mathbb{W}_f \hat{H}_s \tag{7.31}$$

where $\alpha_w = \rho_s/\eta_s$ (m^{-1}) is constant (see (4.17)). The rate of production of live wood dry matter is $G_w = \lambda_w G$, a fraction of which is attributable to elongation of shoots and roots, and the remaining fraction is attributable to the expansion or thickening of stems. Let η and $1 - \eta$, respectively, be the fractions of wood production from expansion (G_{wd}) and elongation (G_{ws}).

As discussed above, leaf and fine-root biomass, and the aggregate cross-sectional area of live wood ordinarily achieve an approximate steady state after a stand closes. However, stem and branch elongation continues until stand senescence (Mäkelä and Valentine 2001). It follows that a dynamic equation for mean pipe length is needed for modelling stand production and growth. In their study, however, Mäkelä et al. (2008b) assumed that for establishing an optimal balance between leaves and fine roots, which themselves are approximately at steady state, it was sufficient to consider a quasi-steady state for wood, where $G_w = \mathbb{W}_w/T_w$, which means that new wood is produced at the same rate that old live wood deactivates.

To achieve quasi-steady state with the model, it was assumed that $\eta = 1$, so that elongation is nil, i.e., $G_{ws} = 0$. Hence,

$$G_w = G_{wd} = \frac{\mathbb{W}_w}{T_w}. \tag{7.32}$$

The elimination of all elongation is botanically unrealistic, but this has no effect on the carbon balance at quasi-steady state, because the missing elongation fraction of the wood production is accounted for by a commensurate increase in the expansion fraction. The resultant constant mean pipe length is denoted by $\hat{H}_s$. Hence, by the pipe model, $\mathbb{W}_w = \alpha_w \mathbb{W}_f \hat{H}_s$. Since $\mathbb{W}_f = G_f T_f$ at steady state and, by definition, $G_i = \lambda_i G$ (i = f, w), (7.32) converts to

$$\lambda_w = \lambda_f \hat{H}_s \frac{\alpha_w T_f}{T_w}. \tag{7.33}$$

which provides the relative allocation of production to live wood.

Finally, mean pipe length at quasi-steady state must be expressed as a function of site quality, as described by the metabolic parameters and/or state variables in a stand. Tree height growth is known to respond strongly to N availability. Mäkelä et al. (2008b) assumed that height growth and, therefore, mean pipe length at quasi-steady state, were proportional to $[N]_f$,

$$\hat{H}_s = c_H[N]_f \tag{7.34}$$

where c_H is a site-specific constant depending on the availability of carbon. This assumption is consistent with previous derivations from the pipe model (Mäkelä 1985; Valentine 1997).

Because both C and N uptake are saturating with dry matter density, the system has a steady state where the production of both foliage and fine roots equals their turnover. At the same time, the export of C and N to wood is occurring at a constant rate. This can be interpreted as a representation of a stand during a period after canopy closure. This formulation makes it possible to work out the steady-state leaf and fine-root masses and the N concentrations that maximise production, G.

The model has seven dynamic variables that need to be solved at the steady state: the three component biomasses, $\mathbb{W}_f$, $\mathbb{W}_r$ and $\mathbb{W}_w$, the foliar N concentration, $[N]_f$, and the three growth allocation coefficients, λ_i. At steady state, the rates of change vanish in the state equations, (7.17)–(7.19) and (7.23), providing four algebraic equations to solve for the unknowns. Two additional equations have been introduced: the condition that the allocation coefficients sum up to unity, and the pipe model relationship between leaf mass and sapwood mass, (7.31), (completed with (7.34) to eliminate mean pipe length at steady state). We thus have six equations and seven unknowns, leaving one free variable. An additional constraint is introduced through optimisation.

In their study, Mäkelä et al. (2008b) solved the steady-state equations for the three component biomasses and allocation coefficients, whereas the tissue N concentration was found by maximising the rate of production, G. Formally, they solved the following optimisation problem:

$$\max\{G([N]_f)\} \tag{7.35}$$

subject to (7.17)–(7.34) with $d\mathbb{W}_N/dt = 0$ and $d\mathbb{W}_i/dt = 0, i = f, r, w$. The solution to this problem requires numerical methods of solving equations and optimisation, and is omitted here.

Mäkelä et al. (2008b) parameterized the model for Norway spruce (*Picea abies*) and Scots pine (*Pinus sylvestris*) under boreal conditions. Similar to those of Hilbert (1990), the results showed an increasing tissue N concentration and increasing leaf to fine root ratio with increasing nitrogen availability in the soil (Fig. 7.8). There was a threshold of nitrogen availability above which the tissue N and leaf biomass saturated, whereas the N uptake rate and fine root biomass peaked as further increase in fine-root biomass would have resulted in no more N uptake but increasing maintenance respiration. In reality, additional constraints, such as requirements of

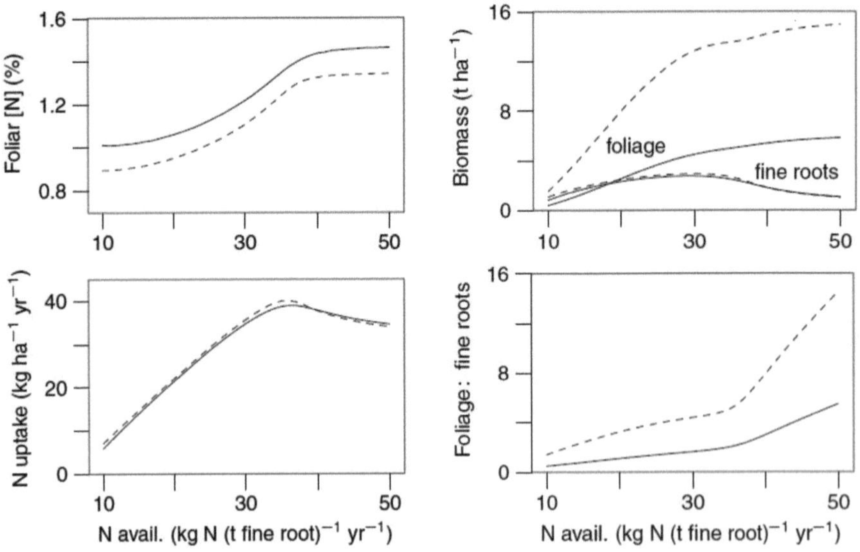

Fig. 7.8 Left panels: Foliar nitrogen concentration (% of dry matter) (top) and N uptake (bottom) as a function of N availability to fine roots in Scots pine (solid line) and Norway spruce (dashed line). Right panels: Leaf and fine root mass (top) and leaf mass to fine root mass ratio (bottom) in Scots pine (solid line) and Norway spruce (dashed line). (Redrawn from Mäkelä et al. 2008b)

water and other nutrients, likely prevent the fine-root biomass from decreasing with increased N availability.

The seedling experiments (Ingestad 1980) and optimisation models (Hilbert 1990; Mäkelä and Sievänen 1987) showed decreased carbon allocation to roots with increasing nitrogen availability. The canopy steady-state model by Mäkelä et al. (2008b) reproduces this result, but further predicts that it is not so much allocation to leaves but rather allocation to woody growth that benefits from the increased availability of nitrogen (Fig. 7.9). This is consistent with empirical observation (Litton et al. 2007).

In accordance with the model by Dewar (1996), the NPP:GPP ratio decreases with increased nitrogen availability (Fig. 7.9), as respiration increases linearly with tissue nitrogen content but production only increases in a saturating manner. Interestingly, the leaf-specific rate of photosynthesis is predicted to be relatively constant across different N availabilities, because at lower N availability low tissue N concentrations are compensated by high mean light availability, and at higher N availability the denser canopies become more light-limited (Fig. 7.9). This prediction was also obtained from Dewar's model (Dewar 1996).

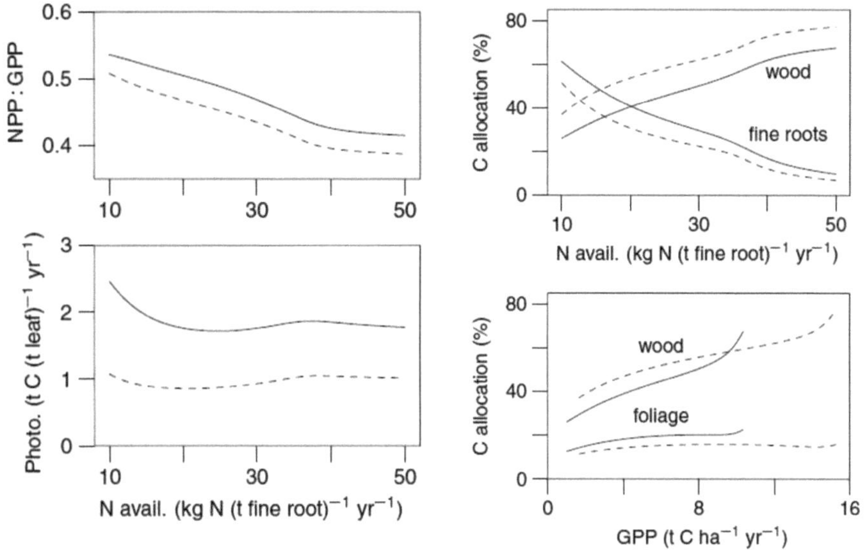

Fig. 7.9 Left panels: The ratio of Net Primary Production to Gross Primary Production (top) and leaf-specific rate of photosynthesis (bottom) as a function of N availability to fine roots in Scots pine (solid line) and Norway spruce (dashed line). Right panels: Allocation of growth to wood, leaves and fine roots as a function of N availability to fine roots (top) and Gross Primary Productivity (bottom) in Scots pine (solid line) and Norway spruce (dashed line). (Redrawn from Mäkelä et al. 2008b)

7.5.5 Dynamic Co-allocation of Carbon and Nitrogen

The above models as well as most other similar approaches applying the evolutionary paradigm to the co-allocation of carbon and nitrogen have focused on either of the two extremes of plant growth, namely, exponential growth (Hilbert 1990; Ågren and Franklin 2003), or closed canopies at steady state (Dewar 1996; Franklin 2007; McMurtrie et al. 2008; Mäkelä et al. 2008b; Dybzinski et al. 2011). This leaves several open questions related to stand dynamics: Is the optimal solution in an open stand similar to the exponential-growth optimum? Does the steady-state response represent the essence of the closed-canopy dynamics? What happens when the system shifts from free growth to closed-canopy growth? How does the degree of plasticity of the adaptive traits affect the optimal strategies at different phases of stand development?

Valentine and Mäkelä (2012) addressed these questions in their extension of the above steady-state model to a dynamic situation of an even-aged stand throughout its rotation. This required relaxing the assumption of nil height growth. Instead, they introduced a height growth model derived from the requirements of C balance (see Valentine and Mäkelä 2005), the pipe model (Shinozaki et al. 1964a), and optimal crown allometry (Mäkelä and Sievänen 1992) (Sect. 7.3). In addition, they derived a second height-growth model from the N balance in analogy to height growth rate based on the C balance, such that the two height-growth rates are the same if the C and N balances are in agreement.

During stand rotation, stand density is considerably reduced by mortality due to crowding. To account for this, Valentine and Mäkelä (2012) defined the mortality rate of trees on the basis of leaf and fine root density, and assumed that crown rise was driven by crowding in space (after Valentine et al. 1994b) (Sect. 2.4).

As discussed in Chap. 6, trees in closed stands compete for light and other resources, so rapid height growth is a crucial means of survival and reproductive success for species of low or moderate shade tolerance (e.g., Horn 1971). Growing in height as fast as possible while young and reducing height growth to a minimum at maturity has been shown to be an evolutionarily stable strategy for such species (Mäkelä 1985) (see Sect. 7.6). Because height is related to crown size and leaf mass through optimal crown allometry, and these further constrain sapwood area through the pipe model, maximising height growth will not lead to unrealistic structures, such as trees thinner than could be mechanistically sustainable. On the other hand, maximising total growth when height is a free variable could lead to unrealistically short trees, as height generally poses a carbon cost for trees.

Based on these arguments, Valentine and Mäkelä (2012) chose to look for a strategy that maximises the height-growth rate of the trees at each moment of time, defined at a yearly time step. They optimised photosynthetic N concentration and the co-allocation of C and N to the production of fine-root, leaf, and wood biomass, while accounting for the C costs of respiration and N acquisition from the soil. The model was applied to finding answers to the research questions presented above, with parameter values appropriate for Scots pine (*Pinus sylvestris*) in southern Finland.

Regarding the closed canopy situation which was reached approximately by age 50 years in all site fertilities considered, the results of the dynamic optimisation were consistent with the steady state optimisation by Mäkelä et al. (2008b) (Fig. 7.10), suggesting that the steady state problem described above can be used as a reasonable proxy. In addition, the dynamic model was able to make projections about the plasticity of traits during the lifetime of trees. An interesting projection was that fine root allocation was considerably higher in young, open stands than in stands after canopy closure (Fig. 7.11). Few data are available for direct comparisons with these predictions, though other modellers have assumed that fine-root allocation is greater in the early phase of stand development (Running and Gower 1991; Bossel 1996).

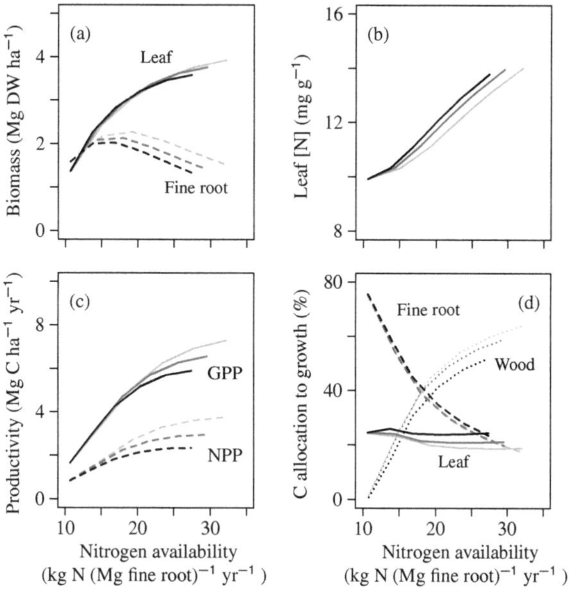

Fig. 7.10 Optimal model solutions over a range of nitrogen availability at 50 (light), 100 (medium) and 150 (dark) yr. Top left: leaf (solid) and fine-root (dashed) biomass; top right: leaf N concentration; bottom left: gross (solid) and net (dashed) primary productivity; bottom right: C allocation to production of fine roots, leaves, and wood. (Redrawn from Valentine and Mäkelä 2012)

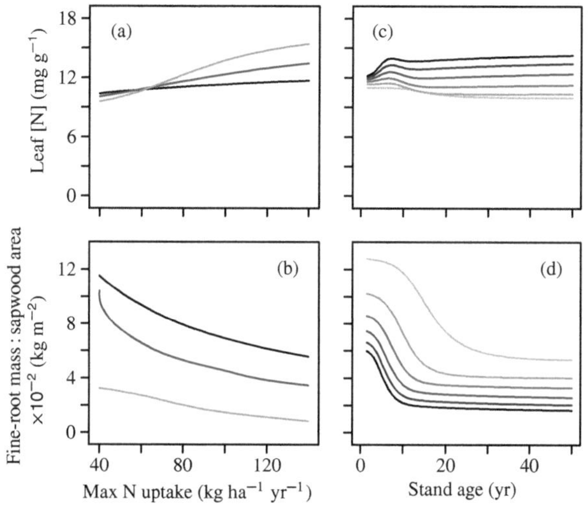

Fig. 7.11 Left: Predicted leaf nitrogen concentration (top) and the ratio of fine-root mass to sapwood area (bottom) as functions of N uptake rate (dark, stand initiation; medium, time of peak height growth; light, stand age 50 yr). Right: Predicted leaf nitrogen concentration (top) and the ratio of fine-root mass to sapwood area (bottom) as functions of stand age for six maximum N uptake rates ranging from 20 to 140 kg N ha^{-1} yr^{-1}. (Redrawn from Valentine and Mäkelä 2012)

7.5.6 Summary

We have presented a development of the functional balance theory from a simple idea to models with increasing detail and increasing number of assumptions. However, all these models share the assumption that the nitrogen and carbon budgets are balanced, and the balance is achieved across different environments through plasticity of plant structure. The key structural variables were (1) the ratio of fine-root mass to leaf mass and (2) the leaf nitrogen concentration.

The main result of the functional balance approach, retained from the simplest to the most complex model, is the dependence of the fine-root mass to leaf mass ratio on the relative availabilities of carbon and nitrogen. In the models that allow for variability of tissue nitrogen concentration, this also depends on the relative availabilities of resources, somewhat dampening the response in the root-leaf ratio.

It is noteworthy that these results were obtained using variable formulations of the evolutionary optimisation problem, where the objective function ranged from relative growth rate to actual NPP and height growth. On the other hand, the assumptions constraining the optimisation problem in relation to resource acquisition and tree structure were similar in all cases. In the dynamic stand model (Valentine and Mäkelä 2012) height growth was explicitly connected with the N and C balances of the tree, resulting in a dependence between maximum height growth and N availability. In the static approach (Mäkelä et al. 2008b), a similar pattern was obtained by maximising NPP at steady state while demanding that the stable height increase with N availability. These results suggest that optimisation problems may be simplified with appropriate choices of constraints and objectives, although it is difficult to provide any general rule for guiding such choices.

Valentine and Mäkelä (2012) also considered the possibility that there was no plasticity in tree structure, such that the environment only affected the growth rate of trees directly. This led to a violation of the functional balance, i.e., depending on the environment and stand developmental stage, either carbon or nitrogen was limiting growth while the other resource was available in excess. This situation is commonly known as the law-of-the-minimum and often assumed in growth models (e.g., Dybzinski et al. 2011). Valentine and Mäkelä (2012) found that the time-courses of height for the optimal-balance and law-of-the-minimum approaches from their model were similar qualitatively, though the optimal balance approach yielded the taller trees at any given point in time.

If the traits crucial for the optimal co-allocation of N and C lack sufficient plasticity, then the law-of-the-minimum solution may be closest to the truth. As a consequence, however, it may be necessary for trees to 'waste' the excess element, e.g., by increasing the rate of respiration (Dewar 2001), or by releasing sugars as root exudates (Hari and Kulmala 2008). In this regard, growth based on agreement between the C and N balances is more efficient, and, we conjecture, more likely from a Darwinian standpoint. We further conjecture that if a model indicates that it is less profitable to achieve agreement between the N and C balances than to allow for some of the resources to be wasted, then it is likely that the model lacks the mechanism to achieve profitable agreement. Moreover, if optimal results suggest wider plasticity than has been observed, then it is likely that additional constraints are at play, which have not been considered.

We should note that all the reviewed studies ignored the dynamics of N cycling in the soil, which may change the solution to some extent especially under nutrient-limited conditions or if carbon is used to prime nutrient release from the soil by accelerating soil organic matter decomposition. The optimisation method would equally well lend itself to situations where the availability of nutrients varies with time, but so far has not been included in most optimisation models (Franklin 2007; Dybzinski et al. 2011; Franklin et al. 2014).

7.6 Evolutionary Games

7.6.1 Evolutionarily Stable Strategies

Game theory is used for analysing evolutionary adaptations in situations where the success of a life strategy of an individual essentially depends on the life strategies of the other individuals in the population. An important example that we already touched upon in the preceding chapters is competition for light. If trees were not shaded by neighbours, their optimal life strategy could be one that produces fairly circular crowns that reach in all directions to capture resources. In reality, however, this strategy would soon be outcompeted by individuals that are able to bring their crowns above the rest of the canopy and cut off the best part of the incoming light.

Shading is a typical example of a case where the selection of traits takes place under asymmetrical competition. This is when competition for resources is decisive, and consequently, natural selection favours forms that would not be optimal in the absence of competitors, nor are they optimal for the community as a whole. Under competition those traits are selected, because they provide a competitive advantage to the individual. This kind of selection leads to what is called an *Evolutionarily Stable Strategy* (ESS), i.e., a life strategy that, when adopted by all individuals of a population, cannot be outcompeted by any other feasible strategy (Maynard-Smith and Price 1973).

The concept of ESS was defined by John Maynard Smith in his classic book "Evolution and the theory of games" as follows (Maynard Smith 1982):

A strategy I is an evolutionarily stable strategy if and only if

(1) when playing I against itself the payoff is larger than when playing any other strategy J against I, or

(2) if playing I against itself gives the same payoff as playing a J against I, then I against I is better than J against itself

In terms of classical game theory, this definition is almost identical to the Nash solution of a game problem. A Nash equilibrium is reached in a game where no cooperation is allowed among the players. Other solution concepts have been developed and are categorized on the basis of the type of collaboration and information flow between the contestants. The ESS is consistent with the individual selection postulated in the theory of evolution.

The first biological applications of game theory were to animal behaviour, such as "the war of attrition" and sex ratios (Maynard Smith 1982). In plants, one of the first applications was to height growth in a static Scotch Auction game (Rose 1978). In a Scotch Auction we consider a situation where all contestants play the same strategy, except for one that decides to play a "cheating strategy". In the case of tree height, all trees are assumed to be equally tall, except for one that "cheats" and grows a bit taller. This will give the cheater an advantage in terms of light capture, but it will also incur a cost in terms of having to construct a taller stem. Maximum tree height can be found at the point where the costs and benefits of cheating balance, i.e., it is no longer profitable to cheat by growing taller than the others. More generally, the ESS in a Scotch Auction is a strategy that no cheater can beat.

7.6.2 *Differential Games*

The Scotch Auction as well as many of the game applications to animal behaviour apply static game theory where the strategies are single parameters, such as body shapes, sex ratios or decisions in single individual-to-individual contests. This presupposes that the payoff from any one contest does not depend on what happened earlier in the game. It is conceivable, however, that maximising a payoff over a lifetime does not involve maximising the same variable at every moment of time. For example the maximum seed production need not be obtained by allocating the maximum amount of resources to seed growth at every moment of time. In optimisation, such problems are treated with dynamic optimisation techniques, one example of which was given in Sect. 7.3.2 where a height growth pattern was found that maximised the expected lifetime production (see (7.3)). When the same approach is applied to game theory, we talk about *dynamic* or *differential game theory*. Instead of a single parameter, it allows us to search for a time function as the ESS.

To define an evolutionary differential game problem, consider a population with N members. Each of these has a time-dependent state, x_i. The population at time t consists of the set $\{x_1(t), x_2(t), \ldots, x_N(t)\}$ of individual states which can differ from each other because of environmental histories or genes. We consider a genetically determined property, u, which can be a time function (such as timing of growth) or a number (such as leaf size). The variation of u is constrained to a set D.

The time development of the individual states depends on the states x_i and life strategies u_i of all the members in the population through a specified function f_i:

$$\frac{\mathrm{d}x_i(t)}{\mathrm{d}t} = f_i[x_1(t), x_2(t), \ldots, x_N(t), u_1(t), u_2(t), \ldots, u_N(t)]$$
$$x_i(t_0) = x_{i0} \tag{7.36}$$

where the latter equation defines an initial constraint and t_0 denotes the initial time.

In the evolutionary game the players are maximising their payoffs that are thought to represent, in some manner, the number of their surviving offspring. Each payoff J_i depends on the states x_i of all the members of the population and thus indirectly on their respective life strategies u_i, and can accumulate over (part of) their life time, T:

$$J_i = K_i[x_i(T), \ldots, x_N(T)] + \int_0^T L_i[x_1(t), x_2(t), \ldots, x_N(t)]\,\mathrm{d}t \tag{7.37}$$

Above, K_i is a payoff to player i that is evaluated at a specified final time of the game, T. For example, K_i could quantify the competitive position reached by an individual tree at the time when seed production commences. If the tree has grown to be larger than its neighbours at this point, but still has enough resources to use towards ample seed production, its payoff K_i is high. The function L_i, on the other hand, quantifies payoffs that accumulate over time. For example, if seed production occurs during most of the lifetime of the tree, $L_i(t)$ could be the rate of seed production at time t. The time dependent payoff could also be modified by the probability of survival, such that later potential payoffs are less important, because it is more likely that the tree has already died (see (7.3)).

The differential game problem is now formulated as follows: For each $i = 1, \ldots, N$,

$$\max_{u_i \in D} J_i \tag{7.38}$$

subject to

$$\frac{\mathrm{d}x_1(t)}{\mathrm{d}t} = f_1[x_1(t), x_2(t), \ldots, x_N(t), u_1(t), u_2(t), \ldots, u_N(t)]$$

$$\vdots$$

$$\frac{\mathrm{d}x_N(t)}{\mathrm{d}t} = f_N[x_1(t), x_2(t), \ldots, x_N(t), u_1(t), u_2(t), \ldots, u_N(t)]$$

$$x_1(t_0) = x_{10}, \ldots, x_N(t_0) = x_{N0}$$

As in the Scotch Auction described above, the ESS of the differential game can be found by looking for a population-level strategy that cannot be beaten by any invading cheater. We therefore say that if the whole population is playing strategy I except for a member j that plays J, then j is playing J against I. This is a special case of "playing the field," where a fraction q of the population is playing the exceptional strategy (Maynard Smith 1982). When $q = 1/N$ the ESS can be interpreted as almost identical with the Nash strategy in classical game theory (Starr and Ho 1969). Strategy I is an ESS if, for $i = 1, \ldots, N$ and for all $I \neq J$, either

$$J_1[u_1(I), \ldots, u_N(I)] > J_i[u_1(I), \ldots, u_i(J), \ldots, u_N(I)]$$

or

$$J_1[u_1(I), \ldots, u_N(I)] = J_i[u_1(I), \ldots, u_i(J), \ldots, u_N(I)]$$

and

$$J_i[u_1(J), \ldots, u_i(I), \ldots, u_N(J)] > J_i[u_1(J), \ldots, u_N(J)] \tag{7.39}$$

The first equation above with a "≥" sign instead of the inequality ">" is identical to the Nash strategy. The methods of solving such problems are beyond the scope of this presentation but can be found in mathematics textbooks and papers (e.g., Luenberger 1979). The necessary conditions for the Nash equilibrium were presented by Starr and Ho (1969).

An example of how to apply differential games to the analysis of ESS in trees was presented by Mäkelä (1985) who studied evolutionarily stable height growth strategies. Other applications of game theory to the plant kingdom include, e.g., height growth (King 1990), carbon allocation (King 1993; Gersani et al. 2001) and crown base height (Lindh 2016). In the following we summarise the problem formulation and results by Mäkelä (1985).

7.6.3 *Height Growth as a Differential Game*

In the preceding chapters we have emphasized the significance of tree height in both competition for light and allocation of growth resources. At least in relatively homogeneous stands in relatively high latitudes where the sunshine is lateral, the height of a tree relative to the whole canopy seems to be an important indicator of survivorship. On the other hand, as the trees grow taller, the proportion of growth allocated to the non-productive woody parts seems to increase, thus decreasing the overall growth potential of the tree (Valentine 1985; Mäkelä 1986).

Mäkelä (1985) presented an analysis of tree height as an ESS strategy in a differential game setting. The height growth pattern was searched for trees growing in a homogeneous canopy surrounded by other members of the same species. This represents a situation occurring after disturbance, such as a forest fire, when pioneer species occupy the site and the individuals compete with each other for resources and space. It is then those individuals that survive to maturity and enter the seed production phase that will produce the maximum number of offspring. The seed production ability at that time depends on the sizes of the productive parts of the tree, especially its leaves. The objective function of the game problem was therefore formulated as maximisation of foliage biomass at a fixed time, T, during the active seed-producing phase in the lifetime of the tree corresponding to the function K_i in (7.37):

$$J = W_f(T) \tag{7.40}$$

For instance in normal, natural stands in the boreal forest, the age of the onset of seed production may be 60–80 years.

The leaf mass at the start of the seed production phase was determined using the carbon balance approach of Chap. 3:

$$\frac{dW_f}{dt} = \lambda_f G - s_f W_f \tag{7.41}$$

where G is growth rate, λ_f is allocation of growth to leaf mass, and s_f is leaf-specific rate of senescence.

To make the model solvable analytically, some simplifying assumptions were made. Firstly, respiration was assumed proportional to photosynthesis, such that

$$G = f_c \sigma_f W_f \tag{7.42}$$

where σ_f is foliage-specific photosynthesis rate which is converted to growth rate by the constant f_c.

Secondly, allocation to wood was assumed to increase linearly with tree height:

$$\lambda_f(t) = 1 - cH(t) \tag{7.43}$$

where $H(t)$ is tree height and c is a constant. Note that according to the pipe model, c is proportional to the amount of sapwood required per unit new leaf mass.

Thirdly, the shading by neighbours was assumed to be derivable from the relative height difference between the subject tree and the mean of its neighbours:

$$\sigma_f = \sigma_{f0} + b[H(t) - \hat{H}(t)] \tag{7.44}$$

where $\hat{H}$ is the average height of the canopy, σ_{f0} is the specific photosynthetic rate when the subject tree is as tall as the average canopy and b is a parameter reflecting the magnitude of the reaction to shade. σ_{f0} was assumed constant.

The set of the possible solutions was restricted by the assumption that there is a physical limit, η, to the maximum height growth rate, so it was required that

$$0 \le \frac{dH(t)}{dt} \equiv z(t) \le \eta \tag{7.45}$$

These assumptions lead to the following game problem (which is a special case of (7.38)): For each $k = 1, 2, \ldots, N$ find a z_k that maximises (in the Nash sense)

$$J_k = W_f^k(T) \tag{7.46}$$

subject to the constraints ($i = 1, 2, \ldots, N$)

$$\frac{\mathrm{d}W_{\mathrm{f}}^{i}(t)}{\mathrm{d}t} = f_{\mathrm{c}}\sigma_{\mathrm{f}}^{i}(t)(1 - cH^{i}(t)W_{\mathrm{f}}^{i}(t) - s_{\mathrm{f}}W_{\mathrm{f}}^{i}(t) \tag{7.47a}$$

$$\frac{\mathrm{d}H^{i}(t)}{\mathrm{d}t} = z^{i}(t) \tag{7.47b}$$

$$\frac{\mathrm{d}\sigma_{\mathrm{f}}^{i}(t)}{\mathrm{d}t} = f_{\mathrm{c}}^{-1}b(z^{i}(t) - (N-1)^{-1}\sum_{\substack{j=1\\ j\neq i}}^{N} z^{j}(t) \tag{7.47c}$$

$$W_{\mathrm{f}}^{i}(0) = W_{\mathrm{f0}}^{i},\ H^{i}(0) = H_{0}^{i},\ \sigma_{\mathrm{f0}}^{i} = \sigma_{\mathrm{f0}}^{i} \tag{7.47d}$$

In the above, the definition of $\hat{H}^i$ as the average height of the remaining stand has been utilised, giving

$$\frac{\mathrm{d}\hat{H}^{i}(t)}{\mathrm{d}t} = (N-1)^{-1}\sum_{\substack{j=1\\ j\neq i}}^{N} z^{j}(t) \tag{7.48}$$

Equation (7.44) has been differentiated with respect to time on this basis.

Mäkelä (1985) found the solution to the game problem using the theory of differential games (Starr and Ho 1969). It turned out that the ESS of this problem was a so-called bang-bang strategy where the height growth function gets values at the boundaries, switching from one boundary to the other at prescribed times. Here, one switching time, τ, was found, with the value

$$\tau = \frac{b - f_{\mathrm{c}}\sigma_{\mathrm{f0}}c}{\eta bc} \tag{7.49}$$

The optimal height growth strategy was hence one with maximum growth, η, up to the switching time τ and zero growth afterwards (Fig. 7.12). The final height reached was therefore

$$H_{\mathrm{max}} = \frac{b - f_{\mathrm{c}}\sigma_{\mathrm{f0}}c}{bc} \tag{7.50}$$

This result illustrates the trade-off between the metabolic cost of height growth on one hand, and its competitive benefits for light capture on the other hand. It is beneficial for the tree to grow as fast as possible in the early part of its lifetime, to avoid becoming overshadowed by neighbours, but at some point the metabolic costs overcome this benefit, implying that no further height growth will improve the photosynthetic gain and hence the seed production.

Mäkelä (1985) further considered the possible evolutionary significance of the structural constraints assumed in the model. Obviously, if a mutant appeared with a faster maximum height growth than the rest of the population, it would benefit

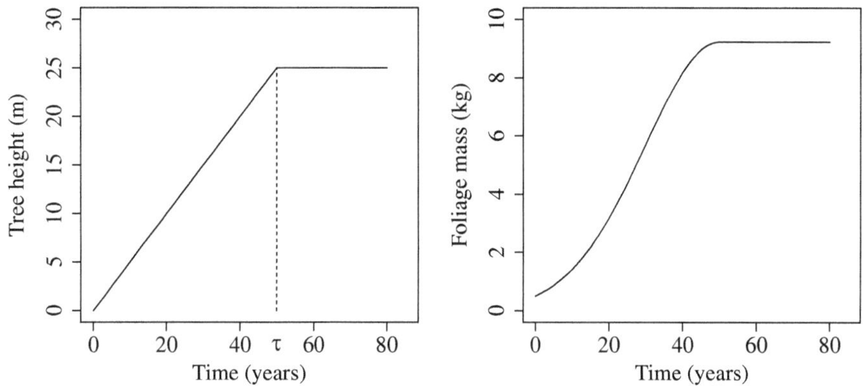

Fig. 7.12 Evolutionarily stable height growth strategy (left panel) and the corresponding foliage development (right panel). The parameter values used here were $\sigma_{f0} = 0.74\,\mathrm{yr}^{-1}$, $f_c = 0.5$, $\eta = 0.5\,\mathrm{m\,yr}^{-1}$, $b = 0.0072\,\mathrm{kg\,m}^{-1}\,\mathrm{yr}^{-1}$, $c = 0.0131\,\mathrm{m}^{-1}$. Initial state $W_{f0} = 0.5\,\mathrm{kg}$, $H_0 = 0\,\mathrm{m}$. (Modified and redrawn from Mäkelä 1985)

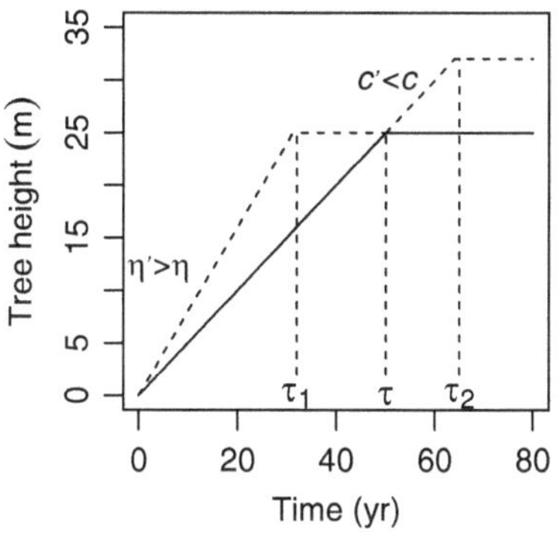

Fig. 7.13 Possible replacements of the ESS strategies subject to modifications in structural parameters. The solid line represents the current ESS for height growth. If a mutant appears that can grow faster than the resident population ($\eta' > \eta$) it will reach the maximum height fast and will outcompete its neighbours, thus making this trait more frequent in the population. On the other hand, if another mutant appeared that could sustain a more slender stem than the resident population ($c' < c$), it could grow taller than its neighbours with the same cost, and would also become more frequent in the population. (Redrawn from Mäkelä 1985)

from reaching above the rest of the canopy when young. This early competitive advantage would provide a long-term benefit in terms of productivity, although the evolutionarily stable maximum height would not be affected (7.50). On the other hand, a reduction in the value of parameter c would represent a switch towards more slender stems, likely requiring some improvement in the mechanical stability or water conductance properties of wood. This would allow for larger maximum tree heights. Within the above framework, no finite ESS can be found to restrict such structural improvements, provided that they are physically possible (Fig. 7.13).

7.6.4 Adaptive System Dynamics

The methods of optimisation and game theory both postulate a proxy of evolutionary fitness, then find the traits that maximise the fitness under the prescribed conditions. An implicit assumption of these methods is, in addition to the fitness proxy and model structure, that the optimum or ESS can always be achieved by mutation and selection.

An alternative method for finding evolutionarily stable strategies is to model the process of selection. Such models of *adaptive system dynamics* explicitly describe the population as consisting of individuals with a variety of possible traits and their interactions with each other and the environment (Geritz et al. 1998; Brännström et al. 2013). These interactions lead to either increasing or decreasing numbers of individuals with particular traits, depending on how the system has been defined. New forms are introduced through "mutations" in the traits. An ESS is reached if a mutation can no longer invade the existing, or resident, population. The dynamics of the model will hence lead to the emergence of "the fittest" *sensu* the model, without explicitly having to define a fitness proxy.

Theoretical analysis of adaptive system dynamics has shown that although an ESS, if adopted by the population, cannot be invaded by any other strategy, the population may not always converge towards the ESS (Geritz et al. 1998; Brännström et al. 2013). This depends on the relative fitness of traits in the neighbourhood of the ESS. Interestingly, it is also possible that several strategies coexist in a stable population that cannot be invaded by mutant traits. These can be strategies that on their own do not satisfy the conditions of an ESS.

Simulating the adaptive dynamics of a population allows us to consider a variety of population structures and environmental conditions, which is not so straightforward if the system has to be defined formally as an optimisation problem or game. As shown by Geritz et al. (1998), one mechanism creating multiple strategies in the stable population is the existence of patches with different environmental conditions within the dispersal area of the population. If the environmental differences are sufficiently small, the system will converge to an ESS corresponding to the mean environment. However, if the variability is large, a branching towards more than one trait value may occur in the population. Similar variability of traits could conceivably be caused by environments that vary in time.

Another cause for multiple traits is an environment where one trait creates a competitive advantage for the other, compared with a one-trait situation. Using the terminology of Maynard Smith (1982) (see Sect. 7.6.1), the payoff of playing I against J is better than the payoff of I against I, and the payoff of J against I is better than the payoff of J against J. Neither strategy is therefore an ESS, although their combination can be evolutionarily stable in the sense that it cannot be invaded by other strategies.

To our knowledge the first application of adaptive system dynamics to tree structure, Dybzinski et al. (2011) studied carbon allocation between roots, leaves, and stems in stands that were assumed to be at equilibrium regarding community size and age distribution. In addition, to simplify the description of competition for light, the model used the perfect-plasticity approximation for the spatial arrangement of tree crowns (Strigul et al. 2008, see Sect. 6.6.2). These assumptions meant that an explicit description of interactions between size classes was not necessary in the model. The results were very similar to those obtained by Mäkelä et al. (2008b) and Valentine and Mäkelä (2012) using a conventional optimisation approach (see Sect. 7.5). The model predicted an ESS where carbon allocation to fine roots increased with decreasing nitrogen availability, and there was a trade-off between root allocation and stem growth, with increasing allocation to stems when nitrogen availability was increasing.

Lindh et al. (2014) used adaptive system dynamics to study the conditions under which developing a taproot would be an evolutionarily stable strategy in trees. The study considered populations of trees where mutual shading could reduce the evolutionary success of individuals, and where the life strategy choices were to allocate growth to either longer stems or deeper taproots. The benefit from the taproot was to reduce mortality in environments where water was limiting growth. Both tree height and taproot length also inferred a cost related to their construction and maintenance. The study found that the ESS depended on the environment, with increased drought mortality leading to an ESS with taproot while in a less drought-prone environments the taproot was not included in the ESS. In medium environments this model also resulted in a mixed-trait stable equilibrium where trees with and without taproots coexisted.

7.7 Summary and Outlook

The above sections have highlighted some of the applications of evolutionary optimisation to tree structure and consequent carbon allocation. Another field where optimisation techniques have been widely used is in eco-physiological models to balance different metabolic processes, for example, to model stomatal control which regulates both carbon uptake and transpiration (Cowan and Farquhar 1977; Hari et al. 1986; Mäkelä et al. 1996; Dewar et al. 2018). In all these applications the role of optimisation has been to provide additional constraints to plant and ecosystem behaviour, so as to avoid increasing the complexity of the model through additional submodels based on more detailed physiology.

Although several studies have applied optimisation to analyse plant structure and growth allocation, to date few – if any – model applications to practical situations include embedded optimisation routines. This may be because optimisation is relatively time-consuming computationally. Instead, summary models have been created that incorporate the result of optimisation through the overall patterns of tree form or life strategies predicted to be evolutionarily stable by the optimisation model. A wide-spread example is the use of environment-sensitive root-shoot ratios as an application of the functional balance theory. Another similar case is the crown allometry model (7.5) implied by the optimal height growth strategy of Sect. 7.3.2.

The examples covered in this chapter illustrate the fact that there is no well-defined, unique fitness proxy – or objective function – to be applied in the eco-evolutionary problem, nor is it always clear which traits should be assumed adaptive and which should be taken to constrain the problem. Theoretically, different choices may lead to consistent results, because different subsystems of complex systems may be partially independent of each other, allowing for constraining one subsystem while analysing the interrelationships of the other (Ahl and Allen 1996). Above, this conclusion was corroborated in practice, as similar results could be obtained with different fitness proxies when optimising co-allocation of carbon and nitrogen at quasi-steady state (Sect. 7.5.4) or over a full rotation (Sect. 7.5.5). The latter case can be taken as an example of sequential or hierarchical optimisation where either a constraint or the objective function is based on another evolutionary optimisation problem (Sect. 7.6.3).

Eco-evolutionary optimisation can be regarded as an example of an organising principle that constrains system behaviour (Franklin et al. 2020). Other organising principles include, e.g., perfect plasticity (Sect. 6.6.3) and perfect aggregation (Sect. 6.6.2). While such principles are derivable from biological theory, they cannot be reproduced by reductionist approaches. On the contrary, organising principles reduce complexity in models by providing additional constraints that replace the need to dive into more detail at ever lower hierarchical levels (see Sect. 1.5). The need for organising principles helping reduce complexity without losing biological soundness has recently been increasingly recognised, e.g., in the context of global dynamic vegetation models aiming for robust global predictions in a changing climate (Franklin et al. 2020).

Nevertheless, no matter how strong the biological theory behind eco-evolutionary modelling, each model still relies on unique hypotheses about plant dynamics, stable constraints and fitness proxies. This puts a strong emphasis on the need to treat the results, not as theoretically deduced facts, but as hypotheses that must be carefully evaluated against empirical observation.

7.8 Exercises

7.1 Derive the result of (7.10) from the functional balance assumption, i.e., that N uptake equals its use in growth.

7.2 Show that (7.12) maximises the relative growth rate under the assumptions made in (7.8)–(7.10).

7.3 `R` code for simulating the OptiPipe model (Valentine and Mäkelä 2012) called "`OptiPipe.R`" is provided in the electronic supplementary material of this chapter. Download the model and explore its behaviour by varying its parameter values and inputs.

7.4 Consider the different fitness proxies used in the examples of this chapter. What do they have in common and how do they differ? How could such differences be justified and what problems might such ambiguity infer?

Chapter 8
Predicting Stand Growth: Parameters, Drivers, and Modular Inputs

In this chapter we consider different methods of estimating the inputs to the tree and stand growth models presented in this book. How does the selected method depend on the specific questions we want to ask with the model? To gain insights into this, we first outline some general ideas and theory about linking models with data. We then illustrate input quantification for model applications by introducing different methods of parameterisation for the core model presented in Chap. 5. These include empirical parameterisation of a reduced form of the model from standard forestry data, estimation of structural parameters from direct measurements, and using submodels in a modular system to derive time-dependent or structure-dependent parameters from environmental drivers and/or model state.

8.1 Introduction

Any model that is constructed to analyse real-world situations—for example, to make growth predictions, or to search for the best course of management actions—should predict reality, not only in its pattern of behaviour, but also with acceptable accuracy. Forest models, in particular, need to predict realistically the outcomes of any relevant management actions across different environments: How does stocking density affect stand growth and mean diameter development? How does the residual stand behave after harvest removals? Which harvest schedules produce the best economic revenue? How does forest production vary regionally? Is climate change going to increase or decrease productivity and carbon sequestration? The realism is obtained by linking the model to measurements.

A real-world situation is specified for a model by its inputs. The inputs of a dynamic model include: (a) the initial values of the state variables, and (b) the values of *drivers* and *parameters* over the entire simulation time. Drivers are typically variables that quantify the environment, such as climate, but they can also include

A. Mäkelä, H. T. Valentine, *Models of Tree and Stand Dynamics*,
https://doi.org/10.1007/978-3-030-35761-0_8

control actions, such as forest management. Parameters, on the other hand, are envisaged as reflecting some internal characteristics or traits of the system, and they are usually assumed constant over time. While drivers can simply be taken as given, e.g., from meteorological measurements, model parameters must be estimated from appropriate data sets before the model can be used for prediction, i.e., for simulation of the time-courses of the state variables.

In this chapter we consider different methods of estimating the inputs to the tree and stand growth models presented in this book. How does this depend on the specific questions we want to ask with the model? To gain insights into this, we first outline some general ideas and theory about linking models with data. We then illustrate input quantification for model applications, using the core model presented in Chap. 5 as an example.

8.2 Linkages Between Models and Data

The linkages between data and model essentially depend on the type of model (see Sects. 1.5, 1.6, and 1.7). Emprical growth and yield models usually assume that climate and site properties remain stable during the time of interest (Skovsgaard and Vanclay 2008). The models therefore do not have explicit external drivers, and any environmental impacts are embedded in model parameters. The parameter values are determined for each site by fitting the model to data with statistical methods. The fitting uses data on independent and dependent variables at the same level of organisation (Fig. 8.1).

Process-based, or mechanistic models, by contrast, derive growth from processes that are defined mechanistically at a lower level of resolution (lower level of hierarchy) than model outputs. Environmental drivers affect the process rates of

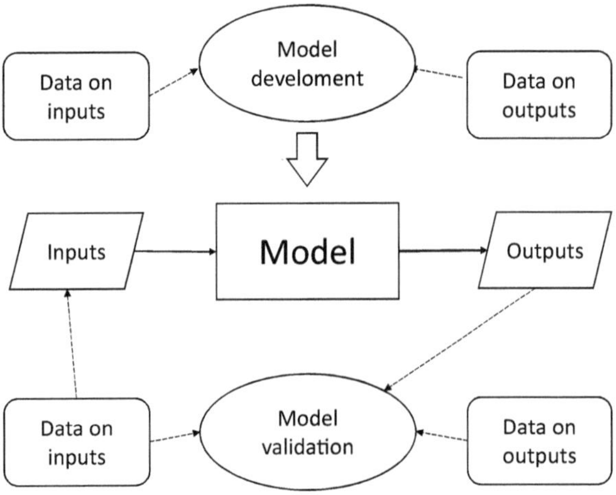

Fig. 8.1 Model-data relationships in parameter estimation of purely empirical models. An empirical model uses data on independent and dependent variables to find the best statistical fit with observations. The model, including its parameter estimates, is evaluated against a similar data set, including observations of independent and dependent variables

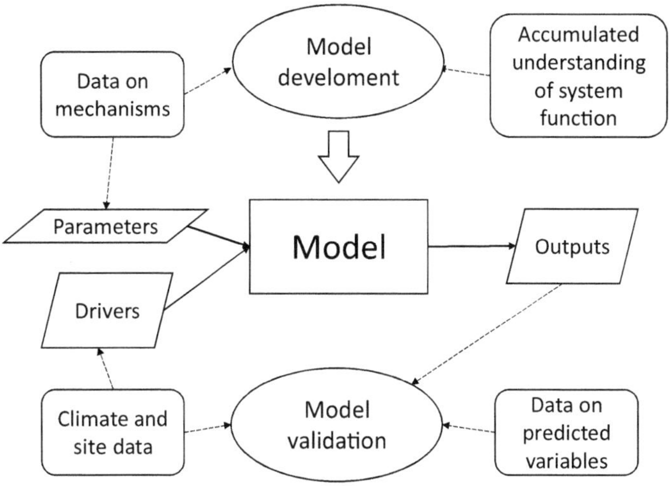

Fig. 8.2 Model-data relationships in parameter estimation of purely mechanistic models. A mechanistic model uses independent inputs on drivers and parameters, estimated on the basis of their physical definition, to produce the outputs, while the model structure has been formulated on the basis of previously accumulated understanding of the functioning of the system. Comparing the outputs with data provides an independent test of the model

the model, and its parameters should be directly quantifiable on the basis of their definition. They should therefore be fully independent of model outputs (Fig. 8.2).

The models discussed in this book derive tree and stand growth from material balances controlled by physiological processes under some structural constraints. They thus belong to the class of mechanistic models.

However, as our main scale of focus has been individual tree growth at an annual time resolution, the models do not have any direct environmental drivers, such as radiation, temperature or rainfall. Instead, these impacts are embedded in quantities that appear as independent input parameters in the model. Similarly, we have assumed some rather straightforward structural regularities expressed as ratios or allometries at the whole-tree level. The model presented in Chap. 5 has 20 parameters that describe (1) mean annual metabolic rates and (2) tree structure (Table 8.1).

As noted above, model parameters should usually be defined as quantities constant in time, whereas drivers vary in time independently of the system. Channelling environmental impact in a model through its parameters would not make sense if the environment-sensitive parameters had not been defined such that they remain relatively constant in time. This is indeed possible, as long as the model incorporates the essential dynamic properties of the system at the hierarchical level of interest, because the lower level detail can generally be regarded as random noise that does not alter system dynamics. We illustrated this general result of *hierarchy theory* (see Sect. 1.5) in the case of mean annual leaf-specific photosynthesis rate in Sect. 3.2 (Fig. 3.1). As already noted in Sect. 1.5, hierarchy theory also suggests that indeed,

Table 8.1 Structural (S) and metabolic (M) parameters of the core model presented in Chap. 5. The indices s, b and t refer to stem, branches and coarse roots, respectively

Symbol	Type	Units	Definition
η_i	S	kg m^{-2}	Pipe model coefficient, $i = \text{s, b, t}$
ϕ_i	S	–	Form factor, $i = \text{s, b, t}$
ρ_i	S	kg m^{-3}	Wood density, $i = \text{s, b, t}$
z	S	kg C m^{-2}	Exponent in crown allometry
β_A	S	$\text{m}^2\,(\text{kg DW})^{-1}$	Coefficient in crown allometry
α_r	S	–	Ratio of fine-root to leaf biomass
σ_f	M	$\text{kg C (kg DW)}^{-1}\,\text{yr}^{-1}$	Leaf-specific rate of photosynthesis
a_σ	M	–	Reduction factor of photosynthesis
r_{M_f}	M	$\text{kg C (kg DW)}^{-1}\,\text{yr}^{-1}$	Specific maintenance respiration rate of leaves
r_{M_r}	M	$\text{kg C (kg DW)}^{-1}\,\text{yr}^{-1}$	Specific maintenance respiration rate of fine roots
r_{M_p}	M	$\text{kg C (kg DW)}^{-1}\,\text{yr}^{-1}$	Specific maintenance respiration rate of wood
T_f	M	yr	Life span of leaves
T_r	M	yr	Life span of fine roots
Y_G	M	kg DW (kg C)^{-1}	Growth efficiency

the best method of model construction is to confine the model to only one level below the focal level of interest, because models including mechanism at several nested levels necessarily become complex and difficult to handle, hypersensitive to small changes in lower level parameter values, and tend to lose their mechanism as noise (Landsberg 1986; Reynolds et al. 1993).

Even if the parameters were generally fairly constant over time under normal conditions, we still need to consider the variability of parameters in space, for example in different climates and sites. When climate change impacts are considered, a trendlike change in appropriate parameter values will have to be accounted for. In this case, the parameters become time-dependent and may therefore be treated as generalised driving variables to the model. We therefore need to define methods for the estimation of model parameters in variable environments. This may require the application and scaling up of more detailed physiological models that use environmental drivers explicitly at faster time scales, and also including such models as independent sub-modules in the simulation (Fig. 8.3). It is important for such parameter estimation that the links between the detailed models and the growth model are defined appropriately.

Another class of parameters in mechanistic models, and particularly in the models of this book, are parameters defining tree structure. These can be determined directly from dedicated structural measurements. Alternatively, they may also be considered as scaled-up results of more detailed models of tree structure (Mäkelä 2003).

According to hierarchy theory, some inputs to our scale of interest come from the hierarchical levels above, when these pose constraints to the modelled system level. An important class of such constraints is provided by evolutionary optimi-

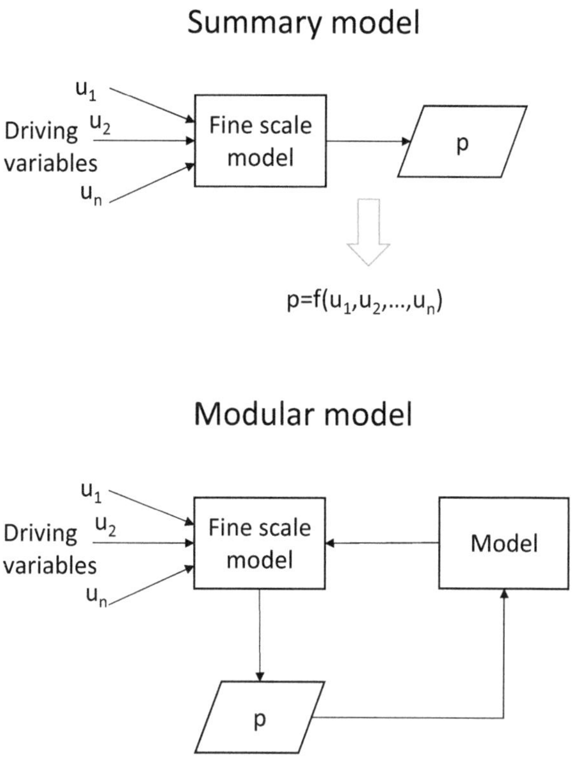

Fig. 8.3 Incorporating finer scale effects in a dynamic model. **Top:** A summary model—aka emulator—is developed to describe the effect of driving variables on model input parameter *p* by means of simulations with a finer scale model that cover the appropriate range of variability of the inputs. In this case, *p* is assumed independent of the focal system. **Bottom:** The input parameter *p* depends (weakly) on the state of the focal system. The changes in state are accounted for by a feedback from the main model to the finer scale model. In both cases the scales have been clearly separated and both can be considered independently

sation (Chap. 7). Once the evolutionary optimisation problem has been solved, the resulting functional or structural rules can be utilised as constraints to a mechanistic model. Typically, these would provide rules for optimal structures or balances between the metabolism of different elements, which can be either universal or environment-sensitive. Central results for the growth models of this book include the optimal co-allocation of carbon and nitrogen, which implies confined leaf–fine-root biomass ratios, and the optimal crown allometry that constrains the relationship between basal-area at crown base and crown length.

In the following sections, we present different ways of determining the parameter values of the core model presented in Chap. 5. First, we present a method of converting the model to a traditional growth and yield model that can be parameterised using statistical fitting (Valentine and Mäkelä 2005; Valentine et al. 2013). Secondly, we discuss the determination of structural parameters, and thirdly, we review a regional parameterisation of the model (Mäkelä et al. 2016). Finally, we present a modular approach to incorporating finer-scale structural effects to describe timber properties (Mäkelä 2002; Mäkelä and Mäkinen 2003).

8.3 Empirical Estimation of the Core Model

Most of the models of tree and stand dynamics used in forestry and forest ecology derive from empirical methods. The choice of variables in such models is often limited to those measured in forest inventories or research trials. Historically, stem diameter, specifically dbh, has been the most commonly measured attribute of a tree. Consequently, most empirical tree-level models use either dbh or the corresponding basal area, B, as the principal predictor of other attributes, traits, and rates of interest, including the rate of change in dbh itself. In our process-based approach, however, we recognise that the cross-sectional area at any point on a stem accumulates as a function of crown development acropetal to that point. Accordingly, our mean-tree model is formulated so that cross-sectional growth—and consequently the growth of dbh—is a response to crown-length dynamics. In this regard, measurements of dbh are of vital importance for testing the accuracy of a model's predictions.

The core of our process-based model comprises four differential equations—(6.22), (6.23), (6.24), and (6.26)—whose solutions, respectively, comprise time-courses for the four state variables that summarise the size and spatial confines for an average tree in a stand, namely, tip height (H), spacing (X), crown-base height (H_c), and basal area (B). With sufficient data, this set of differential equations can be fitted with statistical procedures. By sufficient data we mean annual or periodic remeasurements of each of the four state variables, with measurements procured both before and after the onset of closure and crown rise.

The tip height, crown-base height, and basal area equations were introduced by Valentine and Mäkelä (2005) for use by both process-based and empirical modelers. For example, empirical versions of the core model could be used to project future inventories of standing crops and to evaluate forest management options, while the process-based version with elaborations could be used to investigate issues related to forest processes, evolutionary theory, and climate change. The spacing equation, in its present form, provides for self-thinning, as explained in Sect. 6.5.1. Together, the core equations have come to be known as the *Bridging model*.

As we show below, each of the four equations of the core model can be calibrated more or less independently. Nonetheless, one could attempt to estimate some of the parameter values of the set of equations simultaneously. For example Valentine et al. (1997) estimated four parameters of a stand-level model (called Pipestem) with an iterative procedure that involved (a) numerically integrating the model starting with measurements of the starting values and guesses of the parameter values, and (b) calculating a cost function comprising weighted sums of squares of residuals of the state variables. New proposed values of the parameters were generated with a numerical search algorithm and the whole procedure was repeated until the cost function converged to an apparent minimum. The numerical search algorithm was based on the downhill simplex method of Nelder and Mead (1965). Green et al. (1999) used the method of Bayesian synthesis to analyse the uncertainty in inputs, parameter values, and predictions of the Pipestem model. In a similar vein,

Van Oijen et al. (2013) used Bayesian methods to evaluate and calibrate six different stand-growth models, including one called *Bridging*, which was built around our core model. Various options for estimation and evaluation of models comprising a set of differential equations are included in an R package called FME. Soetaert and Petzoldt (2010) provide an overview of the package with examples.

8.3.1 Considerations for Fitting

To fit the differential equations of our core model, we need to consolidate the parameters of the height-growth model. We start with (5.26), the process-based model of height growth rate, $\mathrm{d}H/\mathrm{d}t$ (m yr^{-1}), that we derived in Chap. 5:

$$\frac{\mathrm{d}H}{\mathrm{d}t} = \frac{L_\mathrm{c}}{z\left(W_\mathrm{f} + W_\mathrm{r} + W_\mathrm{p}\right) + \beta_1 W_\mathrm{f} L_\mathrm{c}} \left(Y_\mathrm{G}\left(P - R\right) - \frac{W_\mathrm{f}}{T_\mathrm{f}} - \frac{W_\mathrm{r}}{T_\mathrm{r}}\right) \tag{8.1}$$

We first convert the model into a form that can be fitted to data. To allow for an effect of hydraulic limitation on the rate of photosynthesis, P (kg C yr^{-1}), we assume that the specific rate of photosynthesis, σ_f, decreases as a function of crown length, whence $P = \sigma_\mathrm{f}\left(1 - \alpha_\sigma L_\mathrm{c}\right) W_\mathrm{f}$ (see Sect. 3.4.3). Substituting the following structural and functional relationships from Chap. 5,

$$\begin{aligned}
(P - R_\mathrm{M}) &= \left[\sigma_\mathrm{f}\left(1 - \alpha_\sigma L_\mathrm{c}\right) - r_{\mathrm{M_f}}\right] W_\mathrm{f} - r_{\mathrm{M_r}} W_\mathrm{r} - r_{\mathrm{M_w}}\left(\beta_1 H + \beta_2 H_\mathrm{c}\right) W_\mathrm{p} \\
W_\mathrm{r} &= \alpha_\mathrm{r} W_\mathrm{f} \\
W_\mathrm{p} &= (\beta_1 H + \beta_2 H_\mathrm{c}) W_\mathrm{f} \\
H &= H_\mathrm{c} + L_\mathrm{c},
\end{aligned}$$

into (8.1) and consolidating the many parameters into 5 aggregate parameters provides

$$\frac{\mathrm{d}H}{\mathrm{d}t} = g_1 L_\mathrm{c}\left(\frac{g_2 - g_4 H_\mathrm{c} - L_\mathrm{c}}{g_3 + g_5 H_\mathrm{c} + L_\mathrm{c}}\right) \tag{8.2}$$

where

$$g_1 = \frac{Y_\mathrm{G}\left(r_{\mathrm{M_w}}\beta_1 + \alpha_\sigma \sigma_\mathrm{f}\right)}{(1 + z)\beta_1} \tag{8.3}$$

$$g_2 = \frac{\left(\sigma_\mathrm{f} - r_{\mathrm{M_f}} - r_{\mathrm{M_r}}\alpha_\mathrm{r}\right) - \left[(1/T_\mathrm{f}) + (\alpha_\mathrm{r}/T_\mathrm{r})\right]/Y_\mathrm{G}}{r_{\mathrm{M_w}}\beta_1 + \alpha_\sigma \sigma_\mathrm{f}} \tag{8.4}$$

$$g_3 = \left(\frac{z}{1 + z}\right)\frac{1 + \alpha_\mathrm{r}}{\beta_1} \tag{8.5}$$

$$g_4 = \frac{r_{\mathrm{Mw}}(\beta_1 + \beta_2)}{r_{\mathrm{Mw}}\beta_1 + \alpha_\sigma \sigma_{\mathrm{f}}} \tag{8.6}$$

$$g_5 = \left(\frac{z}{1+z}\right)\frac{\beta_1 + \beta_2}{\beta_1} \tag{8.7}$$

Crown length equals tree height before any onset of crown rise, in which case (8.2), reduces to:

$$\frac{\mathrm{d}H}{\mathrm{d}t} = g_1 H \left(\frac{g_2 - H}{g_3 + H}\right), \quad H_{\mathrm{c}} = 0 \tag{8.8}$$

The asymptote of (8.8), calculated by setting $\mathrm{d}H/\mathrm{d}t = 0$, is $H_{\max} = g_2\,(\mathrm{m})$. Substituting $H - H_{\mathrm{c}}$ for L_{c} in (8.2) yields $H = g_2 + H_{\mathrm{c}}(1 - g_4)$, so the maximum tree height changes with the crown-base height and may exceed g_2.

The parameter values assigned to the mean-tree model in Table 6.1, together with $\sigma_{\mathrm{f}} = 2.5$ and $\alpha_\sigma = 0.015\,(\mathrm{m}^{-1})$, provided: $g_1 = 0.03233\,(\mathrm{yr}^{-1})$, $g_2 = 27.39\,(\mathrm{m})$, $g_3 = 1.055\,(\mathrm{m})$, $g_4 = 0.4783$, and $g_5 = 0.8042$.

Correlations among g_3, g_4, and g_5 may cause a parameter identification problem that would preclude finding the global minimum of a cost function when fitting the model to data with, say, R package FME. Note, for example, that $g_5 = [(\beta_1 + \beta_2)/(1 + \alpha_r)]\,g_3$, and in the absence of hydraulic limitation (i.e., $\alpha_\sigma = 0$), $g_5 = [z/(z+1)]\,g_4$. Should a parameter-identification problem arise, one might assign fixed values to g_3 and g_5 — based on independent information from sampling or prior information.

The use of difference equations is common in forestry. It is possible to integrate (8.2) over a time interval t to $t + \Delta t$ if the crown-base height, H_{c}, is zero or if we assume that $H_{\mathrm{c}} > 0$ remains constant over that interval (see the appendix of Valentine and Mäkelä (2005) for details). The integration, however, results in a nonlinear difference equation that must be solved iteratively to obtain $H_{t+\Delta t}$. As an alternative, we suggest a straightforward approximate solution of the height-growth model, viz.,

$$H_{t+\Delta t} \doteq H_t + g_1 L_{\mathrm{c},t}\left(\frac{g_2 - g_4 H_{\mathrm{c},t} - L_{\mathrm{c},t}}{g_3 + g_5 H_{\mathrm{c},t} + L_{\mathrm{c},t}}\right) + \epsilon_{t+\Delta t} \tag{8.9}$$

This approximate model, which retains the possible identification problem, can be fitted, for example, with the non-linear least squares function nls in R or Bayesian methods.

The other three equations of the core model are driven by the height growth rate. Equation (6.24),

$$\frac{\mathrm{d}H_{\mathrm{c}}}{\mathrm{d}t} = \max\left\{0,\, S_{\mathrm{r}}\left(\frac{\mathrm{d}H}{\mathrm{d}t} - \beta_{\mathrm{x}}\frac{\mathrm{d}X}{\mathrm{d}t}\right)\right\},$$

is the time derivative of the crown-rise model, (2.69),

$$H_{c,\,t+\Delta t} = \max\left(0,\, H_{c,\,t},\, H_{t+\Delta t} - \alpha_X - \beta_X X_{t+\Delta t}\right)$$

The numerical switch, S_r, smooths the transition from no crown rise to active crown rise (see Sect. 6.5.2). Our approach has been to first fit the model and then add the numerical switch, with assigned parameter values, to smooth the transition from no crown rise to active crown rise.

The average spacing among trees at time $t + \Delta t$ is modeled by

$$X_{t+\Delta t} = \begin{cases} X_t, & \text{if } H_{c,\,t} < k_X H_X \\ X_t + \dfrac{u}{\beta_X}\left(H_{t+\Delta t} - H_t\right), & \text{if otherwise} \end{cases} \tag{8.10}$$

We consider increases in spacing that arise from self-thinning. As explained in Sects. 2.4 and 6.5.1, the first case of this model indicates that there is no change in spacing until the average crown-base height, H_c, rises to height $k_X H_X$. We recall that H_X is the average tree height when crown rise commences. The second case, where $H_{c,\,t} \geq k_X H_X$, is a linear model that indicates that the spacing among trees increases with further height growth. The value of the parameter, β_X, obtains from the crown-rise model, and the control parameter, u, obtains from fitting this linear model with data from actively self-thinning stands. The corresponding differential equation is (6.20). However, we choose to use

$$\frac{\mathrm{d}X}{\mathrm{d}t} = S_X \frac{u}{\beta_X}\frac{\mathrm{d}H}{\mathrm{d}t}, \tag{8.11}$$

where, as explained in Sect. 6.5.1, S_X is a numerical switching function that smooths the transition between no self-thinning and active self-thinning as crown-base height, H_c, approaches and passes height $k_X H_X$.

Basal area, B, is the cross-sectional area of tree stem at breast height, 1.3 m. If breast height is between the tip and base of the crown, then from (5.42), we deduce that

$$B = \beta_A (H - 1.3)^z, \qquad 1.3 \geq H_c \tag{8.12}$$

Below the crown, the growth rate of basal area, by our theory, is

$$\frac{\mathrm{d}B}{\mathrm{d}t} = \frac{\rho_s z \beta_A}{\rho_m}(H - H_c)^{z-1}\frac{\mathrm{d}H}{\mathrm{d}t}, \qquad 1.3 < H_c \tag{8.13}$$

To approximate basal area growth over a discrete time step, Δt, we suggest using

$$\bar{H}_c = \left(H_{c,\,t+\Delta t} + H_{c,\,t}\right)/2 \tag{8.14}$$

then

$$B_{t+\Delta t} \doteq B_t + \frac{\rho_s}{\rho_m}\beta_A\left[(H_{t+\Delta t} - \bar{H}_c)^z - \left(H_t - \bar{H}_c\right)^z\right], \qquad 1.3 < H_{c,\,t+\Delta t} \tag{8.15}$$

These models account for the difference in density between juvenile and mature wood by inclusion of the ratio of juvenile to mature wood bulk densities, ρ_s/ρ_m (see Sect. 5.3). Bulk density—dry weight per unit wet volume—can be obtained from increment cores. These same parameters apply to the more general model of cross-sectional growth, (5.42). A simpler version of basal area growth model applies where: (a) the crown-base height is above breast height and (b) crown length remains approximately constant after stand closure (see Valentine et al. 2012; Bravo-Oviedo et al. 2013).

8.4 Estimating Structural Parameters

Forest mensurationists have devised protocols, fabricated special instruments, and adapted digital technology to achieve measurements of an assortment of forest-related variables (see, e.g., Kershaw et al. 2017). Recent mensuration texts also include explanations of popular sampling methods, many devised by forest mensurationists, for estimating those attributes of trees, stands, and the forest ecosystem for which exhaustive measurement is impractical or impossible. These methods are particularly useful for measuring the state variables—tree height, crown height, and stand density—and the response variables—dbh, basal area, volume, and biomass—of the core model.

Sampling methodology adapted from other fields can be applied to estimate structural parameters and dimensions of trees. In particular, Jessen (1955) devised randomized branch sampling (RBS) to estimate fruit abundance on individual orchard trees. RBS also has been used for estimation of leaf mass (Valentine and Hilton 1977, et seq.), and adapted for estimation of aboveground tree biomass, volume, and mineral content (Valentine et al. 1984; Gregoire et al. 1995). Gregoire and Valentine (2008, Chapter 13) describe the theory and protocols of RBS in detail with examples and diagrams. They also provide an estimator for average stem length. The RBS method is easiest to apply to trees with opposite or alternating branching, but protocols for efficient randomized sampling of conifer crowns are also available (see, e.g., Valentine et al. 1994a; Schlecht and Affleck 2014).

The use of randomized branch sampling to estimate total leaf mass or average active pipe length or both results in the selection of small branches from which leaf mass is measured. The same small branches also can be used for the estimation of some of the structural parameters of the core model.

We recall that β_1 and β_2 [kg active wood m^{-1} (kg leaf)$^{-1}$] are composite parameters constructed from the form factors ϕ_b, ϕ_s, and ϕ_t and the specific pipe lengths l_b, l_s, and l_t for b, branches; s, active stem; and t, coarse roots, whence,

$$\beta_1 = \frac{\phi_b}{l_b} + \frac{\phi_s}{l_s} + \frac{\phi_t}{l_t}$$

$$\beta_2 = -\frac{\phi_b}{l_b} + \frac{1-\phi_s}{l_s}$$

Optionally, we can substitute η_i/ρ_i for l_i, $(i = b, s, t)$, where η_i is the pipe-model ratio and ρ_i is the bulk density. To calculate the specific stem length, the pipe model ratio, and the bulk density from a sample branch, we need (a) the dry weight of the leaves and (b) a bolt of known length, perhaps 10 cm of longer, cut from the basal section of the sample branch that supports all the leaves. From the bolt we measure: (1) the cross-sectional area for the calculation of the pipe-model ratio, (2) the dry weight per unit wet length for the calculation of the specific stem length, and (3) the dry weight per unit wet volume, which, by definition, is the bulk density.

To calculate the form factor, ϕ_b, we need an estimate of $\hat{H}_b$, the average length of the active pipe within the branches off the main stem. We obtain one observation of branch pipe length by measuring the distance from a leaf along the connecting internodes to junction with the main stem.

The form factor on the main stem above crown base, ϕ_s, determines the value of z, as indicated in Sect. 5.2.3, that is,

$$z = \frac{1}{\phi_s} - 1, \qquad 0 < \phi_s < 1$$

Alternatively, the value of z determines the value of ϕ_s. Let D_c be the diameter at the crown base so that $D_c^2 \propto A_c \propto L_c^z$. Then, given a measurement of the stem diameter, D', at, say, distance $L' = L_c/k$ $(k > 0)$ from the crown tip, we obtain

$$z = 2\,\frac{\ln(D_c/D')}{\ln(L_c/L')} \tag{8.16}$$

Then,

$$\phi_s = \frac{1}{z+1}$$

Measurements of roots are hindered by soil, rocks, and other impediments. Soil removal by hydraulic excavation has worked well in some studies of roots (e.g., Stout 1956; Richardson and ZuDohna 2003). Otherwise, digging around the tree stump (and tap root, if any) exposes first-order coarse roots for measurement of aggregate cross-sectional area, bulk density, and dry weight per unit length. Estimation of the specific pipe length for coarse roots requires the measurement or an estimate of the tree's total leaf dry weight and the sum of the dry weights per unit length of the first-order coarse roots. Likewise, the pipe-model ratio requires the total leaf dry weight and the aggregate cross-sectional area of the first-order coarse roots.

Kalliokoski et al. (2008) were able to expose sample roots by digging and the use of a pnueumatic Soil-Pick tool that produces a super-sonic jet of air that exposes root branches without damage down to 2 mm in thickness. Their ability to expose and measure whole coarse-root lengths suggests that the estimation of average-pipe lengths of tree roots is feasible.

Albaugh et al. (2006) obtained stand-average estimates of coarse-root biomass by excavating square pits with horizontal areas of 1 m^2. The excavations afforded estimates of aggregate coarse-root biomass kg DW m^{-2}, which, when divided by stand density (trees m^{-2}), provides an estimate of coarse-root biomass per tree. Estimation of α_r, the ratio of fine-root mass (kg m^{-2}) to leaf mass (kg m^{-2}), can be accomplished on a stand-average basis by using core samplers in lieu of digging pits.

8.5 Environment-Sensitivity of Metabolic Parameters

The environment-sensitive metabolic parameters of the model of Chap. 5 include the leaf-specific rate of photosynthesis, σ_f, the tissue-specific rates of maintenance respiration, r_{M_f}, r_{M_r}, and r_{M_p}, the lifetime of leaf and fine-root tissue, T_f and T_r, and the ratio of fine-root mass to leaf mass, α_r.

Establishing the production potential of a growth site and its dependence on climatic and edaphic factors essentially depends on the effect of environmental drivers on these parameters. As noted above, this dependence can be studied using detailed eco-physiological models and measurements, but applying such results to long-term and large-scale analyses requires several input variables at high resolution. Here, we present an example of how the parameters can be determined for regional analysis using summary models and stand-scale data. The approach closely follows Mäkelä et al. (2016).

8.5.1 *Photosynthesis*

A key climatic input to the carbon balance models of Chaps. 3, 4, 5, 6, and 7 is the mean annual specific rate of photosynthesis, σ_f (yr^{-1}), which relates the photosynthetic production of a tree to its active leaf mass:

$$P = \sigma_f W_f \tag{8.17}$$

On the other hand, stand-level photosynthesis can be expressed using the Light Use Efficiency approach based on the Lambert-Beer equation:

$$\mathbb{P}_g = \mathbb{P}_0[1 - \exp(-k a_{SL} \mathbb{W}_f)] \equiv \mathbb{P}_0 f_{APAR} \tag{8.18}$$

where we have denoted the proportion of photosynthetically active radiation absorbed by the canopy by f_{APAR}, and $\mathbb{P}_0$ (kg C m^{-2} yr^{-1}) is the rate of gross photosynthesis of a stand with complete light interception and $\mathbb{W}_\mathrm{f}$ is the stand-level leaf mass. This relates to the leaf-specific photosynthetic rate of the mean tree of the canopy as

$$\sigma_\mathrm{f} = \frac{1}{NW_\mathrm{f}} \mathbb{P}_0 f_{\mathrm{APAR}} \tag{8.19}$$

where N is stand density and $\mathbb{W}_\mathrm{f} = NW_\mathrm{f}$. More generally, information about stand structure will allow us to compute the proportions of stand photosynthesis allocated to trees of different size or location (Chap. 6). The bulk of the environmental sensitivity of photosynthesis is incorporated in $\mathbb{P}_0$, which may depend on various environmental factors, such as the average light available to the tree during the growing season, the length of the growing season, the availability of water and nutrients, and other factors.

The parameter $\mathbb{P}_0$ is a key link of the growth model with eco-physiological information about stand productivity, and it can be estimated using any canopy photosynthesis model by integrating daily or sub-daily rates over the whole season and assuming that $f_{\mathrm{APAR}} = 1$. It can also be estimated from canopy-level data on photosynthesis as provided by eddy covariance measurements.

For growth model simulations, $\mathbb{P}_0$ can be calculated each year using drivers appropriate for the photosynthesis model employed, or alternatively, an emulator (summary model) (see Fig. 8.3) of the faster scale model can be constructed on the basis of an input-output analysis of the model. Here we illustrate these options using a daily Light Use Efficiency (LUE) model, PRELES, that provides $\mathbb{P}_0$ for a given site and year as a function of daily environmental drivers (Mäkelä et al. 2008a; Peltoniemi et al. 2012, 2015). PRELES is a semi-empirical model that was fitted to data from eddy covariance measurements at different sites in boreal and temperate coniferous forests. The model is robust in the sense that a generic parameterisation for all sites has been shown to provide as good or better results than site-specific parameterisations (Minunno et al. 2016).

The drivers of PRELES include the daily total of photosynthetically active radiation (PAR), daily mean temperature and water vapour pressure deficit (VPD) and daily total rainfall. Such data are usually available from meteorological stations, although PAR may have to be calculated theoretically and adjusted with other measurements, such as sunshine hours. Climate models also provide projections of these variables on a geographical grid both as interpolations of existing meteorological measurement networks, and for future climate predictions. However, simulations show that this detail may be unnecessary if projections concern a fairly confined region under current climate only.

Härkönen et al. (2010) demonstrated that $\mathbb{P}_0$ estimated by PRELES can be accurately predicted in Finland by Effective Temperature Sum (ETS), calculated as the sum of degrees by which the daily average temperature exceeds +5 °C (Fig. 8.4). A threshold of +5 °C is a commonly used standard when calculating the ETS in

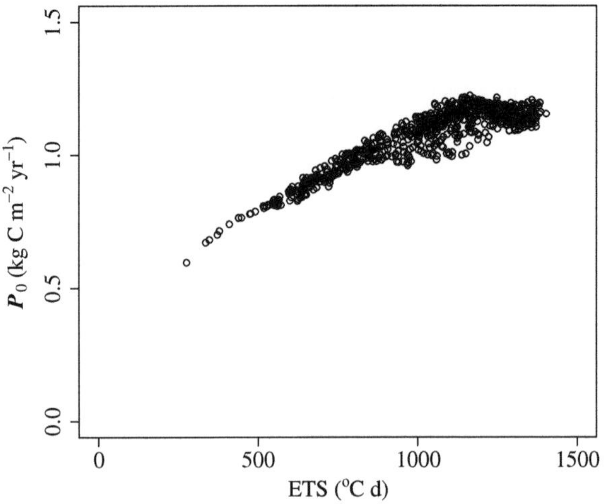

Fig. 8.4 The rate of gross photosynthesis of a canopy with complete light interception ($f_{APAR} = 1$) (kg C m^{-2} yr^{-1}) as a function of Effective Temperature Sum (ETS), calculated as the sum of degrees by which the daily average temperature exceeds +5 °C. The figure depicts results from the PRELES model (Peltoniemi et al. 2015) applied to a 1 km × 1 km grid of interpolated daily weather station data across Finland, averaged over the period 1980–2010

forestry and agriculture in the Nordic countries (so-called thermal growing season). ETS can explain photosynthesis because water availability plays a minor role in the photosynthetic production of conifers in Finland, and because temperature, PAR and VPD correlate strongly with each other (Mäkelä et al. 2006). This suggests that ETS is an appropriate proxy for current climate effects in Finland. Mäkelä et al. (2016) used the following saturating equation:

$$\mathbb{P}_0 = \mathbb{P}_{0_{max}} \frac{E}{E + E_1} \tag{8.20}$$

where E is ETS and E_1 is a parameter determining the rate of saturation towards $\mathbb{P}_{0_{max}}$.

The instantaneous maximum capacity of photosynthesis depends on leaf nitrogen concentration, $[N]_f$, among and between species (Reich et al. 1998), as was already discussed in Sect. 7.5. This is also manifested in $\mathbb{P}_0$, but this impact is small compared with the direct climate effect, especially in conifers (Peltoniemi et al. 2012) (see also Chap. 7). If leaf N concentration can be approximated for a site, a dependence of $\mathbb{P}_0$ on this can be accounted for, e.g., as follows:

$$\mathbb{P}_{0_{max}} = \mathbb{P}_{0_{max,ref}} \frac{[N]_f}{[N]_{ref}} \tag{8.21}$$

where $[N]_{ref}$ is the leaf nitrogen concentration for a reference site and $\mathbb{P}_{0_{max,ref}}$ is the value of $\mathbb{P}_{0_{max}}$ for the reference site.

8.5.2 Respiration

Tissue respiration models describe instantaneous tissue-specific respiration rates primarily as functions of ambient temperature and tissue nitrogen concentration (Ryan 1991, 1995; Ryan et al. 1996). Ryan (1991) presented a summary model for leaf-specific maintenance respiration rate in conifers as follows:

$$r_{M_f} = 27[N]_f \times e^{0.07T_a}, \tag{8.22}$$

where $[N]_f$ is leaf N concentration and T_a is annual mean temperature. A similar equation was given for deciduous trees, with the difference that the duration of leaf area display (D_L, days) also was included:

$$r_{M_f} = 0.059 D_L [N]_f \times e^{0.07T_s}, \tag{8.23}$$

where T_s is average growing season temperature.

However, there is evidence that respiration rates acclimate to prevailing temperatures and tend to be at least loosely controlled by the availability of photosynthates. Whether respiration can be assumed proportional to photosynthesis has been discussed extensively in the literature (Ryan et al. 1996; Dewar et al. 1999; Cannell and Thornley 2000; Mäkelä and Valentine 2001), with the current consensus pointing to a fairly strict connection between the two at least in the long term (Dewar et al. 1999; Högberg et al. 2001; Wertin and Teskey 2008; Van Oijen et al. 2010). Mäkelä et al. (2016) assumed that the dependence of leaf maintenance respiration rate on temperature was similar to that of photosynthesis. However, based on empirical results from the Flakaliden nutrient optimisation experiment in Sweden (Stockfors and Linder 1998), they assumed that the dependence on leaf nitrogen was down-regulated as follows:

$$r_{M_f} = r_{ref} \frac{\mathbb{P}_0}{\mathbb{P}_{0_{ref}}} \sqrt{\frac{[N]_{ref}}{[N]_f}} \tag{8.24}$$

where r_{ref} is the respiration rate when $\mathbb{P}_0 = \mathbb{P}_{0_{ref}}$. Because $\mathbb{P}_0$ was assumed to depend linearly on $[N]_f$ (see (8.20) and (8.21)), the above renders a non-linear dependence of r_{M_f} on $[N]_f$. This way, the specific maintenance respiration rates are affected by both ETS and leaf nitrogen concentration but synchronized with photosynthetic production, as suggested by empirical (Wertin and Teskey 2008), theoretical (Dewar et al. 1999) and modelling (Medlyn et al. 2011) studies.

8.5.3 Tissue Life Span

Tissue life span has been functionally attributed to general tissue activity (Kikuzawa and Lechowicz 2011). It has been considered a trait that has to be balanced with other leaf properties, so as to optimise the costs of constructing and maintaining the leaf with the benefits obtained from its life-time carbon gain (Williams et al. 1989; Kikuzawa 1991). Consistent with the balanced-trait approach, Reich et al. (2014) presented evidence of increasing leaf life span and decreasing leaf nitrogen with decreasing mean annual temperature in geographical transects of several conifers in North America and in Scots pine (*Pinus sylvestris* L.) in Europe. Ťupek et al. (2015) found that leaf life span of Norway spruce (*Picea abies* (Karst.) L.) and Scots pine was linearly correlated with ETS across Finland (Fig. 8.5).

Mäkelä et al. (2016) assumed that the photosynthetic parameter $\mathbb{P}_0$ reflects leaf tissue activity, so

$$T_\mathrm{f} = T_{\mathrm{f,ref}} \frac{\mathbb{P}_{0_\mathrm{ref}}}{\mathbb{P}_0} \tag{8.25}$$

where $T_{\mathrm{f,ref}}$ is the life span when photosynthetic rate is $\mathbb{P}_{0_\mathrm{ref}}$. This way leaf life span will depend on both climate and nitrogen availability (Kikuzawa and Lechowicz 2011; Reich et al. 2014).

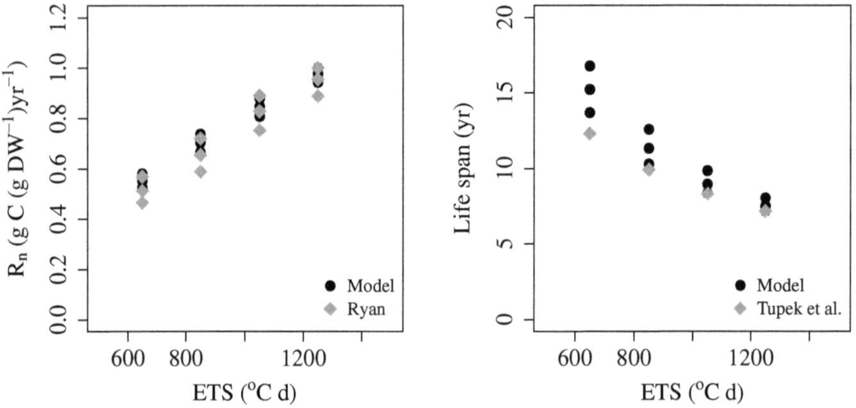

Fig. 8.5 Environmental gradients of metabolic rates in Norway spruce. **Left panel:** Dependence of leaf-specific respiration rate, $r_{\mathrm{M_f}}$, on Effective Temperature Sum (ETS), expressed in degree days—Comparison of model predictions of (Mäkelä et al. 2016) with the summary relationship of (8.22). The rate has been scaled to equal 1 when ETS = 1250 °Cd (R_n). The annual mean temperatures were approximated visually for each ETS from maps provided by the Finnish Meteorological Institute and $[\mathrm{N}]_\mathrm{f}$ was evaluated for three site types for each ETS. **Right panel:** Dependence of leaf life span, T_f, on ETS. Predictions of (8.25) are compared with an empirical model fitted by Ťupek et al. (2015). (Redrawn from Mäkelä et al. 2016)

Fine-root life span of boreal conifers has been shown to increase with decreasing temperature and nitrogen availability (Leppälammi-Kujansuu et al. 2014b) and to increase from south to north (Leppälammi-Kujansuu et al. 2014a). There is also evidence that the dependence of trees on mycorrhizal symbiosis becomes more pronounced in sites with harsher climate or lower fertility, resulting in a carbon cost in the form of root exudates (Litton and Giardina 2008; Näsholm et al. 2013). Leppälammi-Kujansuu et al. (2014a) found almost twice as many ectomycorrhizal root tips per unit fine-root biomass in a Norway spruce stand in southern Finnish Lapland compared to a stand in the south boreal zone. Ostonen et al. (2011) reported a three-fold increase in proportion of the ectomycorrhizal biomass of the total fine root biomass from Germany to northern Finland. To account for the opposite trends with site and climate in the carbon requirement for mycorrhiza and other exudates on one hand, and fine-root life time on the other hand, Mäkelä et al. (2016) suggested to keep the fine-root turnover rate constant and independent of site conditions.

8.5.4 *Effects of Growth Site*

Our growth models do not explicitly include nutrient dynamics but assume that the effects of nutrients are reflected in growth through model parameters in the same way as climatic factors. As already noted in Sect. 7.5, a functional balance assumption leads to a dependence of the fine-root to leaf ratio, α_r, on nitrogen availability. The leaf nitrogen concentration introduced above also depends on site fertility. The results from optimality models can be used when parameterising the growth model for nitrogen effects without explicitly including a nitrogen budget, provided that the availability of nitrogen can be estimated.

Empirical studies have gained quantitative insights into the variation of the fine-root to leaf biomass ratio and leaf and fine-root nitrogen concentrations in boreal conifer stands across geographical and site type gradients (Helmisaari et al. 2007; Ostonen et al. 2011). These studies suggest that α_r increases with decreasing site fertility within a climatic zone and from south to north between climatic zones. Similarly, and consistently with the optimal allocation hypothesis, leaf nitrogen concentration, $[N]_f$, decreases as α_r increases (Fig. 8.6). Based on these studies, it would be possible to use measured values for α_r and $[N]_f$ as model input parameters for each of the sites measured. Because such information is generally not available, Mäkelä et al. (2016) proposed a parameterisation of α_r and $[N]_f$ on the basis of a ground-vegetation based site-type classification used in Finland (Cajander 1949). Site types were quantified as real numbers and functions were devised that mapped these numbers to values of α_r and $[N]_f$ as indicated by the measurements of Helmisaari et al. (2007).

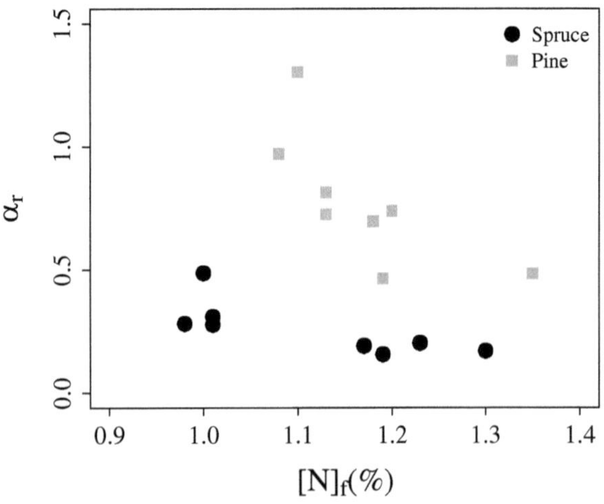

Fig. 8.6 The ratio of fine-root to leaf biomass, α_r, plotted against leaf N concentration, $[N]_f$ in Norway spruce and Scots pine stands in Finland. (Data from Helmisaari et al. 2007)

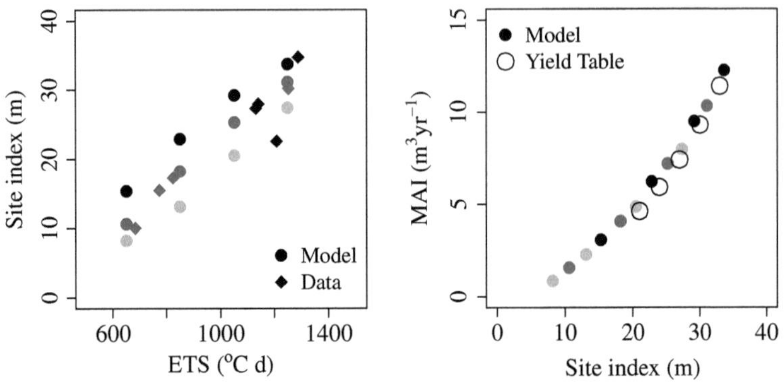

Fig. 8.7 Environmental gradients of productivity in Norway spruce. **Left panel:** Site indices (H_{100}, dominant height at stand age 100 yrs) as a function of Effective Temperature Sum (ETS) in model predictions (Mäkelä et al. 2016) and in data (Helmisaari et al. 2007). The shade of grey indicates site fertility: black: herb rich, dark grey: mesic, light grey: sub-xeric. **Right panel:** Relationship between site index H_{100}, and mean annual increment (MAI) in model predictions (Mäkelä et al. 2016) and in yield tables based on an empirical model (Vuokila and Väliaho 1980). In model simulations, site fertility is indicated with the shade of grey: black: herb rich, dark grey: mesic, light grey: sub-xeric. (Redrawn from Mäkelä et al. 2016)

The quantification of site types and their relationship with plant traits led to a realistic prediction of the geographic distribution of Norway spruce productivity in Finland when photosynthesis, respiration and turnover were predicted with (8.20), (8.21), (8.24), and (8.25) (Fig. 8.7). The assumed dependence of respiration and life span on $\mathbb{P}_0$ and $[N]_f$ yielded environmental gradients that were consistent with other studies (Fig. 8.5).

8.6 Adaptive Adjustment of Structural Parameters

The structural parameters in the core model of Chap. 5 (Table 8.1) are evaluated as mean values of crowns and, by definition, depend on crown shape and the vertical leaf and branch distribution of the tree. Although there is empirical evidence that the means are fairly stable across a variety of tree sizes and stand structures (Ilomäki et al. 2003; Kantola and Mäkelä 2006; Hu et al. 2020) some variability is inevitable. Changes in crown shape and the spatial leaf and branch distributions may occur as responses to ageing or changes in local environment. For example, trees may grow relatively wider crowns after the crown has been released from competition. Trends in branch length relative to crown length may also be related to changes in crown ratio (Kantola and Mäkelä 2006).

Such finer-scale structural changes have been accounted for in the PipeQual model that was developed to investigate the impacts of forest management on stem structure and timber quality (Mäkelä 2002; Mäkelä and Mäkinen 2003; Kantola et al. 2007). Tree growth in PipeQual is defined at the whole-tree level and follows the approach of Chap. 5 (module TREE). In addition, the model includes a module of the vertical structure of stem and crown, based on branch whorls (module WHORL). A third level of detail is defined when the branch whorls are further split into individual branches (module BRANCH). The full model employs a hierarchical, top-down approach where the coarser-scale model sets constraints for the finer scale but also receives feedbacks from it (Fig. 8.8).

Simulation of growth in PipeQual employs all the modules in turn. The principal growth engine is the TREE module which operates with constant structural parameter values. After calculating the total annual growth, the new state of the tree is input to the WHORL module, and the whorl-level information is updated using

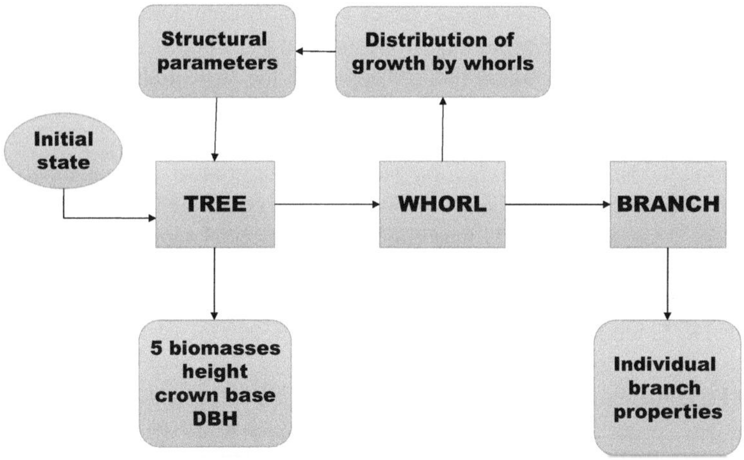

Fig. 8.8 Modular structure of the PipeQual model

this input and whorl-based structural rules. These include, for example, a vertical foliage density distribution and a vertical distribution of the specific pipe length, which has been found to vary somewhat with distance from the tree top (Schneider et al. 2011). This creates a new vertical structure of the stem which is then used for updating the structural parameters that are fed back to the tree level for next year's growth simulation. The state of the whorls is input to the BRANCH module, where individual branches are updated, using statistical models with whorl- and tree-level information as independent variables. The individual branch information does not affect the calculations of tree growth, but it is utilised in forming a non-symmetrical three-dimensional structure for the stem.

The interactive structural parameters are related to (1) the ratio of leaf mass to active pipe area at crown base, (2) form of active pipe volume in the woody parts, (3) form of disused pipe volume, and (4) active pipe turnover rate due to the rise of the crown. The updating method is the same for each parameter. It involves calculating the true value of the parameter in WHORL, then estimating it in terms of the equations used by TREE. For example, the active pipe mass in the stem above the crown base, W_c^T is defined in TREE as the product of crown length, L_c, active pipe area at crown base, A_s, stemwood density, ρ_s, and the form coefficient, ϕ_c (see (4.16) and (4.20)):

$$W_c^T = \rho_s \phi_c A_s L_c \tag{8.26}$$

In WHORL, the corresponding active pipe mass, W_c^W is computed whorl by whorl, and ϕ_c is evaluated from this using the state variables of TREE:

$$\phi_c = \frac{W_c^W}{\rho_s A_s L_c} \tag{8.27}$$

The interaction of the two modules causes a slight discontinuity in the differential equations of TREE at the beginning of the calculation for each year. PipeQual uses a numerical method that makes the tree approach the structural changes gradually (Mäkelä 1999). However, all the parameters determined in this way are conservative, and the use of the method therefore seems justified.

The modular structure used in PipeQual allows for gradual, age and environment related structural changes to be accounted for more realistically than with constant parameters. This is particularly important for the practical application of the model to assess forest management alternatives that account for timber quality. While the finer-scale modules allow us to describe the structural variability more realistically than the TREE module by itself, the top-down approach guarantees that the system remains within realistic limits in the long term.

The PipeQual model has been parameterized for Scots pine (Mäkelä et al. 2000) and Norway spruce (Kantola et al. 2007; Kalliokoski et al. 2016). It has been utilised in economic assessment of forest management options from different perspectives (Hyytiäinen et al. 2004; Cao et al. 2010; Niinimäki et al. 2012), where the detailed structural description has been of significance for the assessment of product value (Fig. 8.9).

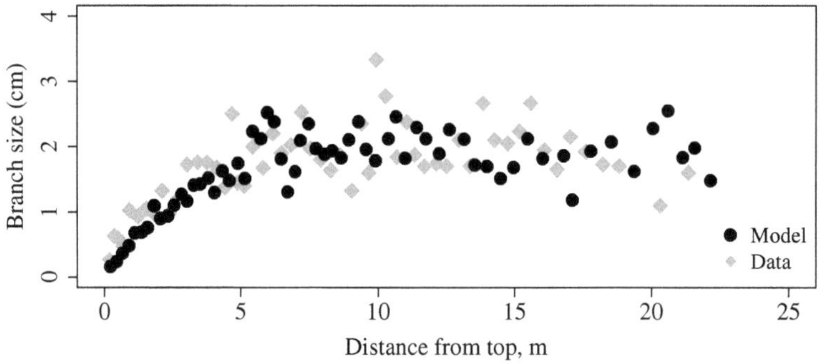

Fig. 8.9 Simulated and measured distribution of mean branch diameter per whorl of an old dominant Norway spruce tree growing in a dense stand. (Re-drawn from Kantola et al. 2007)

8.7 Summary

In this chapter we have introduced different methods of parameterisation for the core model of Chap. 5. The empirical parameterisation, the Bridging model (Sect. 8.3), strongly relies on the structural relationships binding the dimensions and biomasses of the tree together, such that tree dynamics essentially follow from those of tree height and crown-base height only. This allows for the model to be presented in a reduced form, where its functional and structural parameters aggregate into only 5 constants that are identifiable from standard forestry data. Regarded in this way, the model bears resemblance to the descriptive models discussed in Chap. 2.

While the aggregation of parameters facilitates empirical fitting and application of the model, at the same time it provides little information about the original parameter values, as these are combined with each other and cannot be disaggregated without additional information. This means that the aggregated parameterisation will not allow us to derive the dependent variables, such as leaf or fine-root biomasses, from the reduced number of state variables (see Sect. 6.5.2). On the other hand, if all the parameters could be identified on the basis of their definition, the aggregated parameters and all model variables could readily be computed. For this, more data and more attention to eco-physiological and structural detail is required.

As seen above and in Chap. 4, the structural parameters of the model can be determined from measurements of tree architecture and biomass components, and they have proven to be reasonably constant in time. Estimating the metabolic parameters, on the other hand, relies on eco-physiological measurements and models that represent lower-level dynamics compared with the core model. We have argued that, in a stable environment, these parameters can be regarded as effective mean values and thus constant in time without losing any essential dynamics of the model. Nevertheless, variability between regions and sites as well as any trends with environmental change must be accounted for.

We illustrated above how the quantities defined as constant parameters in the core model could actually be made into time-dependent functions, either as time-dependent parameters (Sect. 8.6) or as environmental drivers (Sect. 8.5). This can be done by including lower-level processes as additional modules in the model. The advantage of the modular structure is that the basic dynamic behaviour of the original model is retained, with modifications implied by the adjustment of its parameter values. This allows for more transparency and easier parameter estimation than in models that integrate different level processes without any clear hierarchical structure.

When parameterising forest models, we often have detailed data available at few sites with intensive ecological measurements, whereas standard forestry data is usually available for long time periods from forest management trials, and for extensive regional sets from national forest inventories. In the following chapter we will explore how these different data types can be combined in parameter estimation. Calibrating our core model essentially boils down to determining starting ranges for all parameters from some intensive measurements, then comparing the results to standard forestry data to make sure that the aggregated reduced parameter set leads to growth dynamics that agree with the data.

8.8 Exercises

8.1 Using the parameter values in Table 8.2 as default, analyse the effect of metabolic and structural parameters on maximum tree height without crown rise (Sect. 8.3.1). How does the maximum height change if you modify each of the parameter values by 10%?

8.2 It was concluded in Sect. 8.3.1 that maximum height will be larger if crown base has risen. On the other hand, when the stand is dense causing crown rise, there is also mutual shading between trees, such that the leaf-specific rate of photosynthesis

Table 8.2 Default parameter values

Symbol	Value	Units
σ_f	2.2	$kg\,C\,(kg\,DW)^{-1}\,yr^{-1}$
r_{Mf}	0.3	$kg\,C\,(kg\,DW)^{-1}\,yr^{-1}$
r_{M_r}	0.3	$kg\,C\,(kg\,DW)^{-1}\,yr^{-1}$
r_{M_w}	0.03	$kg\,C\,(kg\,DW)^{-1}\,yr^{-1}$
T_f	3	yr
T_r	1	yr
Y_G	1.5	$kg\,DW\,(kg\,C)^{-1}$
a_σ	0.001	m^{-1}
α_r	0.5	–
β_1	1.4	–
β_2	−0.4	–

is reduced compared with an open-grown tree. How much would σ_f need to decline to provide the same maximum height as in the open-grown tree if we assumed that at the maximum height, crown base height H_c was (a) 5 m, (b) 10 m, (c) 15 m?

8.3 Study the dependence of σ_f on stand leaf mass, $\mathbb{W}_f$ (8.19). Use $\mathbb{P}_0 = 1.5\,\text{kg C m}^{-2}$, $a_{SL} = 14\,\text{m}^{-2}\,(\text{kg DW})^{-1}$, $k = 0.2$.

Chapter 9
Calibration

In this chapter we first introduce some basic concepts central to model calibration, then give examples of sequential methods and Bayesian calibration as used in ecological modelling. Our purpose is to illustrate the use and benefits of these methods in comparison with conventional statistical parameter estimation. Finally, we demonstrate the use of the calibrated models for prediction of forest growth and carbon balance. Readers interested in details and techniques of calibration are referred to the abundant literature on the topic.

9.1 Introduction

Model calibration refers to adjustment of parameter values by means of assimilating new data with previous parameter estimates. The data may comprise observations of some or all of the state, response, and environmental driving variables. Alternative names for the subject matter include *data-model fusion*, *data-model synthesis*, and *inverse modelling*. Data assimilation is used widely in engineering, meteorology, and physical oceanography (Robinson and Lermusiaux 2002).

The theoretical basis of data assimilation is in probability theory, and there especially in the Bayes' Theorem on conditional probabilities. Based on this, data assimilation algorithms provide methods to adjust parameter values on the basis of new data, conditional on all previously available information. While calibration in principle is independent of the type of model, it is of particular use in mechanistic modelling where both model structures and parameter values are derived from prior theoretical and empirical information. The prior information is useful but

Electronic Supplementary Material The online version of this chapter (https://doi.org/10.1007/978-3-030-35761-0_9) contains supplementary material, which is available to authorized users. The videos can be accessed by scanning the related images with the SN More Media App.

A. Mäkelä, H. T. Valentine, *Models of Tree and Stand Dynamics*,
https://doi.org/10.1007/978-3-030-35761-0_9

often extremely uncertain and insufficient for quantitatively accurate predictions (Robinson and Lermusiaux 2002).

Data assimilation algorithms can be classified as either *sequential* or *non-sequential* (Zobitz et al. 2011). The sequential methods are based on continuous monitoring of the process being modelled, such that parameters can be re-calibrated each time new data are brought in. Several methods of sequential calibration derive from the Kalman filter (Kalman 1960), which was originally designed for fitting and analysing systems of linear equations but has been extended to a variety problems. The non-sequential methods often are called data-model (or model-data) fusion methods. They are not iterative in the sense of sequential methods, but utilise all data at hand in a batch.

Both sequential and non-sequential methods are based on ideas presented by the Bayes Theorem (see Sect. 9.4). This allows us to compute a new—calibrated—estimate of parameter values by merging their prior values and new information gained from the process. The Kalman filter was first interpreted in the Bayesian context by Ho and Lee (1964) and this interpretation has since become standard (e.g., Meinhold and Singpurwalla 1983). The calibration of mechanistic ecological models usually applies non-sequential methods based on Bayesian theory (e.g., Green et al. 2000; Poole and Raftery 2000; Van Oijen et al. 2005; Raftery and Bao 2010; Hartig et al. 2012). In non-sequential Bayesian calibration and synthesis, we first assess the probability distributions of the parameters on the basis of prior information, then adjust the parameter estimates within their plausible uncertainty ranges to such values that keep the model outputs within the observed limits (Fig. 9.1).

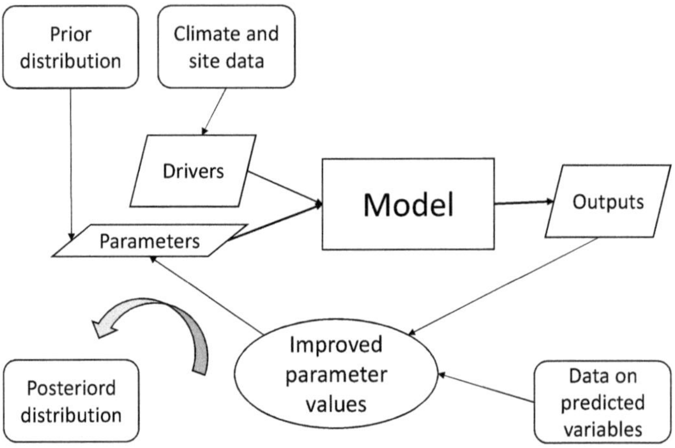

Fig. 9.1 Model-data relationships in parameter estimation of mechanistic models with (Bayesian) calibration. Plausible ranges of the parameter values are first estimated from measurements on the basis of their mechanistic definition for the prior parameter distribution. The parameters are then adjusted iteratively through a systematic comparison of modelled vs. measured output variables. This process yields the posterior parameter distribution that combines the prior information on parameters, and the information from the data on predicted variables. A new data set on predicted values is needed for model testing

Utilising data on model outputs in parameter estimation, Bayesian calibration merges the process-based and empirical estimation methods (Fig. 9.1). The calibrated parameter values are constrained by our prior knowledge but adjusted such that they ascertain a good fit of model predictions to observations. With systematic calibration methods we can also analyse the uncertainties and error margins of the predictions and get an idea of correlations between individual parameter estimates (see, e.g., Richardson et al. 2010; Keenan et al. 2012). From a puristic process-model perspective, the downside of model calibration is that it makes the values of the model parameters dependent on output data. On the other hand, it also makes model predictions more reliable at least in conditions similar to the data. Most importantly, it opens up extensive forest mensuration and inventory data sources for process-based parameter estimation.

In this chapter we first introduce some basic concepts central in model calibration, then give examples of sequential and non-sequential calibration as used in ecological modelling. The purpose is to illustrate the use and benefits of the methods in comparison with conventional statistical parameter estimation. Finally, we also demonstrate the use of the calibrated models for prediction of forest growth and carbon balance. Readers interested in details and techniques of calibration are referred to the abundant literature on the topic (see e.g., Wang et al. 2009; Poole and Raftery 2000; Raftery and Bao 2010; Richardson et al. 2010; Keenan et al. 2011; Zobitz et al. 2011).

9.2 Basics of Sensitivity and Uncertainty Analysis

9.2.1 Sensitivity

In a dynamic model, we can express the dependence of the state variables on its parameters (including initial states if not known exactly) as follows:

$$\frac{\mathrm{d}\mathbf{x}\,(t, \boldsymbol{\alpha})}{\mathrm{d}t} = f\,(\mathbf{x}\,(t, \boldsymbol{\alpha})\,, t, \boldsymbol{\alpha}) \tag{9.1}$$

where $\mathbf{x}$ is a vector consisting of, say, m state variables and $\boldsymbol{\alpha}$ is a vector of n inputs, i.e., the parameters of the model. We denote components of the vectors by x_j and α_i, respectively.

The *sensitivity* of the state to input parameter i at time t, $\mathbf{s}_i(t)$, is defined as the variation in the state relative to a small variation in parameter i when all other parameters are kept at their default values. Mathematically,

$$\mathbf{s}_i\,(t) = \frac{\partial \mathbf{x}(t, \boldsymbol{\alpha})}{\partial \alpha_i} \tag{9.2}$$

If the model is nonlinear, as in most cases, the sensitivity depends on the parameter value itself and may therefore vary when changing its default value.

Sometimes it is more informative to calculate the relative sensitivities of the output components, also called *elasticity*. The elasticity of output j to input i is defined as the relative change of output divided by the relative change in the input parameter:

$$e_{ji}(t) = \frac{\partial x_j(t, \boldsymbol{\alpha})}{\partial \alpha_i} \frac{\alpha_i}{x_j(t, \boldsymbol{\alpha})} \tag{9.3}$$

Computing the elasticity in addition to absolute sensitivity allows us to compare the relative significance of different parameters for model sensitivity.

We can gain insights into the sensitivity function by considering its time derivative, which can be derived from the differential equations of the model. By definition, the time derivative of the sensitivity component s_{ji} is

$$\frac{\mathrm{d}s_{ji}(t)}{\mathrm{d}t} = \frac{\mathrm{d}}{\mathrm{d}t} \frac{\partial x_j(t, \boldsymbol{\alpha})}{\partial \alpha_i} \tag{9.4}$$

The order of differentiation can be changed because the parameters and time are independent of each other, such that

$$\frac{\mathrm{d}s_{ji}(t)}{\mathrm{d}t} = \frac{\partial}{\partial \alpha_i} \frac{\mathrm{d}x_j(t)}{\mathrm{d}t} \tag{9.5}$$

This allows us to solve for the sensitivity functions either analytically or by simulating them in parallel with the state variables, for which the time derivatives are given. For that, it remains to set initial conditions for the sensitivity. Because the initial state is completely determined by the initial values, the initial sensitivity to initial values is $s_{ji}(0) = 1$ and $s_{ji}(0) = 0$ for all other parameters.

As an example to illustrate sensitivity, consider the Mitscherlich model (Eq. 2.49) applied to stand volume growth:

$$\frac{\mathrm{d}x(t)}{\mathrm{d}t} = \alpha_1(\alpha_2 - x(t)); \quad x(0) = \alpha_0 \tag{9.6}$$

where $\alpha_1 = 0.03\,\mathrm{yr}^{-1}$ is a rate constant, $\alpha_2 = 500\,\mathrm{m}^3\,\mathrm{ha}^{-1}$ is the maximum value of $x(t)$ and $\alpha_0 = 10\,\mathrm{m}^3\,\mathrm{ha}^{-1}$ is the initial state. We can now formulate the sensitivity functions for the parameters as follows:

$$\begin{aligned}
\frac{\mathrm{d}s_0(t)}{\mathrm{d}t} &= \frac{\partial}{\partial \alpha_0} \frac{\mathrm{d}x(t)}{\mathrm{d}t} \\
&= -\alpha_1 \frac{\partial x(t)}{\partial \alpha_0} \\
&= -\alpha_1 s_0(t); \quad s_0(0) = 1 \\
\frac{\mathrm{d}s_1(t)}{\mathrm{d}t} &= \frac{\partial}{\partial \alpha_1} \frac{\mathrm{d}x(t)}{\mathrm{d}t} \\
&= \alpha_2 - x(t) - \alpha_1 s_1(t); \quad s_1(0) = 0 \\
\frac{\mathrm{d}s_2(t)}{\mathrm{d}t} &= \frac{\partial}{\partial \alpha_2} \frac{\mathrm{d}x(t)}{\mathrm{d}t} \\
&= \alpha_1(1 - s_2(t)); \quad s_2(0) = 0
\end{aligned} \tag{9.7}$$

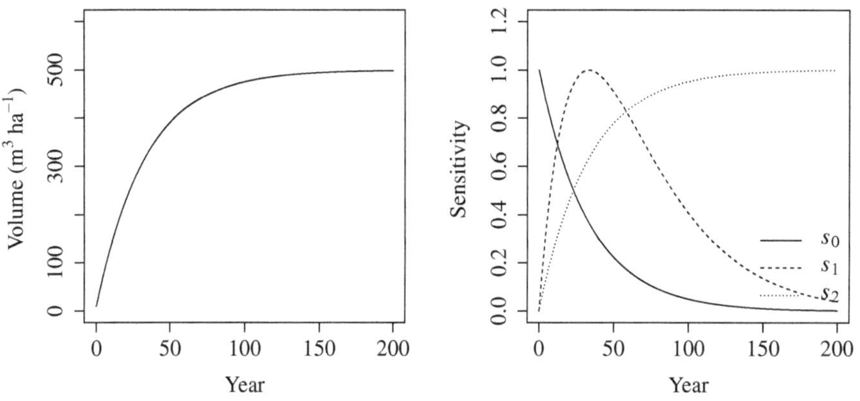

Fig. 9.2 Development of stand volume, x (left panel), and its normalised sensitivity to parameters α_0, α_1 and α_2 (right panel)

Simulations of the model and the sensitivity functions show that the sensitivity to the initial state vanishes with a time constant of $1/\alpha_1$, whereas the sensitivity to the maximum parameter, α_2, increases from zero to unity as x moves from the initial state towards its stable value at α_2. The sensitivity to the rate parameter, α_1, peaks at the point of maximum growth of the state variable x (Fig. 9.2).

An alternative method of estimating the sensitivity functions is purely numerical and is based on the definition of sensitivity, Eq. (9.2). Here, model simulations are first carried out for the default values of the parameters, then the parameters are varied one at a time by a small amount, $\delta\alpha_i$. The sensitivity of state variable x_j to parameter α_i is then estimated as follows:

$$s_{ji}(t) = \frac{x_j(t, \alpha_0, \ldots, \alpha_i, \ldots, \alpha_n) - x_j(t, \alpha_0, \ldots, \alpha_i + \delta\alpha_i, \ldots, \alpha_n)}{\delta\alpha_i} \tag{9.8}$$

The result will be the closer to the theoretical sensitivity, the smaller the value of $\delta\alpha_i$ used in the approximation.

9.2.2 *Uncertainty*

While sensitivity is a property of the model, *uncertainty* is related to the degree of actual knowledge or information we possess of the real system that the model is simulating. The uncertainty of model estimates consists of input uncertainty and structural uncertainty. Inputs include drivers and parameters, and structural uncertainty refers to the gaps of knowledge in model structure that create error in model outputs.

Parametric uncertainty is defined as the probability of error in the parameters and can thus be expressed as probability distributions. When the parametric uncertainty propagates through the model, it creates corresponding probability distributions of the calculated values of $\mathbf{x}$ or any function of those. This is what we call model output uncertainty. It depends on both the sensitivity of outputs to parameters, and the parametric uncertainty.

To illustrate the interaction of model sensitivity and parametric uncertainty, consider the propagation of a small error in the parameters, $\Delta\alpha_i$, to the state variables (i.e., the real value of the parameter is $\alpha_i + \Delta\alpha_i$). This creates a corresponding error in $\mathbf{x}$ which can be approximated if we know the sensitivity of $\boldsymbol{x}$ to the error of α_i:

$$\mathbf{\Delta x}_i(t) = \mathbf{s}_i(t)\Delta\alpha_i \tag{9.9}$$

where $\mathbf{\Delta x}_i$ is the vector of errors in $\mathbf{x}$ due to the error of parameter α_i and the sensitivity is evaluated at the default value α_i. If the parameters are independent of each other, the total error in $\mathbf{x}$ due to all parameters is the sum of those of the individual parameters:

$$\mathbf{\Delta x}(t) = \sum_{i=1}^{n} \mathbf{s}_i(t)\Delta\alpha_i \tag{9.10}$$

We can see from the above that if the sensitivity of the state variables to a parameter is small, even large input errors may be insignificant for model predictions, and *vice versa*, negligible errors may overcome the effect of high sensitivity in parameters. However, even small input errors may have a strong effect on the most sensitive outputs.

When analysing the uncertainty of outputs in a complex model, we have to account for nonlinearities and interactions of parameters. This means that the sensitivity to a parameter is not constant across the parameter space, but depends on all parameter values. If the uncertainty range of parameters is large, it is thus not sufficient to evaluate the sensitivity function at the default parameter value only, as is suggested in (9.10), but the entire uncertainty region of the parameter space has to be walked through. Systematic computational methods of *generalised sensitivity analysis* have been developed for this purpose. One of them, the Morris method (Morris 1991; Campolongo et al. 2007), starts off by defining a metric, μ_i^*, as the mean of (the absolute value of) the sensitivity of an output to parameter i in the uncertainty region of the parameter space. It further utilises the observation that the larger the variance, σ_i, of this sensitivity, the larger the non-linearities and interactions with other parameters. Both μ_i^* and σ_i can be estimated numerically by means of a random walk through the parameter space (Campolongo et al. 2007). They can be used for screening the propagation of parametric uncertainty to the outputs, and ranking the parameters on the basis of their contribution to the overall uncertainty.

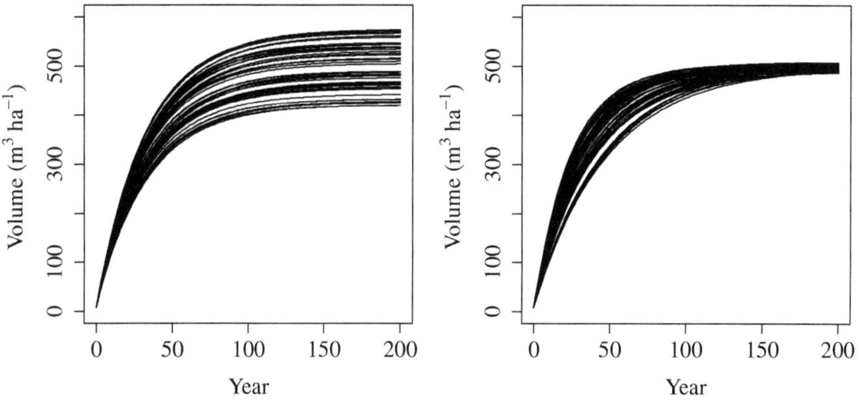

Fig. 9.3 The uncertainty of volume development as described with Eq. (9.6) with different assumptions of parameter uncertainty. The graphs depict 50 simulations where the parameters α_0, α_1 and α_2 were drawn randomly from even distributions. In both panels the initial value is drawn from the range $\alpha_0 \in [8, 10]\,\mathrm{m^3\,ha^{-1}}$. Left panel: $\alpha_1 \in [0.029, 0.031]\,\mathrm{yr^{-1}}$, $\alpha_2 \in [420, 580]\,\mathrm{m^3\,ha^{-1}}$. Right panel: $\alpha_1 \in [0.02, 0.04]\,\mathrm{yr^{-1}}$, $\alpha_2 \in [480, 520]\,\mathrm{m^3\,ha^{-1}}$

Monte Carlo simulations can also be used in a more straightforward way for screening the propagation of the input uncertainty to the output. This is done by simulating the model a large number of times, each time drawing parameter values from their probability distributions. The resulting distribution of the output describes the output uncertainty and can be analysed statistically (Fig. 9.3).

Parameters are *correlated* if the same outputs can be produced by combinations of parameters that correlate with each other. This is often the case in process-based models, where parameters are included for their mechanistic significance, not for their statistical robustness. A simple example is the case where two parameters always occur multiplicatively—any combination of the two parameters that yields the same product will also yield the same outputs. Correlations may create redundant uncertainty if not detected; in this example, the product may be constrained to a narrow interval while both of its components appear to vary widely. Detecting correlations between parameters is crucial for model calibration, as will be elaborated in more detail in the subsequent sections.

9.3 Filtering Methods

9.3.1 Gap Filling Data Streams

In ecology, filtering methods are popular in connection with the method of eddy covariance (e.g. Wang et al. 2009), which measures the rate of exchange of latent and sensible heat, carbon dioxide, and other trace gases between the atmosphere and the biosphere (Baldocchi 2003). One of the principal uses of eddy covariance is to estimate the net rate of movement of carbon to the atmosphere from various kinds of ecosystems, including forested ecosystems. This rate of exchange (carbon flux) is called *net ecosystem exchange* (NEE), which is measured in units of carbon per unit land area per unit time. NEE is closely related to net ecosystem productivity (NEP), which is the difference between gross primary productivity (GPP) and net ecosystem respiration (NER). Any difference in absolute magnitude between NEE and NEP is attributed to the rate of dissolved carbon entering or leaving the ecosystem through the movement of water. If this rate is negligible, then

$$\mathrm{NEE} = \mathrm{NER} - \mathrm{GPP}$$

$$\mathrm{NEP} = \mathrm{GPP} - \mathrm{NER} \tag{9.11}$$

so NEP = −NEE, the atmosphere's loss of carbon is the ecosystem's gain.

With functioning instrumentation and favorable atmospheric conditions, the method of eddy covariance provides a precise estimate of NEE. Since NEE is a temporal rate, it can be integrated with respect to time to provide an estimate of the amount of carbon sequestered by the ecosystem during the period of integration. In practice, total annual sequestration is summed from short-term totals, e.g., half-hourly totals. However, instrumental malfunction or unfavorable conditions—low wind speed, rain, or snow—leave gaps in the NEE record. In this case, an ecosystem-level carbon-balance model, optimised by data assimilation, provides a means to fill in the gaps and to estimate the degree of uncertainty in the resultant estimate of carbon sequestration.

As an example, we consider the analysis of Gove and Hollinger (2006), who assumed that NEE = NER − GPP. During the growing season, GPP (μmol CO_2 m^{-2}) was modeled as

$$\mathrm{GPP} = \frac{A I_0}{K + I_0} \tag{9.12}$$

where I_0 (μmol photons $m^{-2}\,s^{-1}$) is the incident photosynthetic photon flux density (PPFD), and A and K are parameters. Specifically, A is the maximal rate of carbon uptake, and K is the incident PPFD that yields half the maximal rate of carbon uptake. Hence A has the same units as GPP, and K has the same units as I_0.

NER (μmol CO_2 $m^{-2}\,s^{-1}$) was modeled as (after Lloyd and Taylor 1994)

$$\mathrm{NER} = R_{\mathrm{p}} \exp\left(\frac{-E_0}{T + 273.15 - T_0}\right) \tag{9.13}$$

where T (°C) is the ambient temperature, R_{p} (μmol CO_2 $m^{-2}\,s^{-1}$), and E_0 and T_0 (°K) are parameters, though Gove and Hollinger (2006) chose to treat the former as a constant, i.e., $T_0 = 262.2$ °K. In the 'dormant season,' it was assumed that

$$\mathrm{NEE} = \begin{cases} R_{\mathrm{p}} \exp\left(\frac{-E_0}{273.15 - T_0}\right) & \text{if } I_0 < 5 \\ -A & \text{if otherwise} \end{cases} \tag{9.14}$$

The carbon-balance model provided a means to estimate NEE when its measurement by eddy covariance was problematic, which happened 42% of the time. By utilising a particular method of data assimilation—a dual unscented Kalman filter—Gove and Hollinger (2006) minimised the predictive error of their empirical carbon-balance model by assimilating the information in the data streams. They re-estimated the parameter values of the carbon-balance model sequentially on a half-hourly time step when measurements of NEE, I_0, and T were available. The most recently available update of the model was used to fill gaps in the NEE record when only measurements of I_0, and T where available. The analysis accounted for assumed amounts of observation error (noise) in the measurements of the environmental variables and NEE, and provided error bounds for the predictions of GPP and NER and, thus, estimates of half-hourly and annual NEE. The analysis also provided sequential error bounds for the parameter estimates (see Gove and Hollinger (2006) for details).

The model used in this example was very simple, though some physiological underpinnings are evident in both components of the model: ecosystem carbon uptake is a saturating function of PPFD and ecosystem respiration is function of temperature within the domain of physiological activity. However, we also note that some details that one might ordinarily expect, such as the seasonal waxing and waning of the leaf biomass pool and photosynthetic acclimation to changing environmental conditions, are absent from the model. However, omissions of these and other details are of little concern in the data assimilation. The leaf dynamics and acclimation are confounded with the dynamics of other biomass pools, but they are implicitly reflected by the predicted time streams of A, K, R_{p}, and E_0. The simple model combined with the temperature, PPFD, and NEE time streams, is appropriate to meet the modelling objectives: gap-filling and error estimation.

9.4 Bayesian Calibration

Bayesian calibration, model comparison and uncertainty analysis are well suited for process models, because they allow us to make use of all types of data and to deal with uncertainties that are often pronounced with such models (e.g., Poole and Raftery 2000; Raftery and Bao 2010; Reyer et al. 2016). The method can be used for obtaining parameter estimates, measures of uncertainty and correlations between parameters. It starts with estimating the prior probability distributions of parameters. All data on outputs and their probability ranges are also collected, including expert opinion and scattered data points on some variables. These are used to update the parameter distributions using Bayes' Theorem which takes the following form:

$$P(\theta|D) = cP(D|\theta)P(\theta) \tag{9.15}$$

where θ denotes the vector of model parameters and D denotes the data. $P(\theta|D)$ is the probability that the parameter vector takes the value θ when the model output vector has value D; $P(\theta)$ is the prior probability of θ; $P(D|\theta)$ denotes the probability that the vector of the model's predictions equals the measured values D, given that parameters have the value θ; and $c = P(D)^{-1}$ is assumed to be a constant. The result of the analysis is a posterior distribution of the parameters, which can be mapped to the outputs (the same way as in the example of Fig. 9.3) to quantify the predictive uncertainty of the model (Green et al. 1999, 2000; Van Oijen et al. 2005). Different models can further be compared on the basis of the calibration results in Bayesian model comparison (Van Oijen et al. 2013).

The Baysian calibration of simulation models requires the use of advanced numerical tools. A widely used and relatively simply applicable method is provided by Markov Chain Monte Carlo (MCMC) simulation. In Monte Carlo methods, the model is simulated a large number of times with randomly selected values for each parameter (again, see the example of Fig. 9.3). However, in order to modify the parameter distribution to one that makes the model outputs match better with the data, the Markov Chain based algorithm guides the selection of the next parameter vector, such that the parameter distribution gradually moves from the prior to one that agrees best with the data, i.e., the posterior distribution. A large number of iterations is usually required to reach a stable posterior distribution. It is also important to store a large number of parameter vectors after the algorithm has reached a stable region in the parameter space, because the posterior distribution is represented numerically by the stable set of parameter vectors. This means that no additional assumptions are needed for the shape of the parameter distributions (Van Oijen et al. 2005).

The posterior distributions are generally characterised with their means and standard deviations. However, such descriptions alone cannot replace the numerically established distributions, because (a) we have not specified the shape of the distribution, and importantly, (b) the parameters of the posterior distribution are often correlated with each other. If correlations exist, it is crucial to include them in the analysis of predictive uncertainty.

Here we review two examples of Bayesian calibration. One is the study mentioned earlier (Sect. 8.5.1, Minunno et al. 2016) that fitted the PRELES photosynthesis and evapotranspiration model to eddy covariance data. The other study is a calibration of the PREBAS model to data from permanent growth and yield trials and national forest inventories. These studies differ from each other in their motivation, data availability, and role of parameter priors. In the case of PRELES, the calibration used wide priors that were narrowed down in the posterior to make the model predictions closely follow the observations, which were abundantly available from eddy covariance measurements. For PREBAS the situation was the opposite: strong prior information existed for most of the parameter values, but data only covered part of model outputs over part of the normal simulation time. In the former case nonlinear statistical fitting would therefore have been a viable alternative, but Bayesian calibration was chosen because it provided a more straightforward technique for analysing predictive uncertainties, model sensitivity to input uncertainty, and correlation between parameters.

In the following sections we illustrate the key points and critical pitfalls of these estimation procedures.

9.4.1 Calibration of Gas Exchange Model PRELES

Minunno et al. (2016) evaluated the regional applicability of the PRELES photosynthesis model (Sect. 8.5.1, Peltoniemi et al. 2015) using Bayesian techniques. The study included a generalised sensitivity analysis of the model (see Sect. 9.2.2), followed by its calibration to data from multiple boreal coniferous sites. The analysis focused on three research questions: (1) Can a generic set of model parameters be found for all sites? (2) Under what conditions—if any—should the multi-site calibration be used in favour of the site-specific calibration, if both exist for a site? (3) How should data be selected for model calibration to extend model predictions to a site with no prior data?

The outputs of PRELES include daily gross primary production (GPP) and daily evapotranspiration (ET) at the stand scale. Model drivers are daily weather variables, and other inputs include leaf area index and water holding capacity of the soil (Fig. 9.4). Data on all these variables were obtained from ten boreal coniferous forest sites located in Finland and Sweden (Table 9.1). The sites covered a latitudinal

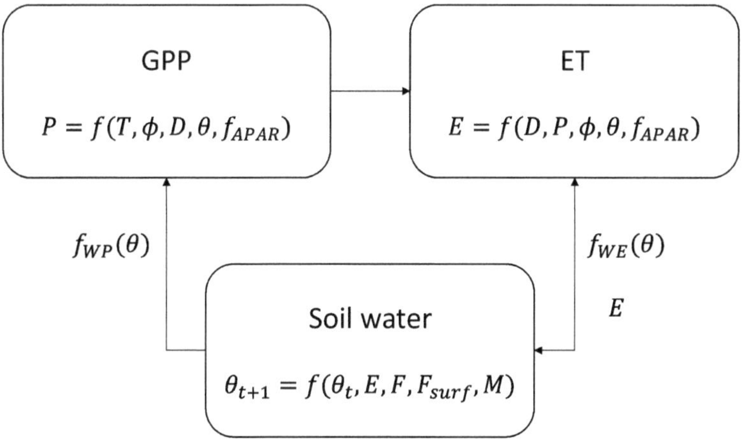

Fig. 9.4 Structure of PRELES (Peltoniemi et al. 2015). Daily canopy photosynthesis, P, is calculated through a multiplicative model from daily mean temperature, T, daily total incident PAR radiation, ϕ, daily mean vapour pressure deficit D, soil water content, θ, and the proportion of absorbed PAR radiation, f_{APAR}. Evapotranspiration, E, consists of transpiration, evaporation from the ground, and evaporation from the leaf surfaces. While transpiration is largely controlled by photosynthesis, the evaporation terms depend mainly on radiation. Soil water is calculated following a bucket model where the water content increases by precipitation, M, and drainage from leaf surfaces, F_{surf}, and decreases through evapotranspiration. The model also accounts for snow

Table 9.1 Sites for PRELES calibration (Minunno et al. 2016). Sites Hyytiälä to Alkkia are in Finland, sites Norunda to Skyttorp are in Sweden. Hyytiälä S is the SMEAR eddy covariance site, Hyytiälä 12 is a temporary measurement site in a 12-year-old stand, and Hyytiälä 75 is a temporary measurement site in a 75-year-old stand. Lat = latitude, Long = longitude, deg = degrees. N = number years of data. Soil = Hp, Haplic podzol; Dfb, drained fertilized bog; Dab, drained afforested bog; Spt, sandy podzolic till. Sp = dominant species: P = Scots pine, S = Norway spruce. LAI = leaf area index (all-sided)

Site	Lat (deg)	Long (deg)	N	Soil	Sp	LAI	Age (yr)
Hyytiälä S	61.51	24.17	11	Hp	P	7.9	40–49
Hyytiälä 12	61.51	24.17	1	Hp	P	7.0	12
Hyytiälä 75	61.51	24.17	1	Hp	P	7.9	75
Sodankylä	67.22	26.38	9	Hp	P	3.8	50–160
Kalevansuo	60.39	24.22	6	Dfb	P	5.7	<40
Alkkia	62.11	22.47	3	Dab	P	9.0	32
Flakaliden	64.07	19.27	7	Spt	S	9.5	43
Norunda	60.1	17.5	5	Spt	P,S	12.7	100
Knottåsen	61.0	16.13	2	Spt	S	7.0	39
Skyttorp	60.07	17.5	1	Spt	P	8.0	NA

Table 9.2 PRELES parameters to be calibrated (Minunno et al. 2016)

Definition	Symbol	Units
Potential light use efficiency	β	$\mathrm{g\,C\,mol\,PPFD^{-1}}$
Delay parameter for ambient temperature response	τ	–
Threshold for state of acclimation change	X_0	°C
Acclimation state maximum	$S_{\max}$	°C
Sensitivity parameter for VPD response	κ	$\mathrm{kPA^{-1}}$
Saturation parameter for light modifier	γ	$\mathrm{mol\,PPFD^{-1}\,m^{-2}}$
Threshold for linear decrease of $f_{\mathrm{W,P}}$	ρ_{P}	–
Transpiration parameter	α	$\mathrm{mm\,(g\,C\,m^{-2}\,kPa^{1-\lambda})^{-1}}$
Parameter adjusting water use efficiency with VPD	λ	–
Evaporation parameter	X	$\mathrm{mm\,mol\,PPFD^{-1}}$
Threshold for linear decrease of $f_{\mathrm{W,E}}$	ρ_{E}	–
Parameter adjusting water use efficiency if soil water is limiting GPP	ν	–

band from 60 to 67 ° N with annual mean temperatures ranging from 0.8 to 7.1 °C, and precipitation from 550 to 850 mm. Daily GPP was calculated from half-hourly NEE according to common practices (Aubinet et al. 2012), and NEE and ET were measured above the forest canopies by the eddy-covariance method (Sect. 9.3.1).

A generalised sensitivity analysis was carried out within the prior uncertainty range of the parameters (see Sect. 9.2.2). The results indicated that model outputs were particularly sensitive to six out of the twelve parameters (Table 9.2), as quantified by the mean-sensitivity metric μ_i^*. The same parameters also showed

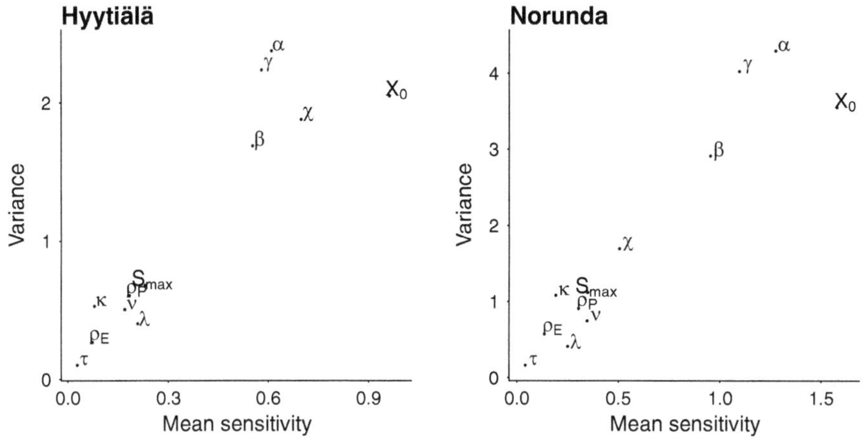

Fig. 9.5 Mean sensitivity (μ^*) of GPP to parameters and its variance (σ) within the prior parameter range in PRELES (Minunno et al. 2016). The relative sensitivities and their variance follow a similar pattern in Hyytiälä, Finland (left panel) and Norunda, Sweden (right panel) even if the absolute sensitivity and its variance are greater in Norunda. (Redrawn from Minunno et al. 2016)

high values of the variance of sensitivity, σ_i^2, which signifies non-linearities and interactions with other parameters. The ranking of sensitivity and the relationship between mean sensitivity and its variance followed a similar pattern at the different sites although absolute values of both varied (Fig. 9.5).

Bayesian calibrations were carried out for each of the sites separately (SS or site-specific) and for pooled data including all sites (MS or multi-site). The SS calibrations were then compared with the MS calibration for each site using different criteria, to find out if the MS calibration was generic enough for use at all sites. The main criteria included assessing the similarity of the posterior parameter sets and the similarity of the output distributions implied by SS and MS, respectively.

The data were highly informative in determining the values of the parameters that were assessed highly influential in the generalised sensitivity analysis (Fig. 9.6). For these parameters, the posterior distributions were much constrained from the priors. The MS calibration was able to constrain even the less important parameters, whereas a lot of uncertainty remained in some of these parameters in the SS calibrations (Fig. 9.6). Parameter estimates across the SS and MS calibrations were consistent for the most influential parameters, whereas differences occurred in estimates of the parameters to which model outputs were less sensitive. The site with the youngest stand (Hyytiälä 12yr), which was measured just 1 year, showed the largest differences in parameter estimates compared to all other sites.

Model performance was evaluated in terms of R^2 and the slopes of the simulated vs. observed GPP and ET, calculated for each calibration at daily time step. The predictions were generated using the maximum a posteriori (MAP, i.e., the modal parameter vector of the posterior distribution) parameter vectors of MS and SS. The

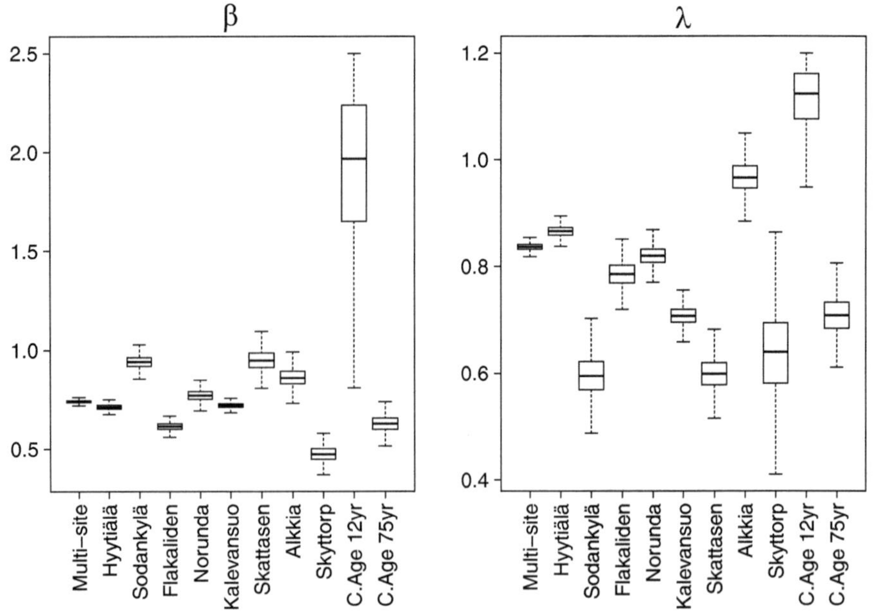

Fig. 9.6 Marginal posterior distributions of PRELES parameters obtained through the multi-site calibration and the site-specific calibrations. Parameter β in the left panel represents parameters with high sensitivity according to the generalised sensitivity analysis, while λ in the right panel is an example of parameters of medium and low sensitivity (see Table 9.2 for definitions of parameters). (Redrawn from Minunno et al. 2016)

variance explained by the model was higher for GPP than for ET, both being in most of the cases higher than 70%; however the model tended to underestimate carbon and water fluxes (slopes lower than 1). In general, after Bayesian calibration, model outputs were characterised by low uncertainty.

In model predictions of GPP and ET for each site, the differences between the multi-site and site-specific calibrations were small in most of the cases (Fig. 9.7).

PRELES also caught the pattern of GPP inter-annual fluctuations for the sites with the long-term datasets (i.e., Hyytiälä SMEAR, Sodankylä, Flakaliden, Norunda and Kalevansuo), and the MS and SS annual predictions were really similar in most cases. Both MS and SS calibrations showed robust performances in predicting the photosynthetic activity of boreal forests also at annual time step, while the model was less accurate in reproducing the annual evapotranspiration (Fig. 9.8).

The results from the site-specific and multi-site calibrations seemed to suggest that the multi-site parameter sets could be used for all sites without a major loss of accuracy. Minunno et al. (2016) investigated this further by means of *Bayesian model comparison* which compares a set of models for the probability that they represent the observed data. The models were defined as the SS and MS calibrations. In the model comparison, the output distributions of GPP and ET created with the posterior distributions of either SS or MS were compared with

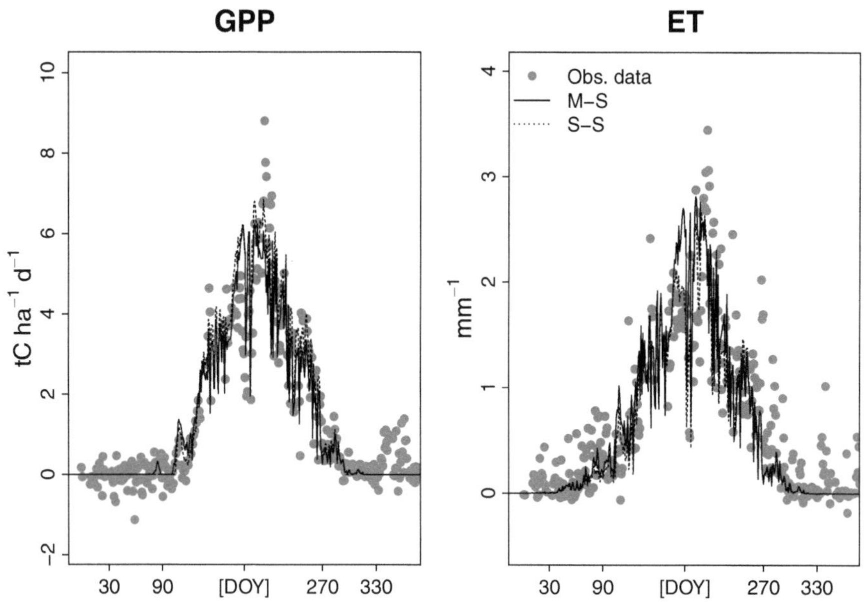

Fig. 9.7 Daily gross primary production (left panel) and evapotranspiration (right panel) at Hyytiälä for a year randomly selected from the dataset. Dots represent the observations, and the lines are PRELES predictions; the dashed line is the output from the site-specific calibrations (S-S), while the continuous lines represent the multi-site calibration (M-S). (Redrawn from Minunno et al. 2016)

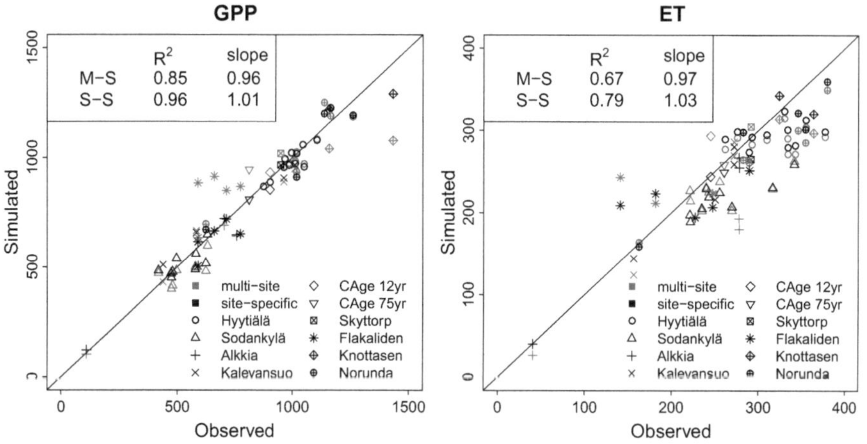

Fig. 9.8 Observed versus simulated annual gross primary production (GPP) and evapotranspiration (ET). Each symbol corresponds to different site; the grey and black colours refer to the multi-site (M-S) and site-specific (S-S) calibration, respectively. The daily observed and simulated data were summed to obtain the annual GPP values in the figure only if the quality flags of the daily GPP measurements were lower than 0.7. (Redrawn from Minunno et al. 2016)

the respective distributions in the observed data, and the probabilities of each of the parameterisations representing the data were worked out, with the assumption that the two models together represented the data perfectly. Two different model comparison analyses were carried out, termed *Forward prediction* and *New site*.

For *Forward prediction*, the data sets for all sites were split into two parts, the first half was used for calibration and the second half for validation. Bayesian model comparison was used to judge whether the SS or MS calibration was more likely to represent the validation data on GPP and ET simultaneously. Interestingly, in four of the ten cases the multi-site calibration turned out to be more likely for the site than its own site-specific calibration. In the *New site* analysis new multi-site calibrations were done where one site was excluded at a time. For each site, all other SS calibrations and the MS calibration without the site itself were then assessed with Bayesian model comparison. Here, the Hyytiälä SMEAR SS calibration had 100% probability of being better than the MS calibration on every site, while the MS calibrations were always better than the other SS calibrations.

The conclusion made from the Bayesian analysis of PRELES was that both the multi-site calibration and the Hyytiälä SMEAR site-specific calibration provided good choices for a general estimation tool for the boreal forest in Northern Europe. This reflects the long measurement record and high quality of the Hyytiälä SMEAR data. It also corroborates the findings that photosynthesis is a generic process derivable from environmental drivers rather than species-specific characteristics (Duursma et al. 2009). However, we must bear in mind that although PRELES is a physiologically motivated model it is not really process-based in its structure. Probably because of this, some of the parameter values were quite different between the sites, although this had little influence on the outputs.

9.4.2 Calibration of Tree Growth Model PREBAS

Minunno et al. (2019) made the dynamic tree growth model of Chap. 5 responsive to environmental variability by using the approach of Sect. 8.5 (but ignoring the variability of foliage nitrogen content), where the photosynthetic production of the stand, as modelled by (8.18), was explicitly calculated with the PRELES model, which was driven by daily weather data. The growth model was taken to describe the mean tree of a stand, and density-dependent processes were slightly modified from those presented in Chap. 5. For example, crown rise was made dependent on the light absorbed by the canopy rather than the mean distance to neighbours, since the former was estimated for the photosynthesis calculation. An upper bound for stand density was estimated using Reineke's density model (Sect. 6.3.2), and absorption of radiation in mixed species stands was simulated as in Sect. 6.4.2, but with species-specific effective extinction coefficients (see (4.52)).

The growth model had 19 parameters (Table 9.3), out of which 17 were calibrated for three European boreal species using Bayesian inference. The two parameters that were not calibrated were the potential photosynthetic production, P_0, that was

Table 9.3 Parameters of the growth model. Parameters marked with $*$ were not calibrated but were assigned values according to the prevailing environment. Note that all state variables were expressed in $kg\,C\,ha^{-1}$ in the calibration

Name	Definition
P_0^*	Maximum annual photosynthetic production of stand ($kg\,C\,m^{-2}\,yr^{-1}$)
m_f	Specific maintenance respiration rate of foliage (yr^{-1})
m_r	Specific maintenance respiration rate of fine roots (yr^{-1})
m_w	Specific maintenance respiration rate of sapwood (yr^{-1})
c	Growth respiration rate (yr^{-1})
T_f	Leaf longevity (yr)
T_r	Fine root longevity (yr)
k_H	Homogeneous extinction coefficient
L_A	Specific leaf area ($m^2\,kg^{-1}\,C$) (all-sided)
s_1	Parameter relating to reduction of photosynthesis with crown length
η_s	Ratio of foliage mass to cross-sectional area at crown base ($kg\,C\,m^{-2}$)
ρ_w	Wood density ($kg\,C\,m^{-3}$)
α_r^*	Ratio of fine roots to foliage
z	Foliage allometry parameter
β_0	Ratio of total sapwood to above-ground sapwood biomass
β_B	Ratio of mean branch pipe length to crown length
β_S	Ratio of mean pipe length in stem above crown base to crown length
χ	Parameter for relating branch length to crown length
c_R	Light level at crown base that prompts full crown rise
N_0	Stand density at which mortality begins when diameter is 25 cm (ha^{-1})

calculated with PRELES using the parameter values estimated in the calibration described above, and the fine-root to foliage mass ratio, α_r, the values of which were specific for site classes and were taken from a previous study (Mäkelä et al. 2016). The rest of the parameters were assigned prior ranges on the basis of the literature.

The calibration data were standard forestry data from from two sources in Finland: (1) long-term thinning experiments and, (2) two consecutive measurement points of permanent sample plots in national forest inventory. The measured variables for each species included mean height, diameter and crown base height, total basal area and volume, stand density and site type (indicative of site index). Both data sets were large (totalling 657 and 223 plot measurements, respectively) and represented a wide area geographically. The long-term experiments covered time spans of up to 84 years and had been designed to study cause and effect relationships by varying the intensity of thinning. In the inventory measurements, the time span (and measurement interval) was only 10 years, but they came from a wider regional distribution, providing a representative sample of average forest variables.

In this calibration the role of the prior parameter distributions was quite significant, because all parameters had a clear physical definition, and relevant value ranges could be found in the literature. In this sense, the calibration also provided a test for model structure, because it was possible that no parameter combinations could be found within the prior distributions that could reproduce outputs close to the observations. However, although the data set was geographically extensive, it only covered part of the output variables predicted by the model. Even if a lot of empirical evidence has been gathered and embedded in the parameters of the growth model about the regularities of tree structure, deviations in these may still reflect in the calibrated parameter values, because no measurements were available, e.g., about foliage and fine root mass.

The thinning experiment data were clearly more restrictive in selecting the posterior parameter ranges than the inventory data. This was reflected in model outputs: the standard deviations of model outputs created with the posterior distribution were generally less than 10% of those created by the prior in the thinning experiment data, compared with 25–40% in the inventory data.

When the two calibrations were tested against each other, it turned out that the thinning experiment calibration was equally well applicable to the inventory data, but the inventory calibration underestimated the stand level measures, basal area and volume, for large basal areas (and volumes) (Figs. 9.9 and 9.10). In accordance with this, the inventory calibration resulted in larger values of parameters related to carbon loss, such as specific maintenance respiration rate, fine root turnover rate and wood density. This was probably because poorer growth sites and sub-optimally managed stands were not well represented in the thinning experiments. On the other hand, the thinning experiment data were more informative for the parameterisation as a whole, as shown by their more restrictive impact on the posterior distribution.

The parameter set from the thinning experiment calibration was used for testing the results against (1) an empirical model of stand growth in Finland (Vuokila and Väliaho 1980) and (2) national statistics of growth in subregions. PREBAS and the empirical model estimates showed similar trends of mean annual increment (MAI) across the country, although PREBAS predicted a somewhat higher MAI than the empirical model especially for spruce. On the other hand, PREBAS estimates of MAI were consistent with the national forest statistics for timber production (Fig. 9.11) (Minunno et al. 2019).

The calibration of PREBAS illustrates how Bayesian inference can make process-based models produce unbiased estimates of output variables while simultaneously retaining consistency of model parameters values with their physical definition. This is important if the models are to be used for decision making, where their sensitivity to both environmental drivers and management actions is a valuable asset.

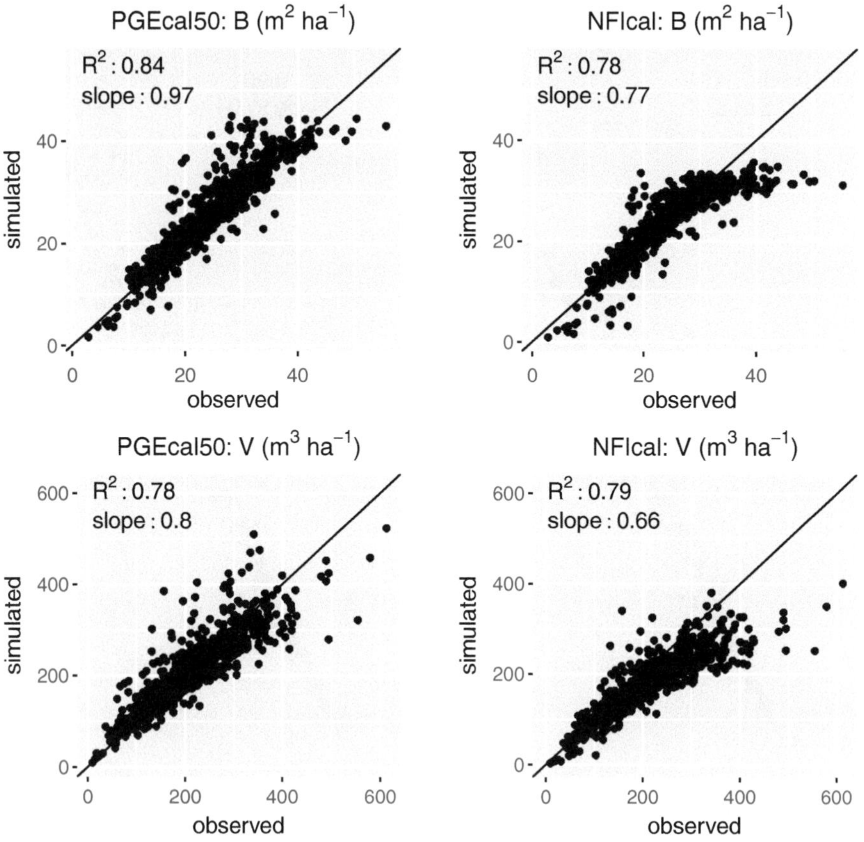

Fig. 9.9 Observed vs. simulated data from thinning experiments of pine stands. The simulations were made with the MAP parameters of a calibration to 50% of the permanent growth experiment (PGE) data and tested against the remaining 50 % of the PGE data (left column, PGEcal50) and against the NFI data (right column, NFIcal). B = basal area, V = volume (Minunno et al. 2019)

9.5 Exercises

9.1 A simulation programme "`Sensitivity.R`" of the Mitscherlich model (9.6) with the related sensitivity functions (9.7) is provided in the electronic supplementary material of this chapter. This code produces the curves in Fig. 9.5. Study the code and make sure that you understand what is going on. Run the programme in R.

9.2 Instead of the Mitscherlich model, consider stand volume development using the logistic model (2.15):

$$\frac{\mathrm{d}x}{\mathrm{d}t} = rx\left(1 - \frac{x}{K}\right) \tag{9.16}$$

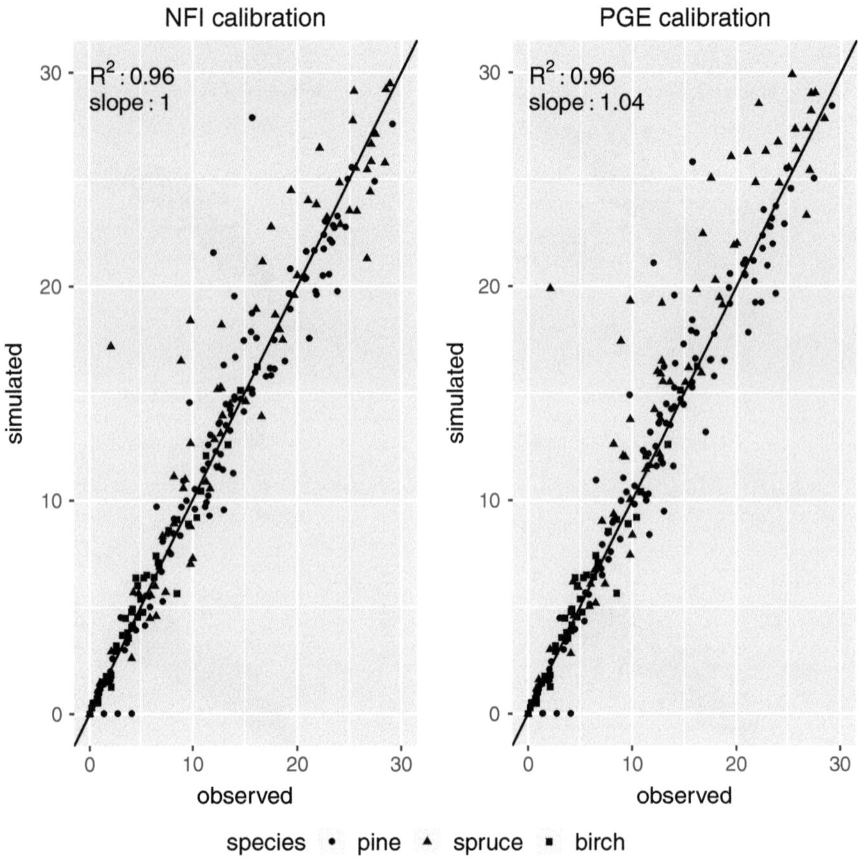

Fig. 9.10 Observed vs. simulated basal area (B) of the inventory dataset. Left panel: the simulated data points were generated using the inventory calibration; right panel: the simulated data were generated using the thinning experiment calibration. Note that thinning experiment calibrations were available only for pine and spruce. When we run the model for stands with birch trees we used the parameter set obtained in the inventory calibration for this species (Minunno et al. 2019)

where we have denoted stand volume by x ($\mathrm{m}^3\,\mathrm{ha}^{-1}$) and r is initial relative growth rate (yr^{-1}) and K is carrying capacity ($\mathrm{m}^3\,\mathrm{ha}^{-1}$). Derive sensitivity functions (9.5) of x with respect to the parameters r and K and initial state x_0.

9.3 Using the R code of Exercise 9.1 as a reference, simulate the state and sensitivity functions of the logistic model with parameter values $r = 0.05\,\mathrm{yr}^{-1}$, $K = 500\,\mathrm{m}^3\,\mathrm{ha}^{-1}$ from initial state $x(0) = 10\,\mathrm{m}^3\,\mathrm{ha}^{-1}$. Plot the sensitivity functions in a normalised form, i.e., with maximum 1 and minimum 0.

9.4 Repeat the analysis of Exercise 9.3 for *elasticity* (9.3) instead of sensitivity. How does elasticity differ from the normalised sensitivity plotted in the previous exercise?

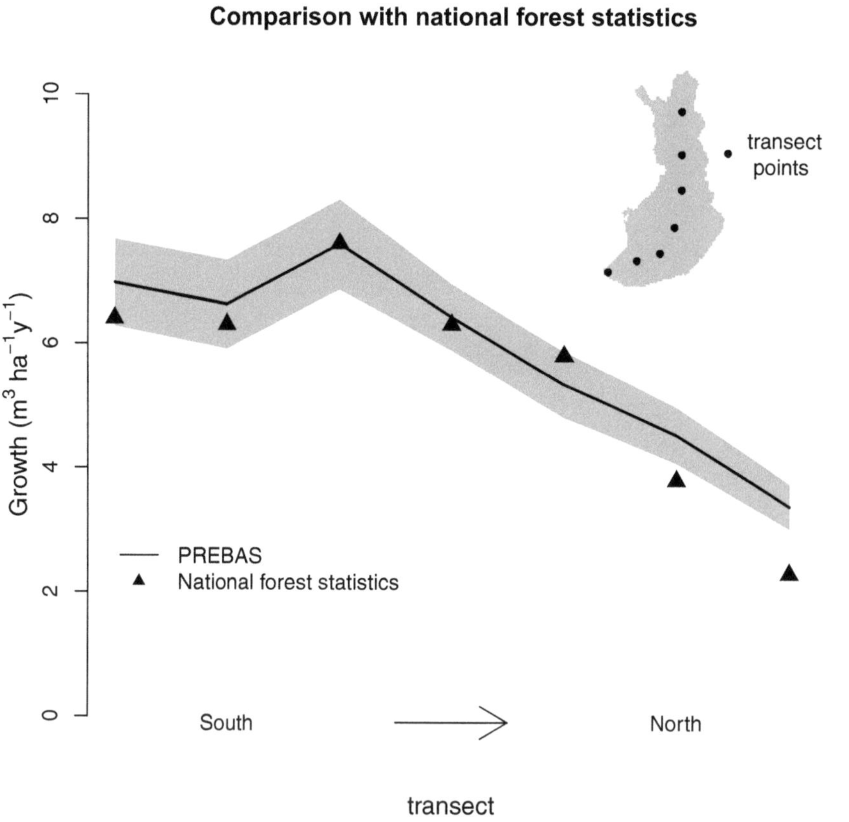

Fig. 9.11 Comparison between Finnish national statistics of forest annual growth (triangles) and PREBAS predictions (black line). The shaded grey area represents model predictive uncertainty. PREBAS estimates were computed for a south-north transect, and national statistics estimates were taken from the regions where the transect points were located and were computed for pine and spruce. Model predictions were the weighted average MAI of pine and spruce at different growth sites, where the weights were obtained from the National statistics on the basis of respective areas covered in each region. (Redrawn from Minunno et al. 2019)

9.5 Instead of deriving the sensitivity functions analytically for the logistic model, calculate them numerically as instructed in (9.8).

9.6 A simulation programme "`Uncertainty.R`" for the Mitscherlich model, (9.6), is provided in the electronic supplementary material of this chapter. This code produces the curves in Fig. 9.3. Study the code and make sure that you understand what is going on. Run the programme in R. How do these curves relate to the sensitivity functions of Fig. 9.5?

9.7 The PRELES and PREBAS models can be found at "github.com/ForModLabUHel/Rprebas_examples". Download the examples of model use and / or the model packages.

Chapter 10
Applications and Future Outlook

In this Chapter we review some of the applications of our modelling approach, with a critical appraisal of the results and an assessment of next steps: what are the key directions of model development at the time of conclusion of this volume? The key applications of the models comprise prediction of stand level growth, production, and economic revenue as affected by different management actions, assessing the regional variability of growth, and as an issue of ever increasing importance, analysing and predicting climate change impacts at the stand scale and also on a larger, geographical scale.

10.1 Introduction

This book has focused on a particular approach to tree and forest growth modelling, centered around a dynamic approach to carbon balance as constrained by the pipe model and other regular, often evolutionarily based structures. The carbon balance approach is shared by most growth models based on physiology, or so-called process-based growth models, where a similar material balance with the constraint of conservation of matter and energy has been applied to other substances as well. We have seen that the simple assumption of carbon balance with the constraint that a unit area can only assimilate a limited amount of carbon alone can lead to quite reasonable approximations of the growth pattern and growth potential of forests. It is noteworthy that such natural constraints are not inherent in so-called empirical growth models. Including environmental factors as limits to growth—if done adequately from the theory perspective and parameterized carefully—will allow for the models to be generalized more widely than any empirical models that, by definition, should only be applied to situations similar to model development data.

A. Mäkelä, H. T. Valentine, *Models of Tree and Stand Dynamics*,
https://doi.org/10.1007/978-3-030-35761-0_10

10.2 Stand-Scale Growth and Production as Affected by Management

Stand-scale forest growth models aim at increasing our understanding of stand dynamics, such as development, competition, and interactions between growth, site and production. One of the key applications of stand-scale models has been the planning of management actions, notably the choice of rotation length and the timing and intensity of harvests. These are influenced not only by stand dynamics but also by the expected economic outcome of the actions. Forest management recommendations in many countries are largely based on results from economic optimisation studies that assess the impact of forest management actions on the economic benefits and costs of these actions, finding the strategies that are the most beneficial for the forest owner, given timber prices, management costs, and interest rates.

Standard optimisation studies in forestry are based on empirical growth models. While these models offer reliable predictions when applied to normal management situations, they may lead to serious problems in optimisation where a solution may be found outside the valid range of model development data or, if restricted, on the border of applicability. This has been taken as an indication of extrapolation leading to unreliable optimisation results (Niinimäki et al. 2012). Models based on ecological resource limitation can avoid at least the most severe consequences of extrapolation.

A key requirement for models assessing the economic outcome of timber production is to include the distribution of trees into different size classes, as this will influence the yield of different assortments. At a more detailed level, the mechanical properties of the stems in each size class affect their usability and value. The pipe model provides a feasible approach to the development of stem structure, including stem shape (see Fig. 5.2) and the development of branchiness into knots in timber (Sect. 8.6). These characteristics have been incorporated in the PipeQual model described in Sect. 8.6 (Mäkelä et al. 2000; Kantola et al. 2007; Kalliokoski et al. 2016). PipeQual depicts stands as collections of tree size classes, such that diameter distributions can also be retrieved, and it utilises the dynamic pipe model to describe the vertical structure of stems, resulting in stems with taper, internal ring structure, and knots that are characterised either as sound or dry.

A frequent method in the analysis of management impacts on stand economics is scenario analysis, where a (usually small) number of management options is compared with each other to find the "best" management strategy (Alegria 2011). The difference between scenario analysis and optimisation is that the latter considers the space of management strategies as continuous, and thus searches for the best strategy, at least in principle, from among an infinite set of alternatives. This set-up requires advanced optimisation tools and often a lot of computation time.

A group of economists led by Prof. Olli Tahvonen at the University of Helsinki, who apply advanced computational optimisation methods, have used the PipeQual model to find the best forest management strategies for different boreal stands

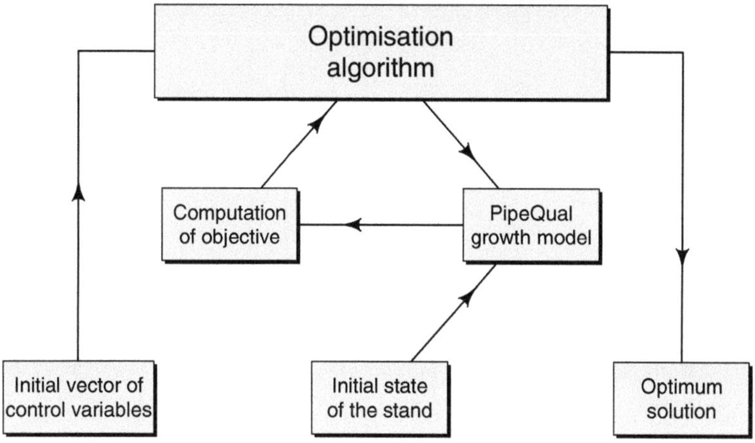

Fig. 10.1 Optimisation procedure for determining cost-effective harvest strategies with the PipeQual model. (Redrawn from Niinimäki et al. 2012)

(Fig. 10.1) (Hyytiäinen et al. 2004; Cao et al. 2010; Niinimäki et al. 2012). They have concluded that PipeQual offers a sound basis for such optimisation, because the validity of the optimisation results does not seem to be restricted to a limited set of states determined by the boundaries of empirical data. The optimisation has also provided a test for the model, as indeed the optimum was being searched from among a wide range of management options, some of which were neither practical nor realistic and led to combinations of model state variables that would rarely be observed in real stands. The model behaviour nevertheless remained logical, which for lack of direct data for all cases, was the only way to assess its performance.

In the optimisation models, the control variables included stand initial density (but not the initial size distribution, which was standardized on the basis of measurements in young sapling stands), the timing of thinnings, the intensity of thinnings in each size class, and the rotation length. In order to reduce the number of decision variables to shorten the computation time, some simplifying assumptions were made. For example, the thinning intensity in different size classes was determined by means of four parameters instead of independent choices in all 10–20 size classes (Fig. 10.2). The number of thinnings during the rotation was also fixed for each optimisation, although different numbers of thinnings were studied separately.

The objective of the optimisation was to maximise the present value of economic surplus over an infinite chain of similar rotations. The gains are formed by the harvested roundwood, with specific prices for sawn timber and pulp of different quality classes, determined by log diameter and branchiness. The related costs are due to logging and depend on the number of trees harvested and the sawlog, pulpwood, and wastewood volumes. All prices and costs are discounted at an interest rate which is assumed constant.

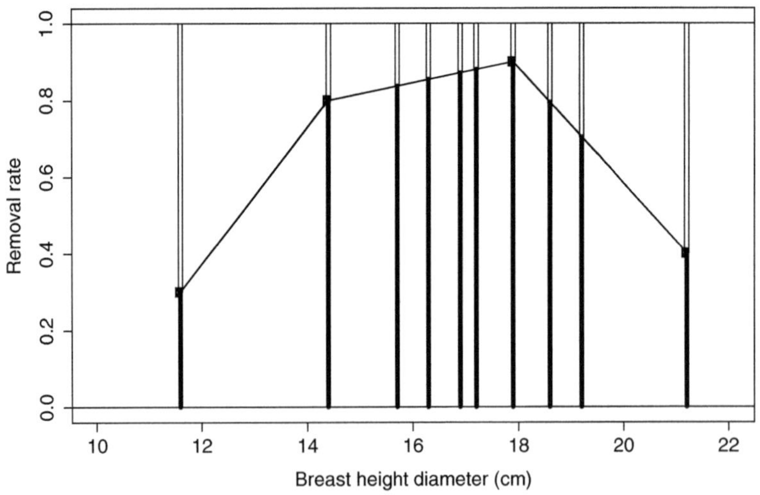

Fig. 10.2 Optimum thinning intensity across tree size classes. The optimum is defined as the four dotted points in the size-class—thinning intensity space. They determine the proportion of trees removed (white area of the columns) and remaining (filled columns) in each size class at each harvest time. (Redrawn from Hyytiäinen et al. 2004)

The optimisation for both Scots pine (Hyytiäinen et al. 2004) and Norway spruce (Niinimäki et al. 2012) stands yielded a pattern of thinning where the early thinnings were conducted as low thinnings and the later thinnings as high thinnings. This was somewhat in contrast with the normal forestry recommendations where low thinnings dominate. The early low thinnings in these studies were attributed to the need of avoiding natural mortality to which the smallest trees are prone. At the later stages of stand development the release of the co-dominants allowed them to grow higher quality logs than the dominants would have done, because they had smaller crown ratios and thus longer bare stems to allocate their growth to. Such differences in growth and resulting wood quality have not been described with empirical growth models.

An interesting outcome of the optimisation for Scots pine was the existence of two distinct peaks in the objective function at different rotation lengths (Hyytiäinen et al. 2004). The first peak represented clear-cutting the stand after the co-dominants had reached sawlog dimensions and some shedding of dry branches had occurred at the butt. The second peak represented a longer rotation when all the trees remaining in the final stock had reached the quality and dimensions required for superior quality butt logs. The choice of optimum was determined by the interest rate and relative prices of superior and regular quality butt logs, particular combinations of these making the two rotation lengths exactly equal in value. Lower interest rates and a higher price difference between normal and superior quality butt log favoured the longer rotation.

The Norway spruce stand analysis resulted in higher stocking densities and shorter rotations than standard management recommendations in Finland (Niinimäki et al. 2012). This was at least partly attributed to the description of wood quality in the model: high densities yielded stems with fewer dry knots and larger high quality butt logs. On the other hand, Norway spruce is fairly shade tolerant, such that the overall growth rate did not notably suffer from the high density. The study also found that optimal thinnings could increase the total timber yield compared with no thinnings, as they reduced the number of trees lost to natural mortality. In contrast, empirical studies have indicated that total yield cannot be increased by thinnings, but these may be based on non-optimal thinning schedules (Niinimäki et al. 2012).

As a word of caution, it is important to note that the results are highly sensitive to tree mortality and relative growth at different stocking levels. Tree mortality, in particular, is difficult to predict by any model (Biegler and Bugmann 2003), and relative growth rates also need to be tested more stringently. Nevertheless, this combination of advanced economic optimisation methods with the process-based, structurally explicit growth model PipeQual has provided new insights into harvesting schedules in boreal forests, and the results have been utilised in the formulation of national forest management recommendations in Finland (Hyytiäinen et al. 2004; Cao et al. 2010; Niinimäki et al. 2012).

10.3 Regional Variability of Growth and Carbon Sequestration

In addition to stand dynamics, such as time development and tree-to-tree interactions, another important application of forest models is to estimate and make predictions of the regional, country level, or even global totals of forest growth and carbon sequestration. This information is important for national forest policies to assess key issues such as sustainable harvest levels, forest potential for climate change mitigation, or the supply of roundwood and biomass to the global markets.

A basis for large-scale empirical estimates of forest production has been provided by ground-based forest inventory networks that have been developed and harmonised world-wide for over a century (McRoberts et al. 2010), and are now being complemented with different satellite-based data streams (Tomppo et al. 2011). Forest inventory data has been widely utilised to describe the current state of forests in country-level forest management planning systems that make projections of forest resources including optimised harvests and impacts of alternative forest policy measures over the next few decades (Lind and Söderberg 1994; Diaz-Balteiro and Romero 1998; Rasinmaki et al. 2009; Yousefpour and Hanewinkel 2014; Eggers et al. 2018). European (Schelhaas et al. 2007; Härkönen et al. 2019) or world-wide (Kindermann et al. 2008) forest production models have also been developed on the basis of inventory data.

Theoretically, forest growth depends on climatic and edaphic conditions and species properties. In geological time, even the edaphic conditions have been largely a function of climate, and together, they have influenced the matching of species with sites. This is the underlying philosophy of the so-called Dynamic Global Vegetation Models (DGVM) that have been developed since the 1990s to predict the responses of vegetation to, firstly, regional climate variability, and secondly, future climatic change (Foley et al. 1996; Cramer et al. 2001; Sitch et al. 2008; Friend et al. 2014; Naudts et al. 2015). Because of their approach, DGVMs operate with potential rather than actual vegetation cover, and even that is often provided at steady state, so they are not directly applicable to forestry projections in the same way as the inventory-based models. However, their main idea of estimating growth from the environment is shared by process-based stand growth models, such as those in this book, though species parameters and soil properties may need to be specified as input (Lasch et al. 2005; Blanco et al. 2007; Gonzalez-Benecke et al. 2016).

One useful application of process-based stand-level models has been to estimate forest productivity in areas that have not been subject to forest inventory (Coops et al. 1998; Waring et al. 2010). This requires data on local climate, soil properties and species parameters on a grid that allows us to simulate the variability of growth over the region in question. Because soil properties and species vary at a much finer scale than climate, a common approach is to use the geographic resolution of climate variables as a basis for the grid, then estimate the proportions of species groups and soil types within each grid box. An estimate of potential growth in each grid box is obtained by combining the information of soil types and species proportionally at the grid box level. Another possibility to reduce the calculation effort, Huang et al. (2017) suggested to use ecoregions instead of gridded cells as the unit of calculation because these could be taken to represent homogenous soil types as well as climate.

In order to predict actual rather than potential productivity, the simulation should be initialised with the current growing stock, including its age distribution and other stand variables that influence production. For a longer-term projection, forest management should also be included in the inputs as it changes the state of the growing stock and therefore has a large impact on future production and other carbon fluxes. This requires that the model can be initialized with data generally available for all forests, i.e., the type of data measured in forest inventories or the multisource inventories that also include variables based on remotely sensed data (Tomppo et al. 2011) (Fig. 10.3).

Process-based models are usually quite complex and their input data needs are more detailed than the inventory type data available on a large scale (Van Oijen et al. 2013). By contrast, the PREBAS model described in Sect. 8.5, which is an application of the core model of Chap. 5, was specially developed to run on a regional scale across Finland using gridded climate data and forest input variables available from National Forest Inventory (NFI). Furthermore, it was calibrated to forestry data from Finnish long-term growth trials and was shown to produce unbiased estimates of volume growth (see Sect. 9.4.2). The model was subsequently applied to 120-year (1980–2100) growth projections in the whole country using multisource inventory data (from 2013) as initial input information. The resolution

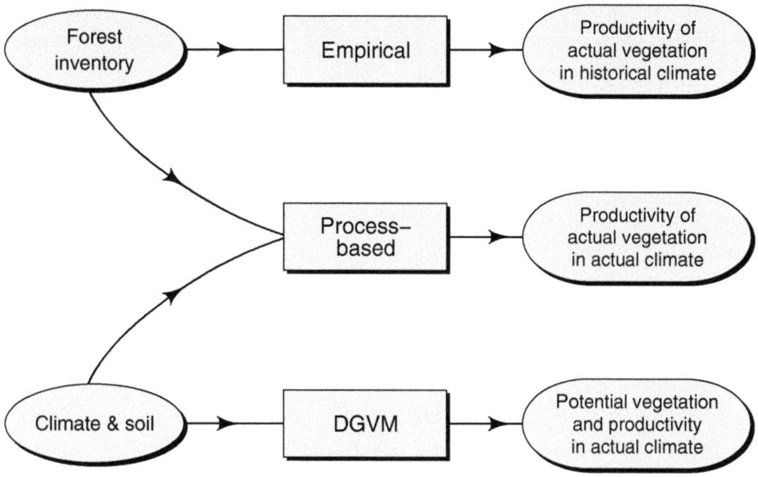

Fig. 10.3 A schematic presentation of data requirements and kinds of model output in different model types intended for large-scale projections of forest growth. DGVM = Dynamic Global Vegetation Model. Forest inventory includes information about site index or site fertility class

of the gridded data is 16 m × 16 m. Weather data representing the current climate was available on a 1 km × 1 km grid, interpolated from meteorological stations using kriging techniques (Venäläinen et al. 2005; Aalto et al. 2013). The multisource data included information about site type that was important for the parameterisation of PREBAS.

Because of the small grid size in the multisource inventory, simulations were only carried out on a sample of grid cells. Finland is divided into 15 forest centres, and 10000 cells were sampled from each forest centre area at the beginning of the simulations. The simulations used information about mean height, breast height diameter and basal area, all separately for Scots pine, Norway spruce, and Silver or Downy birch, soil class, and growth site type. Forest management was applied to the simulated stands using current Finnish management recommendations combined with alternative scenarios of total annual cut. The recommendations defined a thinning schedule and timing of final cut, dependent on site type, basal area, stand mean height, and stand mean diameter (Rantala 2011). However, generally only a subset of stands due for management were actually treated, in order to keep the total annual cut within the forest-centre-specific limits given by the harvest scenario. Each year, stands were randomly selected for management from the subset of stands that should be treated according to recommendations. The simulations were carried out with a super computer of the University of Helsinki, which took about 1 day to simulate the whole country (15 × 10000 grid cells for 120 years).

A comparison of total annual growth with forest statistics at forest centres showed good correspondence between model and data (Fig. 10.4). Cutting level had a big impact on the future development of the growing stock (Fig. 10.5). With the

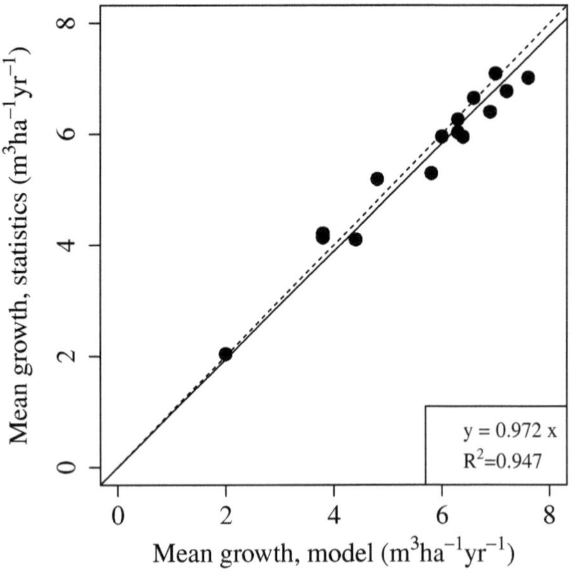

Fig. 10.4 Predicted current annual increment in 15 Finnish forestry regions in comparison with national statistics (Holmberg et al. 2019). The dashed line is the one-to-one line and the solid line is linear regression forced through the origin

largest cutting level there were forest centres where the prescibed cuttings could not be carried out unless harvests were done prematurely relative to management recommendations. These simulations were part of a Finnish model-comparison exercise directed towards decision makers to help them plan forest policy actions to mitigate climate change (Kalliokoski et al. 2019).

10.4 Climate Change Impacts

Environmental factors influence the physiological process rates that determine growth. In this book the process rates have been described as annual effective means, such as the biomass-specific rates of photosynthesis, respiration, turnover or nitrogen uptake. A lot of physiological, empirical, and modelling work has been carried out in the past decades that can be utilised to estimate the dependence of these rates on changing environmental factors, although large uncertainties still remain (e.g., Medlyn et al. 2011).

In addition to the direct environmental effects on process rates, trees will react to changing environments by adapting their structures and balancing different processes in response to the changing relative availabilities of different resources. What paths this adaptation will take cannot be answered simply by analysing the direct effects of the environment on physiological process rates. Trees will likely adapt to the new environment by the same mechanisms that they have adapted to different growing conditions and competitive positions in the past. We have described such adaptive mechanisms by means of eco-evolutionary models (Chap. 7).

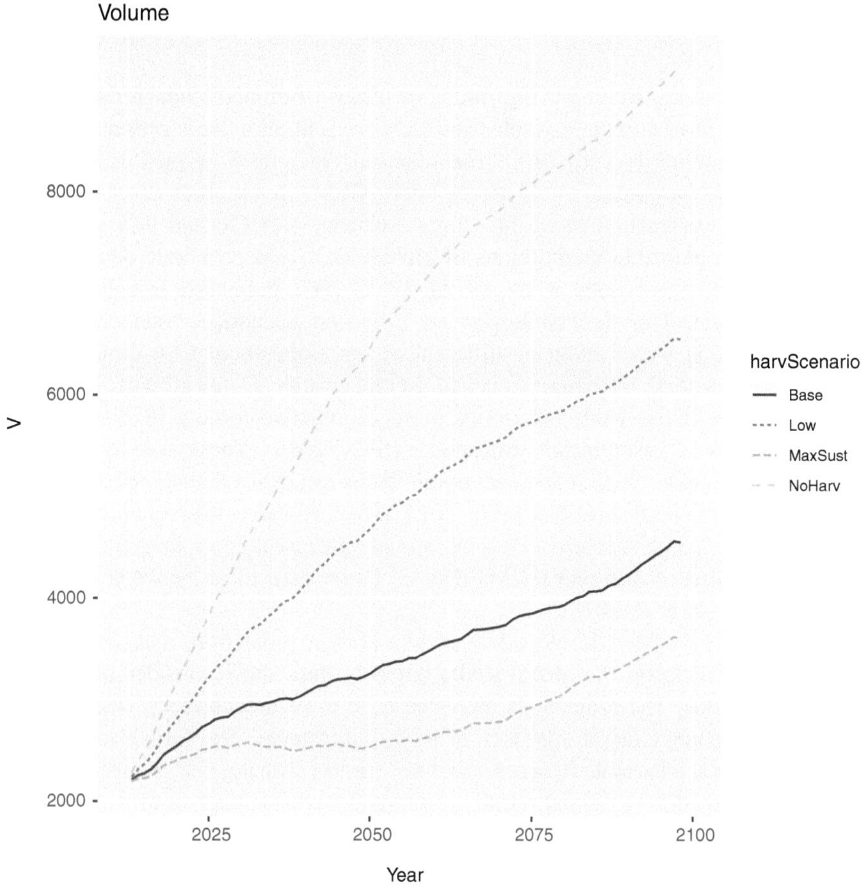

Fig. 10.5 PREBAS predictions of the growing stock in Finland (Mm^3) under different cutting scenarios in current climate conditions. Base = total annual cut as in the past decade (approximately 60 $Mm^3\ yr^{-1}$), Low = 40 $Mm^3\ yr^{-1}$, MaxSust = total annual cut does not exceed total annual net growth according to estimates by Natural Resources Institute Finland, NoHarv = no harvests

Here we demonstrate this issue using our approach to optimal co-allocation of carbon and nitrogen (Sect. 7.5.5) which we apply under the assumption of changing climate. First, we introduce the approach and the representation of climate impacts in the model, then demonstrate the sensitivity of the model to changes in selected metabolic variables, and finally report an exercise where the parameters are changed as expected, according to our best understanding, under some climate change scenarios. To do this, we also briefly introduce the general method of climate impact studies with climate scenarios, and finally, discuss the remaining gaps and uncertainties in the approach.

10.4.1 Climate Scenarios

Climate impacts are studied using future scenarios of climate change that predict the development of climate variables and CO_2 concentration from present day into the future, customarily until 2100. The scenarios have been derived using global climate models (or general circulation models, GCM) in a mutual effort organised by the Intergovernmental Panel on Climate Change (IPCC), and they are based on alternative, plausible assumptions of global social and economic development and related CO_2 emissions in the world. IPCC 2007 published so-called SRES climate scenarios (for Special Report on Emission Scenarios, Nakicenovic and Swart 2000) that were driven by different assumptions about CO_2 emissions. A more recent method is to use so-called Representative Concentration Pathways (RCP) that are defined through the increase in radiative forcing in 2100 caused by all greenhouse gases in the atmosphere (IPCC 2013). These scenarios can be linked with so-called Shared Socioeconomic Pathways (SSP) that describe different global alternative futures (O'Neill et al. 2014, 2017). However, a lot of uncertainty is related to the climate scenarios. Despite mutual agreement about the general trend, differences between different GCMs may be larger than those between scenarios (e.g., Kalliokoski et al. 2018).

The GCMs have been developed to produce climate projections that comply with the current climate on the global scale, but they may show considerable bias at specific locations. The projections therefore need to be downscaled, which can be done using a change factor approach or a bias adjustment. The former means that the change in each variable between baseline (current climate) and changed climate is recorded and added to actual, locally observed variables under current climate. In the latter method, the GCM projections are adjusted to provide unbiased results for a selected baseline period, for example, 1971–2000. Details of the methods can be found, e.g., in Diaz-Nieto and Wilby (2005) and Räisänen and Räty (2013).

10.4.2 Incorporating Climate Impacts in OptiPipe

OptiPipe (Sect. 7.5.5, Exercise 7.3) derives stand growth from the annual fluxes of carbon and nitrogen. The carbon fluxes include photosynthesis, respiration and tissue turnover. The nitrogen fluxes consist of nitrogen uptake and nitrogen loss through turnover. The rates of these processes in the model depend on state variables and metabolic parameters that represent annual effective means. All together, the model has 26 parameters, some of which relate to the metabolic rates and others to tree and stand structure (Table 10.1). Here we use structural parameter values that apply to Scots pine in southern Finland. We estimate the default metabolic parameters for the region and species using results from the PREBAS model calibration described in Sect. 9.4.2 (Table 10.1). Variability of growth due to site fertility is achieved by varying the parameter of maximum nitrogen availability,

Table 10.1 Parameter values used in OptiPipe simulations

Symbol	Value	Unit	Definition
σ_{fM0}	5.3	$kg\,C\,(kg\,DW)^{-1}\,yr^{-1}$	Maximum specific photosynthesis
K_f	4195	$kg\,ha^{-1}$	Leaf biomass at half maximum rate
γ	3.25	$m^2\,kg^{-1}$	Conversion from σ_{fM0} to $\mathbb{P}_0$ (Eq. 10.2)
U_{max}	40–90	$kg\,ha^{-1}\,yr^{-1}$	Maximum N uptake rate
K_r	4000	$kg\,ha^{-1}$	Root biomass at half maximum rate
r_m	24	$kg\,C\,kg^{-1}\,yr^{-1}$	Maintenance respiration per N
T_f	3.685	yr	Leaf longevity
T_r	1.25	yr	Fine root longenvity
f_f	0.6	yr^{-1}	N resorption from leaves
f_r	0.3	yr^{-1}	N resorption from roots
Y	1.54	$kg\,DW\,kg^{-1}\,C$	Growth yield
z	1.86	–	Leaf allometric exponent
N_o	0.0095	$kg\,N\,(kg\,DW)^{-1}$	Structural leaf [N]
N_{ref}	0.001	$kg\,N\,(kg\,DW)^{-1}$	Reference leaf [N]
N_w	0.0007	$kg\,N\,(kg\,DW)^{-1}$	Wood [N]
β_c	0.8	–	Sapwood spec length parameter
β_r	0.6	–	Sapwood spec length parameter
η_s	560	$kg\,DW\,m^{-2}$	Pipe model coefficient
ρ_w	400	$kg\,DW\,m^{-3}$	Wood density
a_σ	0.02	m^{-1}	Reduction of photosynthesis with height
c_σ	0	m^{-1}	Reduction of photosynthesis with crown length
β_3	1.1	–	Self-thinning parameter
β_1	0.3	–	Self-thinning parameter
m_f	0.007	–	Above-ground self-thinning
m_r	0.014	–	Below-ground self-thinning
r_N	0.6	–	Root to leaf

U_{max}, here from 40 to 90 $kg\,ha^{-1}\,yr^{-1}$. This yields stand mean heights at age 100 yrs ranging from about 11 to 27 m and mean annual increments (MAI) over a 100-yr rotation from 1.1 to 6.5 $m^3\,yr^{-1}$. Increasing site fertility increases nitrogen uptake and foliar nitrogen concentration, and decreases the fine-root to leaf ratio in standing biomass (Fig. 10.6). Carbon use efficiency (CUE = NPP/GPP) remains around 50% regardless of site fertility with the default parameters.

The key metabolic parameters that are likely to change under changing climate include (1) maximum rate of photosynthesis per unit leaf mass (σ_{fM0}), (2) specific maintenance respiration rate per unit nitrogen in live tissue (r_m), (3) fine-root and leaf tissue longevity (T_f and T_r), and (4) parameters related to nitrogen availability to trees in the soil (U_{max} and K_r). We use the PRELES photosynthesis model to estimate the photosynthesis parameters in OptiPipe. Because PRELES, as other conventional photosynthesis models, uses the Lambert-Beer law for the dependence of photosynthesis on leaf area, but OptiPipe uses a different type of saturating

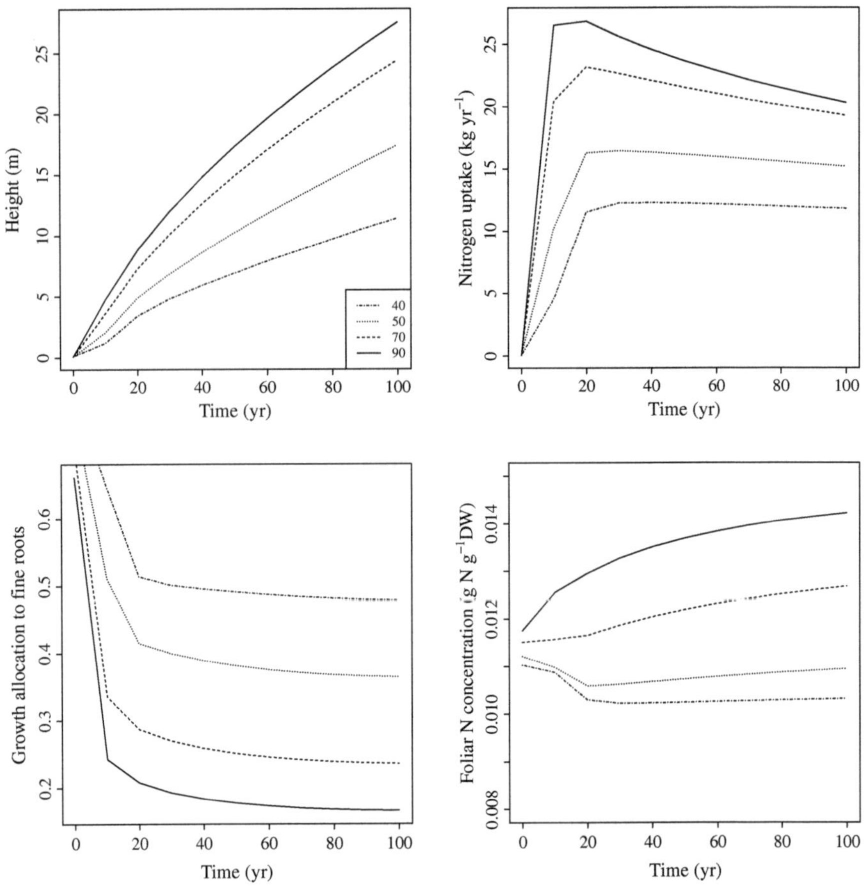

Fig. 10.6 Results from OptiPipe simulations with default parameter values (see Table 10.1) representing Scots pine in southern Finland. The four different curves are simulations with different values of maximum nutrient availability, U_{max}

function, we define a functional correspondence between the two models,

$$P = \sigma_{\mathrm{fM}} K_{\mathrm{f}} \frac{\mathbb{W}_{\mathrm{f}}}{\mathbb{W}_{\mathrm{f}} + K_{\mathrm{f}}} \equiv \mathbb{P}_0 \left[1 - \exp\left(-k a_{\mathrm{SL}} \mathbb{W}_{\mathrm{f}}\right)\right] = \mathbb{P}_0 f_{\mathrm{APAR}} \tag{10.1}$$

This can be achieved by defining a conversion parameter γ such that

$$\mathbb{P}_0 = \sigma_{\mathrm{fM}} \gamma^{-1}; \; f_{\mathrm{APAR}} = \frac{\gamma K_{\mathrm{f}} \mathbb{W}_{\mathrm{f}}}{\mathbb{W}_{\mathrm{f}} + K_{\mathrm{f}}} \tag{10.2}$$

Additionally, σ_{fM} depends on leaf nitrogen but $\mathbb{P}_0$ has been determined in PRELES for an average leaf N concentration. The parameters have therefore been calibrated for $[N] = 0.012\,kg\,N\,(kg\,DW)^{-1}$, which is a typical N concentration in pine needles in southern Finland (Mäkelä et al. 2008b). This allows us to estimate γ, K_f and σ_{fM0} that correspond with PRELES results at the annual scale.

Because temperature influences respiration and foliage turnover, these catabolic processes will also change under changing climate, likely counteracting the changes of photosynthesis. Growth is very sensitive to the ratio of photosynthesis to respiration and turnover, and grossly different growth projections can be obtained depending on what is assumed about this ratio. Here we assume that T_f and T_r decrease and r_m increases with increasing temperature sum as in Fig. 8.5.

A key parameter in the nitrogen uptake model of OptiPipe is the maximum rate of nitrogen uptake, $U_{max} = \sigma_{rM} K_r$. Nitrogen becomes available to plants when microbes decompose soil organic matter. The rate of the decomposition process depends on physical and chemical soil properties, the richness of soil fauna and microbial populations, and environmental factors such as soil temperature and moisture. Models of soil organic matter decomposition include equations describing the decay rates that can also be assumed to relate to the release rates of nutrients.

A proper analysis of climate impacts on nitrogen availability would require that the growth model is linked with a dynamic soil model predicting the N release rate when not only the environmental factors but also the nutrient uptake and the litter input by the vegetation are responding to climate change. Here we demonstrate the possible impacts of climate change on nitrogen availability by simply relating N availability to the environmentally controlled soil organic matter decomposition rate. Based on modelling studies (Comins and McMurtrie 1993) and empirical evidence (Högberg et al. 2017), this seems reasonable as a first approximation, because the total soil organic matter pool exposed to microbes is much larger than the pool available to vegetation, such that the possibly increasing rate of growth and litter fall will not have an additional influence on the availability of nutrients at least in the long term. However, we should bear in mind that advances in our understanding of root exudation and priming may render these assumptions outdated (Sulman et al. 2014; Högberg et al. 2017).

Here we estimate the decomposition rate using an environmental relationship estimated for the soil carbon model YASSO (Liski et al. 2005; Tuomi et al. 2009) which describes the dynamics of soil carbon in upland forest soils using a simple first order dynamic decay model with several carbon pools of different residence times. The parameters of the model have been estimated from a large world-wide data set, and their dependence on easily available climate variables has been estimated empirically. The model uses annual mean temperature and annual precipitation as proxies of soil temperature and moisture. The following equation has been found to apply to all decomposition rates:

$$f_T = e^{0.059T - 0.001T^2}\left(1 - e^{-1.858P}\right) \tag{10.3}$$

where T is mean annual temperature (°C) and P is annual precipitation (m).

We adopt this dependence to describe the change in the maximum N availability, $U_{\max}$. We postulate:

$$U_{\max} = U_{\max 0} \frac{f_T}{f_{T0}} \tag{10.4}$$

where $U_{\max 0}$ and f_{T0} are the variables in current climate and $U_{\max}$ and f_T the same in changed climate.

10.4.3 Sensitivity Screening of OptiPipe

Before considering actual climate change impacts, we first demonstrate the effects of the key parameters on stand growth by varying their values arbitrarily. We note that this will be done assuming a steady rather than transient environment, i.e., the parameter values will be changed but considered constant in the new environment. Because climate change is predicted to increase the potential photosynthetic rates in the boreal zone, we start by considering this change alone. A doubling of σ_{fM0} at 20% steps from its current value results in a steady increase in Optipipe wood production (MAI over 100-yr rotation) in all site classes, the increase being both absolutely and relatively larger, the higher the site fertility (Fig. 10.7). With default parameters, the high-fertility sites are carbon rather than nitrogen limited, so an increase in the potential photosynthesis rate reduces this limitation, leading to increased nitrogen uptake. At the same time, foliar nitrogen concentration decreases while the fine-root to leaf ratio increases. Carbon use efficiency (CUE) also increases slightly with increasing N availability as potential photosynthetic rate increases (Fig. 10.7).

What if the catabolic rates also increase along with increasing gross photosynthesis? We get an insight into this impact by co-varying the specific rate of maintenance respiration, r_{m}, with σ_{fM0}. Here we simply apply the same proportional changes in both. In this case, a small increase in the parameters leads to increased site productivity in all site fertility classes, but further increasing the parameters eventually results in a decline in wood production. Wood production peaks before the parameter values are doubled, however, the peak occurs the later, the more fertile the site (Fig. 10.7). The effects on leaf nitrogen content, nitrogen uptake rate, and fine-root to leaf ratio are in the same direction as in the previous case, however, the effect on CUE is the opposite as it declines slightly with increasing metabolic rates (Fig. 10.7).

Finally, we consider the possibility that nitrogen availability, $U_{\max}$, also increases at the same pace as the other parameters. In the simulations, wood production now increases considerably. A huge proportional increase is obtained for the low-productivity site where MAI increases more than six-fold. On the other hand, the higher productivity sites are not able to utilise the increase of potential productivity for increases larger than 50–60%, yielding a less than double maximum increase in

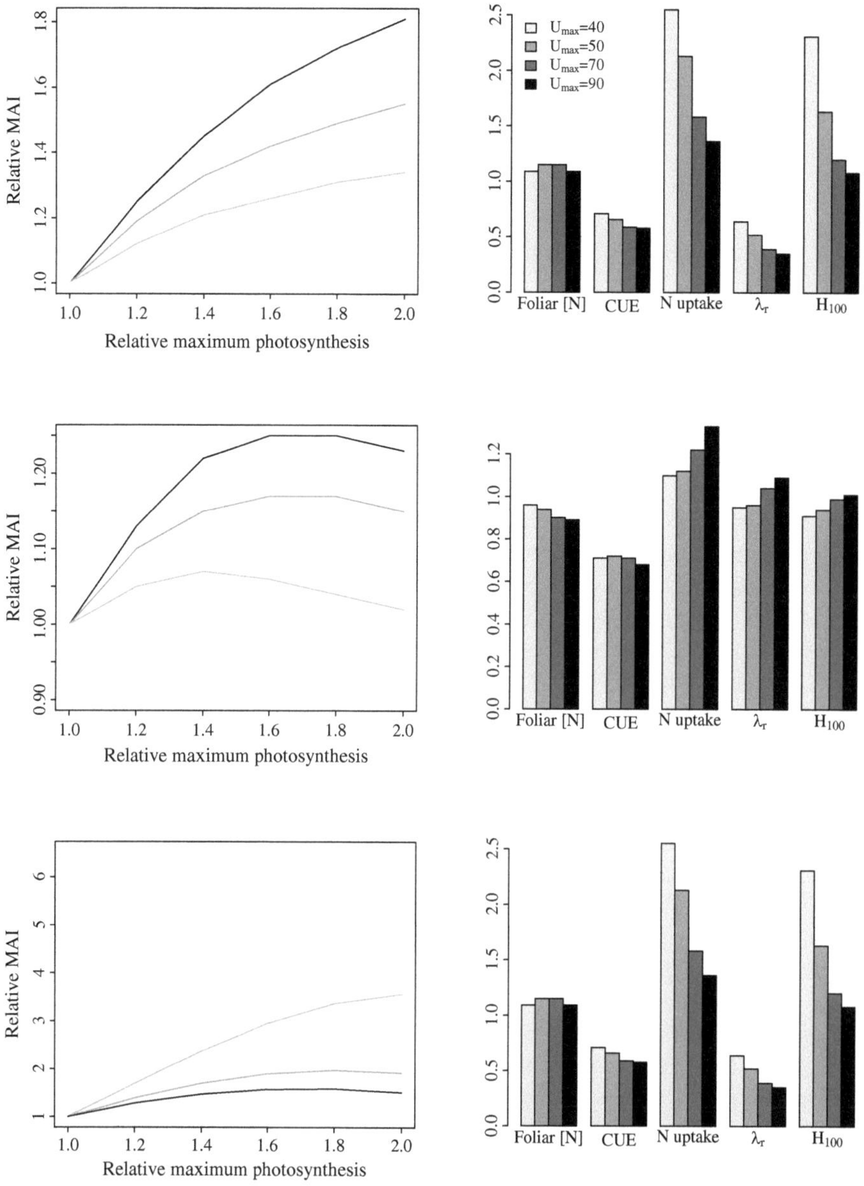

Fig. 10.7 Sensitivity screening of OptiPipe with different parameter values and combinations. Top row: Maximum leaf-specific photosynthetic rate (σ_{fM0}) increases. Middle row: in addition, specific rate of maintenance respiration (r_{m}) increases proportionally the same amount as σ_{fM0}, bottom row: in addition, maximum nutrient availability ($U_{\max}$) increases proportionally the same amount. The left panels show the changes in mean annual increment (MAI) over the 100-year rotation relative to the reference case shown in Fig. 10.6. The right panels show relative changes from reference to double σ_{fM0} in the temporal mean values of foliar N concentration, carbon use efficiency (CUE), N uptake, fine root allocation (λ_{r}) and mean height at 100 yrs (H_{100})

MAI (Fig. 10.7). The increased availability of nutrients makes all sites less nutrient-limited, increasing their leaf nitrogen content and decreasing the fine-root to leaf ratio, whereas CUE decreases considerably. Trees in the lowest fertility site, in particular, are able to benefit from the increased N availability, where the increase in nitrogen uptake is especially large (Fig. 10.7).

10.4.4 Analysis of Climate Change Impacts with OptiPipe

The sensitivity analysis with OptiPipe demonstrates that the simulated climate response depends not only on the alterations in individual process rates, but also on the changes of different rates relative to each other. In order to quantify these relative changes, we need to be able to quantify the environmental effects on those key processes. Here we report a calculation exercise where the five metabolic parameters of OptiPipe (σ_{fM0}, r_{m}, T_{f}, T_{r}, $U_{\max}$) were evaluated under environmental drivers derived from existing climate change scenarios, using the environmental response models of the metabolic parameters as defined in Sect. 8.5.

For climate variables we used SRES scenarios obtained from selected climate models that took part in the Coupled Model Intercomparison Project (CMIP) 3 (Meehl et al. 2007) but also compared these with the more recent representative concentration pathways (RCP) defined by CMIP.5 (Kalliokoski et al. 2018). The downscaling was done using the change factor approach where the baseline weather data came from the Finnish Meteorological Institute and covered the period 1971–2000. Climate projections were originally calculated for a regular 10 km x 10 km grid covering Finland (Kalliokoski et al. 2018). We selected a location in southern Finland to yield daily values of atmospheric CO_2 concentration, mean temperature, precipitation, global radiation and vapour pressure deficit for SRES scenarios B1, A1B and A2 for the period 2000–2100 (Kalliokoski et al. 2018). For the OptiPipe analysis, we chose a climate model that approximately represented an average of the ensemble with respect to both temperature and precipitation change. We evaluated the averages of the climate-dependent parameters of OptiPipe at 30-year intervals during the simulation period, i.e., at 2010, 2025, 2055 and 2085, for all scenarios. This yielded 12 parameter combinations, out of which only six were essentially different (Table 10.2). We simulated OptiPipe for 100 years with each of these parameter combinations, keeping the parameters constant for each simulation.

The relative change in the 100-year mean wood production was positive for all site fertilities for four out of the six cases simulated, whereas the two larger-change cases yielded growth losses for the poorest site type. The increase of production peaked in all site fertilities when σ_{fM0} had increased 30–50% from its current value. This indicates that the assumed increase of respiration and tissue turnover rates was not compensated by the increased availability of nitrogen. The maximum increase of volume growth saturated at about 25% for the most fertile growth sites (Fig. 10.8). These growth changes were accompanied by decreasing foliar nitrogen, CUE and fine-root to leaf ratio but increasing nitrogen uptake rate.

Table 10.2 Parameter values used in OptiPipe simulations under changed climate. CO_2 is in ppm and P_0 is $g\,C\,m^{-2}\,yr^{-1}$, other units as in Table 10.1. The parameter values are representative of climate variables under SRES scenarios B1, A1B and A2 under 30-year periods, the centre of which is indicated in the column headings (Kalliokoski et al. 2018). These scenarios are broadly similar to the following RCP scenarios: RCP 4.5 (B1), RCP 6.0 (A1B), and RCP 8.5 (A2)

	Initial	All	B1	A1B, A2	A1B	A2
Variable	2010	2025	2055	2055	2085	2085
CO_2	380	420	500	550	650	720
σ_{fM0}	5.3	6.2	7.0	7.4	8.3	8.9
r_m	24	28	31	33	37	40
T_f	3.685	3.302	3.069	2.919	2.668	2.529
T_r	1.25	1.12	1.04	0.99	0.90	0.86
f_T/f_{T0}	1	1.11	1.19	1.25	1.35	1.41

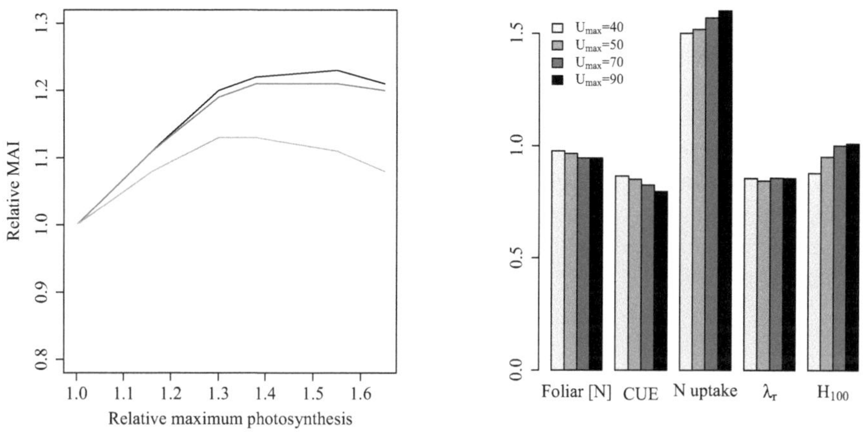

Fig. 10.8 Results of OptiPipe simulations with parameter values estimated using climate change scenarios. The left panel shows the changes in mean annual increment (MAI) over the 100-year rotation relative to the reference case shown in Fig. 10.6. The right panel shows relative changes from reference to maximum σ_{fM0} in the temporal mean values of foliar N concentration, carbon use efficiency (CUE), N uptake, fine root allocation (λ_r) and mean height at 100 yrs (H_{100})

10.4.5 Some Uncertainties in Analysing Climate Change Impacts

The above analyses are obviously largely stylised and qualitative, however, they demonstrate that combined effects of different processes may be complex and unexpected. The overall result that even large increases in primary productivity may not lead to increased volume growth is in line with empirical evidence from Free Air Carbon Enrichment (FACE) experiments indicating that nitrogen availability may severely limit growth (Hyvönen et al. 2007; Norby and Zak 2011). In our example the increased rate of maintenance respiration and tissue turnover also influenced volume production under climate change. Although there is general consensus that increasing temperatures will lead to increases in these catabolic rates, we may have overestimated the temperature effects by simply making a space-for-time substitution, i.e., assuming that the regional variation of these processes can be interpreted to be climate-related and therefore to be transferrable to temporal climate change in one location.

In our simulations the increase of nitrogen availability due to increased decomposition rates was not sufficient to keep up with the increased photosynthesis rate and therefore implied increased N limitation. The sensitivity screening showed that if N availability increased in pace with potential photosynthesis, this could enhance growth significantly. Already some early modelling studies (Comins and McMurtrie 1993) indicated that if soil decomposition processes are modelled using simple first-order decay processes—as is the case in the majority of existing soil models—climate change will lead to nitrogen deficits. More recent empirical and theoretical developments include the idea that root exudates excreted by plants may serve as additional fuel for microbes to enhance the rate of decomposition, so if excess sugars become available due to increased ambient CO_2, this could increase the amount of exudates and thus the release rate of nutrients for plant growth (e.g., Högberg et al. 2017). The argument is akin to the idea of trees allocating more carbon to fine roots under conditions of high CO_2, and low N availability, such as incorporated in OptiPipe. However, the details of plant-symbiont interactions and their effects on growth potential under climate change remain to be revealed and put into fully functional models.

10.5 Quo Vadis?

In this book we have presented an approach to forest modelling that derives from ecological theory and research findings, yet aims for models that are sufficiently simple in structure to lend themselves readily to practical application. A key for keeping the approach simple was, in the spirit of hierarchy theory, to confine the model to a selected level of resolution—individual trees over their lifetime at an annual resolution—while treating higher-resolution phenomena as effective means from external (or linked) modules, and allowing for constraints from lower-resolution phenomena. An important type of constraint was one derived using eco-evolutionary optimisation.

It was important for the applicability of the models that we were able to bridge empirical and process-based data assimilation methods that combine research-based ecological measurements with standard forestry data. The main significance of the standard forestry data is its abundance both spatially and temporally: most countries have collected standard forestry measurements for decades or centuries in growth and yield experiments and national forest inventories. However, this situation is now changing rapidly with the advent of air-borne measurement methods such as LiDAR, drones, and different varieties of satellite sensors. These technologies provide data on the distributions of leaf area, canopy height, and other structural measures at a very high spatial resolution, while routine measurements such as diameter continue, though perhaps with less sampling intensity. Process-based models are able to utilise these data much more readily than traditional empirical models (e.g., Härkönen et al. 2013). This paves the way for versatile airborne monitoring of forests, not only their states, but also their growth, yield, and carbon balances.

Here we have presented the *status quo* of the basic theory and methods of our approach, but of course new information is constantly emerging that requires us to adjust and improve the models. The biggest challenge for any forest model currently is to make reliable predictions under climate change. Although a process-based model should, in principle, be more readily able to do this than an empirical model, we may have overlooked some processes that are not essential under current climate but which may become crucial when the climate changes. Two major phenomena that deserve some attention here are: (1) changes in sink-source balance due to changes in the annual cycle, and (2) changes in the C:N balance due to different above-ground and below-ground responses to environmental change. Below we review the issues and consider methods for incorporating their implications in the type of models presented in this book.

10.5.1 Emerging Eco-Physiological Issues

Sink-source balance Growth models based on the carbon balance are supply-limited as a rule, i.e., it is the balance of photosynthesis and respiration that determines the overall level of growth. In most models, until very recently, this has been taken to mean that all net assimilated carbon is used in growth on an annual basis, which is equivalent to the implicit assumption that the levels of stored carbon do not vary significantly between years. While this assumption seems to comply with the variability of mean growth rates observed between comparable stands (Grace et al. 1987), regionally (Coops et al. 1998; Mäkelä et al. 2016; Härkönen et al. 2019; Minunno et al. 2019), or even globally (Exbrayat et al. 2018), it does not so readily apply to the year-to-year variability observed in stand growth, which seems to be more closely related to temperature and precipitation than photosynthesis (Henttonen et al. 2014). This has been taken as evidence for so-called sink limitiation, i.e., that drivers actually controlling the rate of structural growth are more significant for growth than the level of net photosynthesis (Körner 2003).

If sink limitation really was decisive for determining long-term total growth, then in order to maintain mass balance, there would have to be mechanisms that lead to either reduced photosynthesis or increased respiration under conditions reducing growth. The observed disparity between year-to-year variabilities of growth and photosynthesis does not support this proposition. It seems more reasonable to assume that sink limitation is only important in the short term and is buffered by the variability in the non-structural carbon (NSC) storage (Sala et al. 2012). However, a trend-like change in the environment may lead to situations where growth could not keep up with the increasing potential rates of photosynthesis because, for example, they could not extend the growing season indefinitely, such that some down-regulating mechanisms not included in current models would become important. This would call for an explicit inclusion of the NSC storage as a state variable, as has been done in recent advances in eco-physiological modelling of the sink-source relationships (Schiestl-Aalto et al. 2015, 2019; Guillemot et al. 2015, 2017).

C:N balance We already touched upon the possible changes in the C:N balance due to different above-ground and below-ground responses to environmental change in Sect. 10.4 (Bond-Lamberty et al. 2018). The key issue here is that, while the response of photosynthesis to environmental drivers is quite well understood, we know much less about the response of forest soils and nutrient availability under a changing climate. Nutrient availability is not reacting to immediate environmental variables only, but is a dynamic system itself, dependent on its history and changing with inertia. Importantly, soils develop in interaction with the plants growing on them, and not all of these interactions are well understood under current conditions, let alone under changing climate (Doetterl et al. 2018).

The major interactions between soils and vegetation include nutrient uptake from soils to plants and litter fall from plants to soils. In addition, the role of root exudates, i.e., carbon compounds excreted by the roots directly to the soil or to symbionts (Näsholm et al. 1998; Schimel and Weintraub 2003; Fontaine et al. 2004), has been gaining more attention recently (Stocker et al. 2016; Terrer et al. 2016; Högberg et al. 2017). That litter fall virtually returns the nutrients taken up by the plants, but after a delay, also puts down the photosynthesized carbon not respired previously by the plants. The litter carbon fuels soil microbes that decompose plant material, releasing both nutrients to the soil and carbon to the atmosphere. The carbon in the exudates is much more readily available to microbes and is therefore assumed to accelerate (or *prime*) their functioning and hence nutrient release (Näsholm et al. 1998; Fontaine et al. 2004; Högberg et al. 2017). Without going into more detail we merely state that the big question with relation to climate change is: Is the increased uptake of carbon by trees going to increase or reduce the microbial activity that releases nutrients for the trees (Näsholm et al. 2013; Högberg et al. 2017)?

10.5.2 Trends in Mainstream Methods

It is the sound ecological basis of the models that should make it possible to build on them to advance the approach as new research findings become available. However, accounting for new processes generally seems to create a need for more complex models with more state variables and higher temporal and spatial resolution. Following the spirit of this book, we must bear in mind that our focus is not on the new processes as such, but on their implications on stand productivity in the long term. The key question is then, how do we improve the biological realism of the models without making them more complex than is absolutely necessary? Here we have especially discussed applications of hierarchy theory and eco-evolutionary optimisation as methods of streamlining and simplifying models. The need for overall organising principles that help simplify models yet keep them biologically meaningful has recently been urged among the dynamic vegetation modelling community, as a means for achieving more robust and more easily quantifiable models (Franklin et al. 2020). Other organising principles that are now emerging and seemingly on their way to mainstream vegetation modelling (Franklin et al. 2020) include perfect aggregation (Sect. 6.6.2), the perfect plasticity approximation (Sect. 6.6.3), and maximum entropy production (Dewar et al. 2010).

How do we employ these organising principles in practise? To highlight this, we consider the problem of incorporating the implications of the above emerging issues into our long-term, tree-level approach.

A hierarchical approach would seem useful for linking within-season and long-term processes. A within-season carbon-balance model (e.g., Schiestl-Aalto et al. 2015, 2019; Guillemot et al. 2015, 2017) could serve as a means for iterative refinement of time-dependent parameters in the core model, allowing for temporary deviations from structural constraints due to within-season sink limitation and variable NSC concentration. The constraints could still be valid on average in the longer term, which could be achieved, for example, by treating them as set values—possibly environment dependent—of an acclimation process (Mäkelä 1999). Alternatively, an optimum linkage between the two hierarchical levels could be derived. The significance of the hierarchical approach is that, while allowing for deviations, the higher level constrains the system to follow a robust pattern, which could be lost due to model complexity and parameter ambiguity in a fully bottom-up approach.

Regarding the C:N interactions in plants and soils, we have already demonstrated the applicability of the eco-evolutionary approach for co-allocation of carbon and nitrogen in trees (Sect. 7.5.5). An important addition to the approach presented here is the inclusion of fungal symbiosis. It has already been considered in an optimisation framework, where the plant can allocate carbon either to fine roots or fungal symbionts depending on the cost-effectiveness of the respective nutrient uptake (Thornley and Parsons 2014; Franklin et al. 2014). Furthermore, Franklin et al. (2014) considered the symbiosis using a game theory approach to describe the competition between fungal species. However, all these models considered nitrogen availability as a simple site-specific input parameter (Mäkelä et al. 2008b; Valentine and Mäkelä 2012; Thornley and Parsons 2014; Franklin et al. 2014). It would seem crucial for advancements in modelling C:N interactions in plants and soils that soil models were developed that explicitly include microbial functions in relation to litter decomposition and nutrient release (Meyer et al. 2010).

10.5.3 Trends in Application

Tree growth is but a part of whole ecosystem functioning. Tree and stand models are applicable in integrated modelling systems that combine ecosystem processes, with the objective of drawing light to pressing issues, such as ecosystem carbon sequestration, biodiversity development, and economic implications of different forest management systems and policies. The forest models for such applications need to be both environment and management sensitive, and they must be sufficiently simple to allow for efficient large-scale analysis. The models presented here have already been linked with forest soil carbon models (Holmberg et al. 2019), and are being adapted to analyses of biodiversity and nutrient leakage from watersheds. At the same time, model development towards more reliable responses under climate change is underway. In all this work, we continue to rely on hierarchical treatment of the systems, application of eco-evolutionary principles, and efficient model-data fusion anchoring the theories to empirical evidence.

Solutions to Exercises

Exercises of Chapter 1

1.2 Similar to tree growth, the change in total carbon content of the ecosystem can be written as

$$\text{change} = \text{gain} - \text{loss}$$

where gain is gross primary production (GPP) and loss is ecosystem respiration:

$$\text{change} = \text{GPP} - \text{autotrophic respiration} - \text{heterotrophic respiration}$$

If we denote total ecosystem carbon content by $\mathbb{W}_\mathrm{C}$, the above can be written as the following differential equation:

$$\frac{\mathrm{d}\mathbb{W}_\mathrm{C}}{\mathrm{d}t} = \mathbb{P} - \mathbb{R}_\mathrm{A} - \mathbb{R}_\mathrm{H}$$

where $\mathbb{R}_\mathrm{A}$ and $\mathbb{R}_\mathrm{H}$ denote autotrophic and heterotrophic respiration, respectively.

1.3 Here we define the rate of change of soil carbon content, denoted by $\mathbb{W}_\mathrm{S}$, based on the assumptions given. The input of carbon to the soil comprises senescent material from the tree stand, which have been assumed to be proportional to the foliage, fine root and wood components of the stand (see the McMurtrie and Wolf model in Sect. 1.10.1). The outflow of carbon consists of heterotrophic respiration, which we have assumed to be proportional to the soil carbon content. Thus,

$$\frac{\mathrm{d}\mathbb{W}_\mathrm{S}}{\mathrm{d}t} = [\mathrm{C}][s_\mathrm{f}\mathbb{W}_\mathrm{f} + s_\mathrm{r}\mathbb{W}_\mathrm{r} + s_\mathrm{w}\mathbb{W}_\mathrm{w}] - r_h\mathbb{W}_\mathrm{S}$$

A. Mäkelä, H. T. Valentine, *Models of Tree and Stand Dynamics*,
https://doi.org/10.1007/978-3-030-35761-0

where [C] is the carbon content of dry matter and r_h is the soil specific heterotrophic respiration rate. – Soil models used in applications employ a similar linear dependence of heterotroph respiration of soil carbon, but soil carbon is usually divided into several fractions with variable characteristic respiration rates (e.g., Liski et al. 2005).

Exercises of Chapter 2

2.1 We start with Eq. (2.6),

$$\frac{\mathrm{d}W}{\mathrm{d}t} = \mu_0 W \exp(-at)$$

and rearrange this so as to separate the variables W and t on different sides of the equation, then integrate:

$$\int_W^{W_{\max}} \frac{\mathrm{d}W}{W} = \int_t^{\infty} \mu_0 \exp(-at)\mathrm{d}t$$

In order to introduce $W_{\max}$ into the equation, we have constrained the above integral at the end points, where time goes to infinity and W approaches $W_{\max}$. Carrying out this integration yields

$$\ln(W_{\max}) - \ln(W) = -\frac{\mu_0}{a}[\exp(-\infty) - \exp(-at)] = \frac{1}{a}\mu_0 \exp(-at)$$

Rearranging the left-hand side and combining with Eq. (2.6) yields

$$\ln\left(\frac{W_{\max}}{W}\right) = \frac{1}{aW}\frac{\mathrm{d}W}{\mathrm{d}t}$$

which can be expressed as Eq. (2.7):

$$\frac{\mathrm{d}W}{\mathrm{d}t} = aW \ln\left(\frac{W_{\max}}{W}\right)$$

2.2 We start with Eq. (2.20):

$$\frac{\mathrm{d}W}{\mathrm{d}t} = \mu W (W_{\max} - W)$$

We shall derive Eq. (2.21) from the above. For that, let us consider the transformation of variables $z = 1/W$. This implies

$$\frac{dW}{dt} = \frac{d}{dt}\frac{1}{z} = -\frac{1}{z^2}\frac{dz}{dt}$$

Inserting the transformation into (2.20) gives

$$-\frac{1}{z^2}\frac{dz}{dt} = \mu\frac{1}{z}\left(\frac{1}{z_{max}} - \frac{1}{z}\right)$$

where $z_{max} = \frac{1}{W_{max}}$. Multiply both sides with $-z^2$:

$$\frac{dz}{dt} = \mu\left(1 - \frac{z}{z_{max}}\right) = \mu\left(\frac{z_{max} - z}{z_{max}}\right)$$

Inserting the original variable W back we have

$$\frac{dW^{-1}}{dt} = \frac{\mu}{z_{max}}(z_{max} - z) = a\left(W_{max}^{-1} - W^{-1}\right)$$

where we have used the definition $a = \mu W_{max}$ given in the text. This completes the derivation.

2.3 The impact of parameters on the solution can be studied by calculating H_t for a series of times t with any programming or calculation tool, such as Excel or R. Once a starting value H_0 is given, the next value can be calculated iteratively from the previous one. The impacts of parameters can be seen by plotting the result against time.

Doing this, we find that varying parameter a has an impact on the maximum growth rate and its time of occurrence (top plots of Fig. S.1), while parameter c affects the distribution of growth rate over time. Smaller values of c lead to larger peaks of growth in the beginning combined with a faster slow-down subsequently (middle plots of Fig. S.1), but the timing of the peak is unaffected. Changing H_{max} influences the asymptotic height but also has a small impact on the timing of maximum height growth (bottom plots of Fig. S.1).

While the model can obviously be fitted to time-series or chronosequence data using nonlinear regression, information about the qualitative impact of parameters on model predictions is helpful for an initial assessment of parameter values. Because models may have multiple parameter sets that provide an equally good fit to data, a qualitative understanding of the impact of parameter values on the solution is important for constraining the feasible set of parameter values.

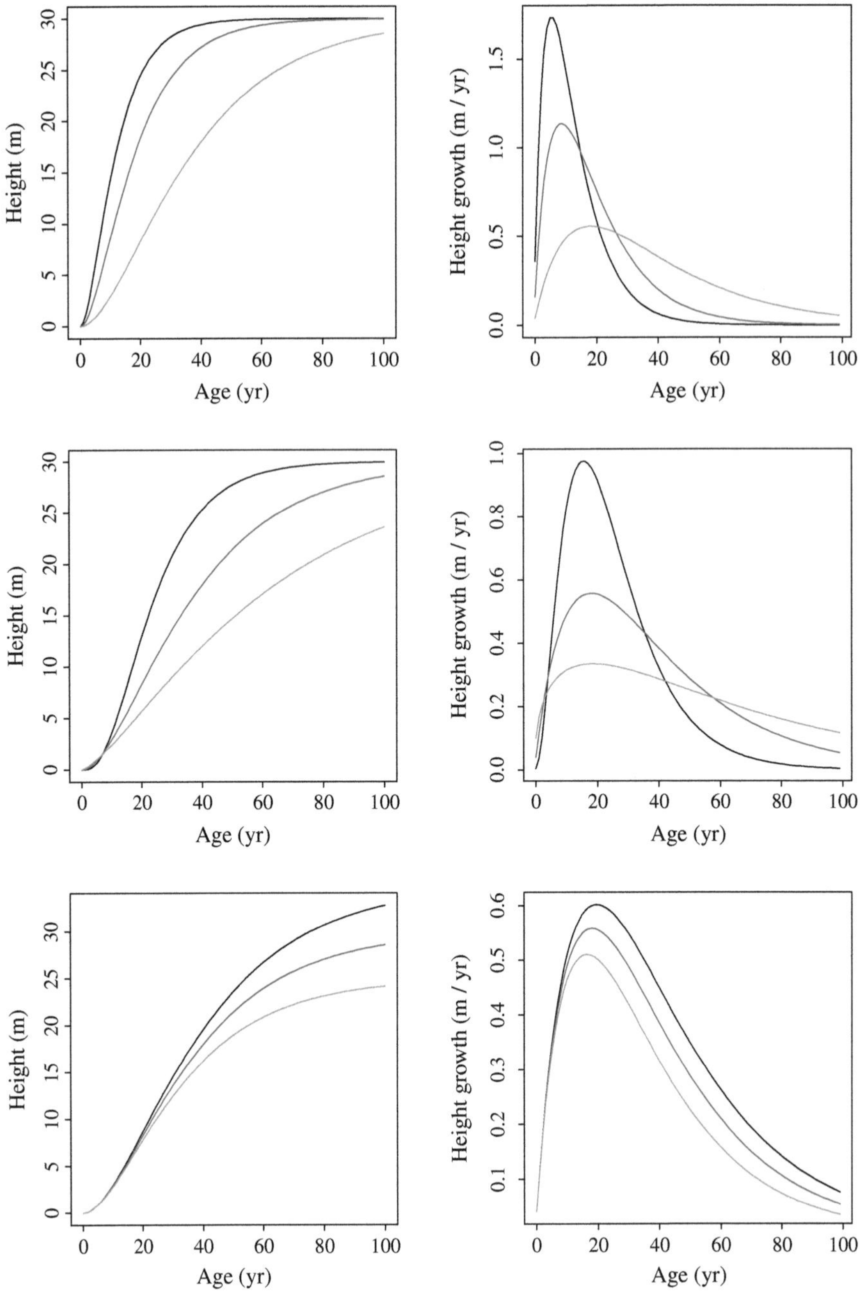

Fig. S.1 Height and height growth according to the Bertalanffy model. Top: parameter a varies, $c = 0.5$, $H_{max} = 30\,\text{m}$. Black line, $a = 0.6$, dark grey line, $a = 0.4$, light grey line, $a = 0.2$. Middle: parameter c varies, $a = 0.2$, $H_{max} = 30\,\text{m}$. Black line, $c = 0.3$, dark grey line, $c = 0.5$, light grey line, $c = 0.7$. Bottom: parameter H_{max} varies, $a = 0.2$, $c = 0.5$. Black line, $H_{max} = 35\,\text{m}$, dark grey line, $H_{max} = 30\,\text{m}$, light grey line, $H_{max} = 25\,\text{m}$

Exercises of Chapter 3

3.2 Using the parameter values of Table 3.1, we first evaluate the slope of the right-hand side of (3.21):

$$\frac{s_f}{\lambda_f}\left[\frac{r_f\lambda_f}{s_f}+\frac{r_r\lambda_r}{s_r}+\frac{r_w\lambda_w}{s_w}+\frac{1}{Y_G}\right]\mathbb{W}_f$$
$$=\frac{0.3}{0.2}\left[\frac{0.2 \times 0.2}{0.3}+\frac{0.2 \times 0.15}{0.9}+\frac{0.01 \times 0.65}{0.05}+\frac{1}{1.3}\right]\mathbb{W}_f$$
$$= 1.599\mathbb{W}_f$$

Similarly, the left-hand-side of the equation is

$$\mathbb{P}_0[1 - \exp(-ka_{SL}\mathbb{W}_f)] = 1.2 \times [1 - \exp(-2.8\ \mathbb{W}_f)]$$

Secondly, we consider increasing $\mathbb{P}_0$ to 1.4 and increasing all turnover rates by 20%. This gives 1.829 for the slope of the straight line.

The equations can be drawn either by hand or by using any appropriate programming tool, such as Excel or R (Fig. S.2). The equilibrium foliage biomass is found at the intersection of the straight line and the exponential. Increasing the slope, such as we did by increasing the turnover rates, will decrease the equilibrium foliage mass, whereas increasing photosynthetic productivity (i.e., the level of the exponential function) will increase the equilibrium.

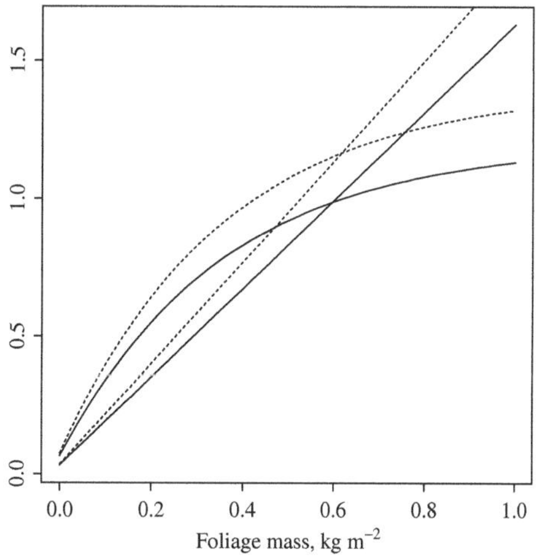

Fig. S.2 The component curves of Eq. 3.21. Solid lines with default parameters as in Table 3.1, dashed lines with turnover rates increased by 20% and $\mathbb{P}_0$ increased from 1.2 to 1.4

3.3 Insert Eqs. (3.20) into (3.11) for leaf mass, i.e., $i = \mathrm{f}$:

$$\frac{\mathrm{d}W_\mathrm{f}}{\mathrm{d}t} = \lambda_\mathrm{f} Y_\mathrm{G}\left[\sigma_\mathrm{f} W_\mathrm{f} - r_\mathrm{f} W_\mathrm{f} - r_\mathrm{r}\frac{\lambda_\mathrm{r} s_\mathrm{f}}{\lambda_\mathrm{f} s_\mathrm{r}} W_\mathrm{f} - r_\mathrm{w}\frac{\lambda_\mathrm{w} s_\mathrm{f}}{\lambda_\mathrm{f} s_\mathrm{w}} W_\mathrm{f}\right] - s_\mathrm{f} W_\mathrm{f}$$

Then insert (3.18) into the above and equate the rate of change to zero to determine the equilibrium (steady state) foliage:

$$0 = \lambda_\mathrm{f} Y_\mathrm{G}\left[P_0[1-\exp(-ka_\mathrm{SL} W_\mathrm{f})] - r_\mathrm{f} W_\mathrm{f} - r_\mathrm{r}\frac{\lambda_\mathrm{r} s_\mathrm{f}}{\lambda_\mathrm{f} s_\mathrm{r}} W_\mathrm{f} - r_\mathrm{w}\frac{\lambda_\mathrm{w} s_\mathrm{f}}{\lambda_\mathrm{f} s_\mathrm{w}} W_\mathrm{f}\right] - s_\mathrm{f} W_\mathrm{f}$$

Rearranging this yields:

$$P_0[1-\exp(-ka_\mathrm{SL} W_\mathrm{f})] = \frac{s_\mathrm{f}}{\lambda_\mathrm{f}}\left[r_\mathrm{f}\frac{\lambda_\mathrm{f}}{s_\mathrm{f}} + r_\mathrm{r}\frac{\lambda_\mathrm{r}}{s_\mathrm{r}} + r_\mathrm{w}\frac{\lambda_\mathrm{w}}{s_\mathrm{w}} + \frac{1}{Y_\mathrm{G}}\right] W_\mathrm{f}$$

This is (3.21).

To derive (3.22), we differentiate (3.17) with respect to W_f:

$$\frac{\mathrm{d}}{\mathrm{d}W_\mathrm{f}}\left(P_0[1-\exp(-ka_\mathrm{SL} W_\mathrm{f})]\right) = P_0[-(-ka_\mathrm{SL})\exp(-ka_\mathrm{SL} W_\mathrm{f})]$$

When W_f approaches zero, the exponential approaches 1, such that the tangent at zero is a straight line with slope $P_0 k a_\mathrm{SL}$. The slope of the straight line has to be smaller than this tangent in order for the two curves to intersect for positive values of W_f. Thus, we require that

$$P_0 k a_\mathrm{SL} W_\mathrm{f} \leq \frac{s_\mathrm{f}}{\lambda_\mathrm{f}}\left[r_\mathrm{f}\frac{\lambda_\mathrm{f}}{s_\mathrm{f}} + r_\mathrm{r}\frac{\lambda_\mathrm{r}}{s_\mathrm{r}} + r_\mathrm{w}\frac{\lambda_\mathrm{w}}{s_\mathrm{w}} + \frac{1}{Y_\mathrm{G}}\right] W_\mathrm{f}$$

This is (3.22).

3.5 From the theory of linear differential equations (e.g., Luenberger 1979) we know that the solution can be expressed in terms of eigenvectors and eigenvalues of the transition matrix and initial values of the state vector. Consider differential equation (3.12):

$$\frac{\mathrm{d}}{\mathrm{d}t}\begin{bmatrix} W_\mathrm{f} \\ W_\mathrm{r} \\ W_\mathrm{w} \end{bmatrix} = \mathbf{A}\begin{bmatrix} W_\mathrm{f} \\ W_\mathrm{r} \\ W_\mathrm{w} \end{bmatrix}$$

Denote the eigenvectors of matrix $\mathbf{A}$ as $\mathbf{e}_i = [e_{1i}\ e_{2i}\ e_{3i}]^T$ and the corresponding eigenvalues by ν_i. The solution can then be written as follows, by component:

$$W_i(t) = z_1(0)e_{1i}\exp(\nu_1 t) + z_2(0)e_{3i}\exp(\nu_2 t) + z_3(0)e_{3i}\exp(\nu_3 t)$$

The eigenvalues and eigenvectors are found by solving the following eigenvalue equation:

$$\mathbf{Ae} = \nu\mathbf{e}$$

In this case, this equation yields

$$\mathbf{A} = \begin{bmatrix} (Y_G\lambda_f(\sigma_f - r_{M_f}) - s_f)e_1 - Y_G\lambda_f r_{M_r} e_2 - Y_G\lambda_f r_{M_w} e_3 \\ Y_G\lambda_r(\sigma_f - r_{M_f})e_1 - (Y_G\lambda_r r_{M_r} + s_r)e_2 - Y_G\lambda_r r_{M_w} e_3 \\ Y_G\lambda_w(\sigma_f - r_{M_f})e_1 - Y_G\lambda_w r_{M_r} e_2 - (Y_G\lambda_w r_{M_w} + s_w)e_3 \end{bmatrix} = \begin{bmatrix} \nu e_1 \\ \nu e_2 \\ \nu e_3 \end{bmatrix}$$

This is essentially a set of three equations with four unknowns, i.e., the three eigenvector components and the eigenvalue. Because the eigenvectors are defined as directions in vector space, we can choose their length by fixing one of the components e_i, and the rest of them are defined in proportion to this fixed component. Ordinarily, equations like this can be solved with numerical methods, such that are available, for example, in R or other numerical packages. However, here we solve the equations "by hand".

If we rearrange the components of the vectors by moving the $s_i e_i$ components to the right-hand side we have

$$\mathbf{A} = \begin{bmatrix} (Y_G\lambda_f(\sigma_f - r_{M_f}))e_1 - Y_G\lambda_f r_{M_r} e_2 - Y_G\lambda_f r_{M_w} e_3 \\ Y_G\lambda_r(\sigma_f - r_{M_f})e_1 - Y_G\lambda_r r_{M_r} e_2 - Y_G\lambda_r r_{M_w} e_3 \\ Y_G\lambda_w(\sigma_f - r_{M_f})e_1 - Y_G\lambda_w r_{M_r} e_2 - Y_G\lambda_w r_{M_w} e_3 \end{bmatrix} = \begin{bmatrix} (\nu + s_f)e_1 \\ (\nu + s_r)e_2 \\ (\nu + s_w)e_3 \end{bmatrix}$$

We now observe that if we multiply the top equation by λ_r and the second equation by λ_f the left-hand sides of the equations are equal. Subtracting the second equation from the first then results in

$$0 = \lambda_r(\nu + s_f)e_1 - \lambda_f(\nu + s_r)e_2$$

such that e_2 can be expressed in terms of e_1:

$$e_2 = \frac{\lambda_r(\nu + s_f)}{\lambda_f(\nu + s_r)} e_1$$

Using the same method for the first and third equation we can similarly solve for e_3:

$$e_3 = \frac{\lambda_w(\nu + s_f)}{\lambda_f(\nu + s_w)} e_1$$

Inserting these into the first equation and rearranging yields the following third order equation for the eigenvalue ν:

$$
\begin{aligned}
\nu^3 = \nu^2 & \left[Y_G \lambda_f (\sigma_f - r_{M_f}) - Y_G \lambda_r r_{M_r} - Y_G \lambda_w r_{M_w} - (s_f + s_r + s_w) \right] \\
& + \nu Y_G \lambda_f (\sigma_f - r_{M_f})(s_r + s_w) \\
& - \nu \left[Y_G \lambda_r r_{M_r} (s_f + s_w) + Y_G \lambda_w r_{M_w} (s_f + s_r) + (s_f s_r + s_f s_w + s_r s_w) \right] \\
& + \left[Y_G \lambda_f (\sigma_f - r_{M_f}) s_r s_w - Y_G \lambda_r r_{M_r} s_f s_w - Y_G \lambda_w r_{M_w} s_f s_r - (s_f s_r s_w) \right]
\end{aligned}
$$

Again, numerical methods exist for solving the above equation. Here we just work out the left-hand side and right-hand side separately to find the eigenvalues graphically (Fig. S.3). From the figure we conclude that the equation has three roots, $\nu_1 = 0.1314$, $\nu_2 = -0.3867$, and $\nu_3 = -0.923$. Inserting these values for the eigenvector components, we find that there are three eigenvectors, each

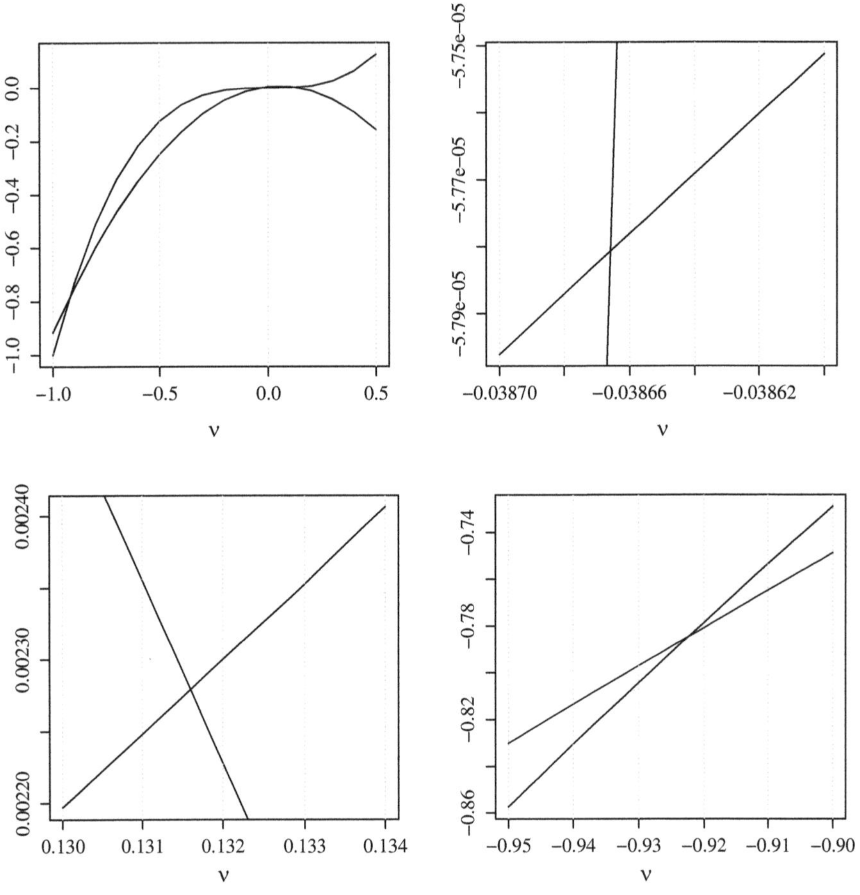

Fig. S.3 Graphical solution of the eigenvalue equation of the differential equation (3.12). Top left: ν^3 and the second-order right-hand-side of the eigenvalue equation showing the three roots of the equation. Bottom left: zooming in to the largest eigenvalue, i.e., the largest value of ν of the three intersection points of the two curves. Top right: zooming in to the middle eigenvalue. Bottom right: zooming in to the smallest eigenvalue corresponding to one of the eigenvalues:

$$\mathbf{e}_1 = \begin{bmatrix} 1 \\ 0.31 \\ 7.73 \end{bmatrix} \mathbf{e}_2 = \begin{bmatrix} 1 \\ -0.13 \\ 0.84 \end{bmatrix} \mathbf{e}_3 = \begin{bmatrix} 1 \\ 20.3 \\ 2.32 \end{bmatrix}$$

Because the largest eigenvector, i.e., the dominant eigenvector, is positive, the system shows exponential growth, approaching the direction of the dominant eigenvector $\mathbf{e}_1$ as time increases. The components of $\mathbf{e}_1$ therefore represent the relative proportions of foliage, fine roots and wood under exponential growth. The role of the rest of the eigenvectors reduces with time as the respective eigenvectors are negative.

At the initial state $t = 0$ and the exponential terms of the solution are $\exp(\nu_i t) = 1$, so the initial state can be expressed as a linear combination of the eigenvectors. If we know the initial state, we can solve for the coefficients $z_i(0)$. Here we simply assume that $z_1(0) = z_2(0) = 0.1$, $z_3(0) = 0.01$. The solution is then (approximately):

$$\begin{bmatrix} W_\mathrm{f} \\ W_\mathrm{r} \\ W_\mathrm{w} \end{bmatrix} = 0.1 \begin{bmatrix} 1 \\ 0.31 \\ 7.73 \end{bmatrix} \exp(0.1314t)$$

$$+0.1 \begin{bmatrix} 1 \\ -0.13 \\ 0.84 \end{bmatrix} \exp(-0.3867t) + 0.01 \begin{bmatrix} 1 \\ 20.3 \\ 2.32 \end{bmatrix} \exp(-0.923t)$$

Exercises of Chapter 4

4.1 (a) Volume = $(10\,\mathrm{cm})^3 = 1000\,\mathrm{cm}^3$, mass = $400\,\mathrm{kg\,m}^{-3} \times 1000\,\mathrm{cm}^3 = 0.4\,\mathrm{kg}$, area = $6 \times (10\,\mathrm{cm})^2 = 600\,\mathrm{cm}^2$. (b) The volume, mass and surface area of the second cube are $(30\,\mathrm{cm})^3 = 27{,}000\,\mathrm{cm}^3$, $400\,\mathrm{kg\,m}^{-3} \times 27{,}000\,\mathrm{cm}^3 = 10.8\,\mathrm{kg}$, and $6 \times (30\,\mathrm{cm})^2 = 5400\,\mathrm{cm}^2$, respectively, so the ratios of these to the first cube are 27, 27 and 9, respectively. (c) Denote the edge length of a cube by d and another cube by kd. The volume ratio is $(kd)^3/d^3 = k^3$, the mass ratio is $(\rho kd)^3/(\rho d)^3 = k^3$, the surface area ratio is $6(kd)^2/6d^2 = k^2$.

4.2 The new mass is (a) $2^3 \times 2\,\mathrm{kg} = 16\,\mathrm{kg}$, (b) $2^2 \times 2\,\mathrm{kg} = 8\,\mathrm{kg}$, (c) $2^{2.4} \times 2\,\mathrm{kg} = 10.6\,\mathrm{kg}$.

4.3 When the tree grows, both its crown volume and crown surface area increase. If we assume that crown shape is not changing during growth, then crown volume is proportional to, say, crown length cubed, i.e., L_c^3, and crown surface area is proportional to crown length squared, L_c^2. According to Exercise 4.2c, foliage mass is proportional to $L_c^{2.4}$. We therefore conclude that (a) the ratio of foliage mass to crown volume, $L_c^{2.4-3} = L_c^{-0.6}$, declines as the crown increases, and (b) the ratio of foliage mass to crown surface area, $L_c^{2.4-2} = L_c^{0.4}$, increases at the same time.

4.4 For maximising light capture, an optimal crown would place all its foliage on the surface. This is approximately the case in some broadleaves, such as poplars. Particularly in evergreen species the increase of the crown requires that some foliage is left in branches shorter than the newly growing ones, thus partly filling the crown volume.

4.5 To derive Eq. (4.2), we start from the definition of allometry in (4.1): $y = kx^a$. Taking the time derivatives of both sides we have

$$\frac{\mathrm{d}y}{\mathrm{d}t} = akx^{a-1}\frac{\mathrm{d}x}{\mathrm{d}t}$$

Now divide both sides with y, using (4.1) for the right-hand side, and elaborate:

$$\frac{1}{y}\frac{\mathrm{d}y}{\mathrm{d}t} = ak\frac{1}{kx^a}x^{a-1}\frac{\mathrm{d}x}{\mathrm{d}t} = a\frac{1}{x}\frac{\mathrm{d}x}{\mathrm{d}t}$$

This is Eq. (4.2).

4.6 We use Eq. (4.21) to estimate active pipe area and then (4.10) to derive foliage mass. First, $B = \pi(10/2)^2 = 78.5\,\mathrm{cm}^2$. The crown base area for the first tree is thus $A_c = 0.8 \times 78.5\,\mathrm{cm}^2 = 62.8\,\mathrm{cm}^2$. This corresponds to foliage mass $W_f = 500\,\mathrm{kg\,m^{-2}} \times 62.8\,\mathrm{cm}^2 = 3.14\,\mathrm{kg}$. For the second tree, the crown base area is $A_c = 0.25 \times 78.5\,\mathrm{cm}^2 = 19.6\,\mathrm{cm}^2$, corresponding to foliage mass $W_f = 500\,\mathrm{kg\,m^{-2}} \times 19.6\,\mathrm{cm}^2 = 0.98\,\mathrm{kg}$.

4.7 We use Eq. 4.26 to calculate the stem mass. For the first tree this becomes

$$W_{\mathrm{stot},1} = 370 \times 5 \times 0.00785 \times [0.53(1 - 0.8^2) + 0.4 \times 0.8^2] = 6.49\,\mathrm{kg}$$

The same for the second tree is

$$W_{\mathrm{stot},2} = 370 \times 10 \times 0.00785 \times [0.53(1 - 0.25^2) + 0.4 \times 0.25^2] = 15.16\,\mathrm{kg}$$

Note that if the two stems had the same relative taper, the taller tree's stem mass would be twice as big as the shorter tree's mass. That the taller tree's mass is more than double that of the smaller tree, shows that it has a more shallow taper on average.

Now to the branch mass of the trees. We use Eq. (4.24) and Table 4.1 for this.

$$W_{\mathrm{bt},1} = 1 \times 400 \times 0.8 \times 0.00785 \times 4^{0.8} = 7.61\,\mathrm{kg}$$

$$W_{\mathrm{bt},2} = 1 \times 400 \times 0.25 \times 0.00785 \times 2.5^{0.8} = 1.63\,\mathrm{kg}$$

Total above-ground woody masses are hence 14.10 and 16.79 kg for the shorter and taller tree, respectively.

4.8 The results of the calculations are shown in the table below. All values are per hectare. The important take-home message of this exercise is that the pipe model equations also approximately apply at the stand level in even-aged stands, because they are linear in basal area.

Site	$r_c B$ (m^2)	W_f (tn)	W_{bt} (tn)	W_{tot}(tn)	V_{est}(m^3)
Spruce 1	24.2	15	25	242	721
Spruce 2	20.1	12	22	111	331
Pine 1	9.8	4.4	2300	138	374
Pine 2	9.6	4.3	12	72	194

4.9 The Cantor set is constructed by dividing a line segment to three equal parts, then removing the middle one of these. If we then use a cover that has length 1/3 of the original line segment, we will need two of these to cover the next level of the set. According to Eq. (4.31), the dimension of the set is

$$D = -\frac{\ln 2}{\ln \frac{1}{3}} = -\frac{0.693}{-1.097} = 0.63$$

Note that the same dimension can be found regardless of the level considered. At the third level, the number of segments is 4 and the length of the segment is 1/9, so

$$D = -\frac{\ln 4}{\ln \frac{1}{9}} = -\frac{1.386}{-2.197} = 0.63$$

The Koch curve is constructed by dividing a line segment similarly into three equal parts, but instead of removing one part, two equal length parts are inserted. To cover the next level curve with covers of length 1/3 of the original, we would need 4 covers. The dimension is then

$$D = -\frac{\ln 4}{\ln \frac{1}{3}} = -\frac{1.386}{-1.097} = 1.26$$

Exercises of Chapter 6

6.1 The Yoda rule states that the maximum mean plant biomass, $\bar{W}_{max}$, scales with tree density N with exponent $3/2$. Trees are self-similar if their shape is retained regardless of size. Mass (W) would therefore scale as the linear dimension (x)

cubed, and the area occupied by a tree (A) would scale as the linear dimension squared:

$$W \propto x^3 \;\; ; \;\; A \propto x^2$$

This implies that $x \propto A^{1/2}$, such that $W \propto A^{3/2}$. Assuming full coverage, $N = 1/A$ trees can be fitted in the total area. Similarly, if there are N trees in an area, their maximum size is such that the entire area is covered by the N trees. Thus

$$\bar{W}_{\max} \propto N^{-3/2}$$

This is the Yoda rule. If we regard the rule dynamically, we also need to assume that once the maximum coverage is achieved, some trees will have to die in order to allow for the remaining trees to grow. This will then create the "self-thinning line".

6.2

(a) To apply (6.7), we need to identify the symbols in terms of the values given. Let us choose the incident radiation as $I_0 = 6000\,\mathrm{mol\,m^{-2}\,yr^{-1}}$ and foliage mass as $\mathbb{W}_\mathrm{f} = 5000\,\mathrm{kg\ ha^{-1}}$. The latter converted to leaf area provides LAI as $L = 5000\,\mathrm{kg\,ha^{-1}} \times 14\mathrm{m^2kg^{-1}} = 7\mathrm{m^2m^{-2}}$. Inserting the values into (6.7) we have

$$I_{\mathrm{abs}} = 6000 \times [1 - \exp(-0.25 \times 7)] = 4956\,\mathrm{mol\,m^{-2}\,yr^{-1}}$$

From this, photosynthesis is obtained using the given light use efficiency: $\mathbb{P}_{\mathrm{g}_2} = 4956\,\mathrm{mol\ m^{-2}\,yr^{-1}} \times 0.00027\,\mathrm{kg\,C\,mol^{-1}} = 1.34\,\mathrm{kg\,C\,m^{-2}\,yr^{-1}}$.

(b) Here we apply the light attenuation piece by piece. Let us define the stand thus: the taller layer has leaf area index 4.5, out of which 20% is above the lower layer. This leaves leaf area index 2.5 for the lower layer. We further assume that 10% of this extends below the crowns of the taller layer. The light reaching the top of the lower layer is then $I_1 = 6000\exp(-0.25 \times 4.5 \times 0.2) = 4791\,\mathrm{mol\,m^{-2}\,yr^{-1}}$. Similarly, the light reaching the bottom of the taller layer is $I_2 = 4791\exp(-0.25 \times (4.5 \times 0.8 + 2.5 \times 0.9)) = 1110\,\mathrm{mol\,m^{-2}\,yr^{-1}}$. We can now calculate the photosynthesis in the three layers:

$$\mathbb{P}_{\mathrm{g}_1} = 0.00027 \times 6000 \times [1 - \exp(-0.25 \times 4.5 \times 0.2)]\,\mathrm{mol\,m^{-2}\,yr^{-1}}$$

$$\mathbb{P}_{\mathrm{g}_2} = 0.00027 \times 4791 \times [1 - \exp(-0.25 \times (4.5 \times 0.8 + 2.5 \times 0.9))]\ \mathrm{mol\,m^{-2}\,yr^{-1}}$$

$$\mathbb{P}_{\mathrm{g}_3} = 0.00027 \times 1110 \times [1 - \exp(-0.25 \times 2.5 \times 0.1)]\,\mathrm{mol\,m^{-2}\,yr^{-1}}$$

Here $\mathbb{P}_{\mathrm{g}_2}$ contributes to both layers in the proportion of their leaf areas, such that the taller component gets $(4.5 \times 0.8)/(4.5 \times 0.8 + 2.5 \times 0.9)$ and the lower component gets the rest. Using these portions, the shares of photosynthesis for the two separate layer components are:

$$\mathbb{P}_{\mathrm{g,\,tall}} = 0.326 + 0.612 = 0.94\,\mathrm{mol\,m^{-2}\,yr^{-1}}$$

$$\mathbb{P}_{\mathrm{g,\,low}} = 0.385 + 0.018 = 0.40\,\mathrm{mol\,m^{-2}\,yr^{-1}}$$

We can see from the above that the total amount assimilated does not depend on how the foliage is distributed in the vertical dimension, as long as we use the horizontal homogeneity assumption with the Lambert-Beer equation.

(c) To illustrate the effect of variable specific leaf area, let us assume that instead of 14 everywhere, the mean specific leaf area in the topmost layer is only $10\,\mathrm{m^{-2}\,kg^{-1}}$, and in the lowest layer $18\,\mathrm{m^{-2}\,kg^{-1}}$. The leaf area in the top and bottom would thus be 10/14 and 18/14 times that used in the previous calculations, respectively. Taking this into account the top and bottom layer photosynthesis would be

$$\begin{aligned}\mathbb{P}_{\mathrm{g}_1} &= 0.00027 \times 6000 \times \left[1 - \exp(-0.25 \times 4.5 \times 0.2 \times 10/14)\right] \\ &= 0.241\,\mathrm{mol\,m^{-2}\,yr^{-1}} \\ \mathbb{P}_{\mathrm{g}_3} &= 0.00027 \times 1110 \times \left[1 - \exp(-0.25 \times 2.5 \times 0.1 \times 18/14)\right] \\ &= 0.023\,\mathrm{mol\,m^{-2}\,yr^{-1}}\end{aligned}$$

In other words, the lower layer would benefit from this, but not in any substantial way in this example.

6.3 Because $X(0)$ is the average spacing, i.e., average distance between trees, it is also the maximum crown radius than can be achieved without crown overlap. The coefficient β_{x} converts this distance to the crown length that obtains at crown closure. The parameter α_{x} determines how sensitive the species is to crown closure: the larger the α_{x}, the more crowding and potential crown overlap or crown reduction at the bottom of the crown the species can tolerate.

6.4 The condition for perfect aggregation is that $y = Nx$, where y is a higher level variable and x is the individual mean. This restricts the candidate variables for perfect aggregation. For example, we could try to find an aggregation for basal area, volume, crown projected area, or total biomass. Variables like tree height or crown radius would most likely not show perfect aggregation.

Exercises of Chapter 7

7.1 Functional balance requires that the amount of nitrogen taken up by the roots equals the nitrogen used for growth. With Hilbert's assumptions this means that

$$\sigma_{\mathrm{N}} f_r W = Y_{\mathrm{G}} \sigma_{\mathrm{C}} f_s W$$

In addition we know that $f_s + f_r = 1$. Inserting this in the above and dividing both sides with W yields

$$\sigma_{\mathrm{N}}(1 - f_s) = Y_{\mathrm{G}}\sigma_{\mathrm{C}} f_s$$

Solving this for f_s leads to Eq. (7.10).

7.2 Relative growth rate r is the total growth rate divided by total mass and is given as

$$r = \frac{1}{W} Y_{\mathrm{G}}\sigma_{\mathrm{C}} W_s = \frac{1}{W} Y_{\mathrm{G}}\sigma_{\mathrm{C}} f_s W = Y_{\mathrm{G}}\sigma_{\mathrm{C}} f_s$$

Here both σ_{C} and f_s depend on [N]. We first insert f_s into the equation:

$$r = \frac{Y_{\mathrm{G}}\sigma_{\mathrm{C}}\sigma_{\mathrm{N}}}{\sigma_{\mathrm{N}} + Y_{\mathrm{G}}[\mathrm{N}]\sigma_{\mathrm{C}}}$$

At the optimum, the derivative of r with respect to [N] must vanish:

$$\frac{\mathrm{d}r}{\mathrm{d}[\mathrm{N}]} = \frac{Y_{\mathrm{G}}\sigma_{\mathrm{N}}\left(\sigma_{\mathrm{N}} + Y_{\mathrm{G}}\sigma_{\mathrm{C}}[\mathrm{N}]\right)\sigma'_{\mathrm{C}} - \sigma_{\mathrm{N}} Y_{\mathrm{G}}\sigma_{\mathrm{C}}\left(Y_{\mathrm{G}}\sigma_{\mathrm{C}} + Y_{\mathrm{G}}[\mathrm{N}]\sigma'_{\mathrm{C}}\right)}{\left(\sigma_{\mathrm{N}} + Y_{\mathrm{G}}[\mathrm{N}]\sigma_{\mathrm{C}}\right)^2} = 0$$

where we have denoted the derivative of σ_{C} with respect to [N] by σ'_{C}. We note that for the above equation to hold, the numerator of the quotient must vanish. We may further divide the equation by $Y_{\mathrm{G}}\sigma_{\mathrm{N}}$. This leaves the following requirement:

$$\begin{aligned}\sigma_{\mathrm{N}}\sigma'_{\mathrm{C}} + &Y_{\mathrm{G}}\sigma_{\mathrm{C}}[\mathrm{N}]\sigma'_{\mathrm{C}} - \sigma_{\mathrm{C}} Y_{\mathrm{G}}\sigma_{\mathrm{C}} - \sigma_{\mathrm{C}} Y_{\mathrm{G}}[\mathrm{N}]\sigma'_{\mathrm{C}} \\ &= \sigma_{\mathrm{N}}\sigma'_{\mathrm{C}} - \sigma_{\mathrm{C}}^2 Y_{\mathrm{G}} \\ &= 0\end{aligned}$$

We now insert σ_{C} and σ'_{C}. Note first that

$$\sigma'_{\mathrm{C}} = \frac{\mathrm{d}\sigma_{\mathrm{C}}}{\mathrm{d}[\mathrm{N}]} = \frac{\mathrm{d}}{\mathrm{d}[\mathrm{N}]} \frac{P_m([\mathrm{N}] - n_1)}{[\mathrm{N}] + n_2}$$

$$= \frac{P_m([\mathrm{N}] + n_2) - P_m([\mathrm{N}] - n_1)}{([\mathrm{N}] + n_2)^2} = \frac{P_m(n_2 + n_1)}{([\mathrm{N}] + n_2)^2}$$

Now we have the requirement

$$\frac{\sigma_{\mathrm{N}} P_m(n_2 + n_1)}{([\mathrm{N}] + n_2)^2} = \frac{Y_{\mathrm{G}} P_m{}^2([\mathrm{N}] - n_1)^2}{([\mathrm{N}] + n_2)^2}$$

Multiplying both sides with $([\mathrm{N}] + n_2)^2 / P_m$ yields

$$\sigma_{\mathrm{N}} (n_2 + n_1) = Y_{\mathrm{G}} P_m ([\mathrm{N}] - n_1)^2$$

The above can be solved for [N], yielding

$$[\mathrm{N}] = \sqrt{\frac{\sigma_{\mathrm{N}} (n_1 + n_2)}{Y_{\mathrm{G}} P_m}} + n_1$$

This completes the derivation of Eq. (7.12).

Exercises of Chapter 8

8.1 As noted in Sect. 8.3.1, maximum tree height without crown rise is achieved when $H = g_2$. With the parameter values given in Table 8.2, $H_{\mathrm{max}} = 27.02$ m.

The changes in H_{max} in response to respective changes in each of the parameters can be readily calculated using some database software, such as Excel. The results are reported in Table S.1.

We note that the most important parameter for maximum height is the foliage-specific rate of photosynthesis, σ_{f}, and the second-most important are the woody respiration and the structure parameter, β_1.

8.2 From Sect. 8.3.1 we have the following equation for maximum height

$$H_{\mathrm{max}} = g_2 + (1 - g_4) H_{\mathrm{c}}$$

Table S.1 Changes in maximum height due to 10% changes in parameter values. p^+ and p^- are the increased and reduced parameter values, respectively, H^+ (m) and H^- (m) are the corresponding maximum heights, and H^+ (rel) and H^- (rel) are the corresponding relative changes in maximum height

Symbol	p^+	p^-	H^+ (m)	H^- (m)	H^+ (rel)	H^- (rel)
σ_{f}	2.42	1.98	31.84	22.16	0.178	−0.180
r_{f}	0.33	0.27	26.34	27.70	−0.025	0.025
r_{r}	0.33	0.27	26.68	27.36	−0.013	0.013
r_{w}	0.033	0.027	24.68	29.86	−0.087	0.105
T_{f}	3.3	2.7	27.48	26.47	0.017	−0.021
T_{r}	1.1	0.9	27.71	26.19	0.025	−0.031
Y_{G}	1.65	1.35	28.17	25.63	0.042	−0.052
a_{σ}	0.0011	0.0009	26.89	27.16	−0.005	0.005
α_{r}	0.55	0.45	25.93	28.12	−0.040	0.040
β_1	1.54	1.26	24.68	29.86	−0.087	0.105

where

$$g_2 = \frac{(\sigma_f - r_{M_f} - r_{M_r}\alpha_r) - [(1/T_f) + (\alpha_r/T_r)]/Y_G}{r_{M_w}\beta_1 + \alpha_\sigma\sigma_f}$$

$$g_4 = \frac{r_{M_w}(\beta_1 + \beta_2)}{r_{M_w}\beta_1 + \alpha_\sigma\sigma_f}$$

Inserting these into the equation for H_{max} and assuming that both H_{max} and H_c are known allows us to solve for σ_f in the equation. This gives us the σ_f that would lead to equivalent maximum height with different crown ratios.

To solve for σ_f we define the following new combined parameters:

$$e_1 = r_{M_f} + r_{M_r}\alpha_r + [(1/T_f) + (\alpha_r/T_r)]/Y_G$$

$$e_2 = r_{M_w}\beta_1$$

$$x = H_{max} - (1 - g_4)H_c$$

The first equation can now be written as

$$\frac{\sigma_f - e_1}{e_2 + \alpha_\sigma\sigma_f} = x$$

which can be solved for σ_f, yielding the following:

$$\sigma_f = \frac{e_1 + e_2 x}{1 - \alpha_\sigma x}$$

Inserting the parameter values given in Table 8.2, the maximum height from Exercise 8.1 and the different crown base heights we can calculate σ_f (Table S.2). Here $e_1 = 1.00556$ and $e_2 = 0.042$.

We see from the results that very small changes in σ_f are associated with large changes of crown base height if all other parameters are kept in their default values. This is consistent with what we know about the response of tree photosynthesis to shading. In dense stands crowns acclimate to the low levels of light by, firstly,

Table S.2 Values of σ_f yielding the same maximum height as no crown rise with different levels of crown rise (H_c). Here H_c is the crown base height at maximum height $H_{max} = 27.024$ m (as in Exercise 8.1)

H_c	x	σ_f	$\sigma_f/\sigma_{f,\text{default}}$
0	27.024	2.200	1.000
5	25.595	2.135	0.971
10	24.166	2.071	0.941
15	22.738	2.006	0.912

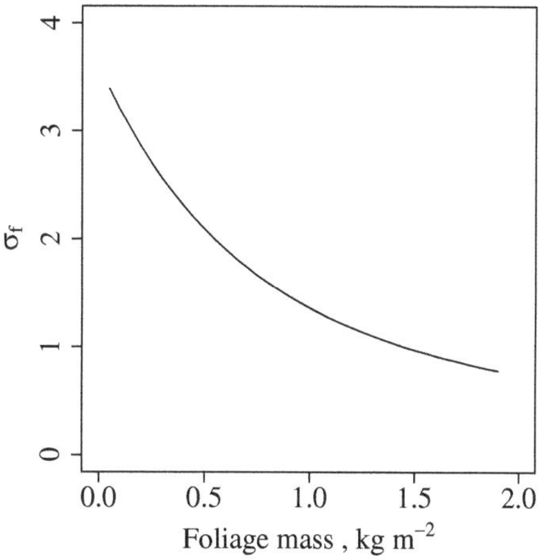

Fig. S.4 Specific rate of photosynthesis, σ_f (kg C (kg DW)$^{-1}$ yr^{-1}), as a function of stand foliage mass in Eq. (8.19)

lifting their crowns upward, such that a greater proportion of the crown remains in full light, and secondly, by many acclimations of the foliage and crown shape to be able to capture light more efficiently.

8.3 Figure S.4 shows σ_f as a function of stand-level leaf mass, $\mathbb{W}_f$. We can see that the decline of σ_f with $\mathbb{W}_f$ is quite fast. Bearing in mind that the core model is rather sensitive to σ_f, as seen in Exercise 8.1, if this whole range was covered during a tree's life time it would cause strong variations in growth rate. However, the observed variability is lower than that predicted with constant parameter values. Firstly, in stands with very low leaf mass or area, the Lambert-Beer equation does not apply as such, because it assumes that all leaf area is homogenously distributed among trees. In reality, leaf area is clustered in crowns and each of these experiences self-shading. This could be quantified by reducing the light extinction coefficient k. Secondly, in a dense stand the specific leaf area increases as an acclimative response to shading. Both of these phenomena mean that the variability of σ_f as a function of stand density is much lower than indicated by Fig. S.4.

Exercises of Chapter 9

9.2 Using the definition of sensitivity function in (9.5), we get the following sensitivity functions with respect to parameters r, K and x_0:

$$\frac{\mathrm{d}s_0(t)}{\mathrm{d}t} = \frac{\partial}{\partial x_0}\frac{\mathrm{d}x(t)}{\mathrm{d}t} = r\frac{\partial x(t)}{\partial x_0} - \frac{r}{K}2x\frac{\partial x(t)}{\partial x_0} = rs_0 - \frac{r}{K}2xs_0; \qquad s_0(0) = 1$$

$$\frac{\mathrm{d}s_r(t)}{\mathrm{d}t} = \frac{\partial}{\partial r}\frac{\mathrm{d}x(t)}{\mathrm{d}t} = rs_r(1 - \frac{2x}{K}) + x(1 - \frac{x}{K}); \qquad s_r(0) = 0$$

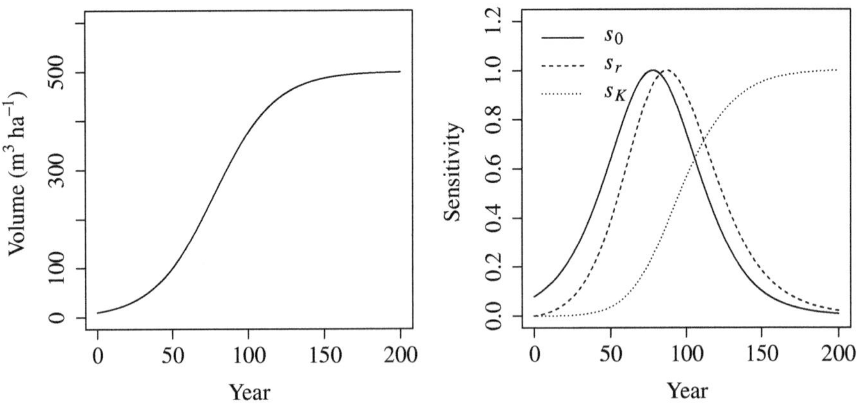

Fig. S.5 Stand volume simulated with the logistic equation (left panel) and related sensitivity functions with respect to parameters r and K and initial state x_0. The sensitivity functions have been normalised to have maximum 1 in the simulation interval

$$\frac{\mathrm{d}s_K(t)}{\mathrm{d}t} = \frac{\partial}{\partial K}\frac{\mathrm{d}x(t)}{\mathrm{d}t} = rs_K - 2\frac{r}{K}xs_K + r\frac{x^2}{K^2}; \qquad s_K(0) = 0$$

Here, the principle of differentiating with respect to each parameter is illustrated with the first equation where the sensitivity is determined with respect to the initial state. The equation does not depend explicitly on the initial state, so we only differentiate the state x itself. In the two other equations, the equation also depends on the parameters. The differential then becomes that of a product, where both (a function of] x and the parameter occur as multiplicative terms.

9.3 The sensitivity functions derived in Exercise 9.2 have been coded in the R programme "`Sensitivity_logistic.R`" available in the electronic supplementary material of this chapter. The results are in Fig. S.5.

9.4 Calculation of elasticity has been coded in the R programme "`Sensitivity_logistic.R`" available in the electronic supplementary material of this chapter. The results are in Fig. S.6. We note that when related to the current state, the sensitivity is largest with respect to initial state. Because the logistic model has a stable equilibrium at $x = K$, the elasticity with respect to K approaches 1 as time increases. The elasticity with respect to the rate parameter r is at its largest at the time when maximum growth rate is achieved.

9.5 The numerical functions of (9.8) have been coded in the R programme "`Sensitivity_numerical.R`" available in the electronic supplementary material of this chapter. The results are in Fig. S.7.

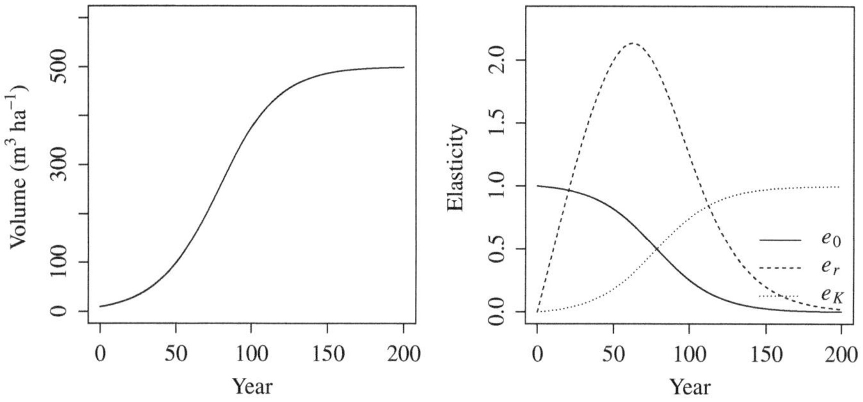

Fig. S.6 Stand volume simulated with the logistic equation (left panel) and related elasticity functions with respect to parameters r and K and initial state x_0

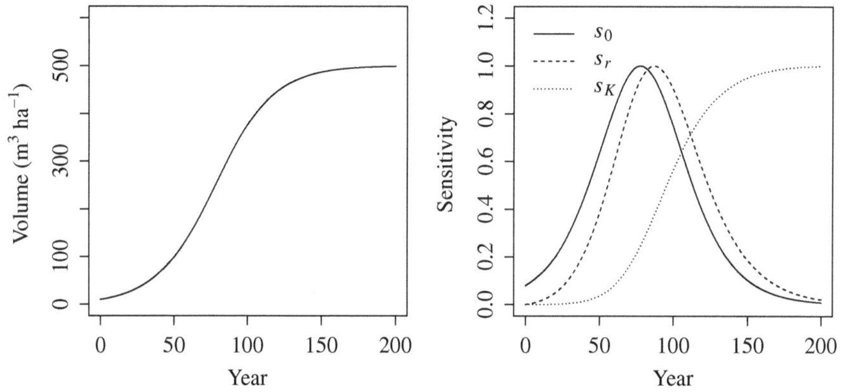

Fig. S.7 Stand volume simulated with the logistic equation (left panel) and related numerically estimated sensitivity functions with respect to parameters r and K and initial state x_0. The parameters were varied by 1 per mille of the default values in the calculation. The sensitivity functions have been normalised to have maximum 1 in the simulation interval

9.6 The left panel of Fig. 9.3 demonstrates high uncertainty about the steady-state parameter, α_2, whereas the right panel shows high uncertainty about the rate parameter, α_1. The sensitivity to α_1 is somewhat higher, so the overall scatter is larger in the left panel.

References

Aalto J, Pirinen PI, Heikkinen J, Venäläinen AI (2013) Spatial interpolation of monthly climate data for Finland: comparing the performance of kriging and generalized additive models. Theor Appl Climatol 112:99–111

Affleck DLR (2006) Poisson mixture models for regression analysis of stand-level mortality. Can J For Res 36(11):2994–3006

Ågren GI (1983) Nitrogen productivity of some conifers. Can J For Res 13:494–500

Ågren GI, Franklin O (2003) Root:shoot ratios, optimization and nitrogen productivity. Ann Bot 92:795–800

Ågren GI, Axelsson B, Flower-Ellis JGK, Linder S, Persson H, Staaf H, Troeng E (1980) Annual carbon budget for a young scots pine. Ecol Bull 32:307–313

Ahl V, Allen TFH (1996) Hierarchy theory, a vision, vocabulary and epistemology. Columbia University Press, New York

Alam A, Kilpelainen A, Kellomaki S (2010) Potential energy wood production with implications to timber recovery and carbon stocks under varying thinning and climate scenarios in Finland. Bioenergy Res 3(4):362–372

Albaugh TJ, Allen HL, Kress LW (2006) Root and stem partitioning of *Pinus taeda*. Trees 20(2):176–185

Alegria C (2011) Simulation of silvicultural scenarios and economic efficiency for maritime pine (*Pinus pinaster* Aiton) wood-oriented management in centre inland of Portugal. For Syst 20:361–378

Allen TFH, Starr TB (1982) Hierarchy: perspectives for ecological diversity. Chicago University Press, Chicago

Assmann E (1970) The principles of forest yield study. Pergamon Press, Oxford

Aubinet M, Vesala T, Papale D (2012) Eddy covariance: a practical guide to measurement and data analysis. Springer, Reading

Baldocchi DD (2003) Assessing the eddy covariance technique for evaluating carbon dioxide exchange rates of ecosystems: past, present and future. Glob Change Biol 9(4):479–492

Bar-Yam Y (2012) Dynamics of complex systems. Perseus Books, Dordrecht/London/New York

Beekhuis J (1965) Crown depth of radiata pine in relation to stand density and height. N Z J For 10:43–61

Berninger F, Coll L, Vanninen P, Mäkelä A, Palmroth S, Nikinmaa E (2005) Effects of tree size and position on pipe model ratios in scots pine. Can J For Res 35:1294–1304

Bertalanffy LV (1957) Quantitative laws of metabolism and growth. Q Rev Biol 32:217–231

Bertalanffy LV (1968) General systems theory. Foundations, development, applications. Revised edition. George Brazilles, New York

A. Mäkelä, H. T. Valentine, *Models of Tree and Stand Dynamics*,
https://doi.org/10.1007/978-3-030-35761-0

Biegler C, Bugmann HKM (2003) Growth-dependent tree mortality models based on tree rings. Can J For Res 33:210–221

Blanco JA, Seely B, Welham C, Kimmins JP, Seebacher TM (2007) Testing the performance of a forest ecosystem model (FORECAST) against 29 years of field data in a *Pseudotsuga menziesii* plantation. Can J For Res 37:1808–1820

Bohlman S, Pacala S (2012) A forest structure model that determines crown layers and partitions growth and mortality rates for landscape scale applications of tropical forests. J Ecol 100: 508–518

Bond-Lamberty B, Bailey VL, Chen M, Cough CM, Vargas R (2018) Globally rising soil heterotrophic respiration over recent decades. Nature 560:80–83

Bontemps JD, Duplat P (2012) A non-asymptotic sigmoid growth curve for top height growth in forest stands. Forestry 85(3):353–368

Borchert F, Slade NA (1981) Bifurcation ratios and the adaptive geopmetry of trees. Bot Gaz 142:394–401

Bossel H (1994) Modeling and simulation. A. K. Peters Ltd, Wellesley

Bossel H (1996) TREEDYN3 forest simulation model. Ecol Modell 90:187–227

Botkin DB, Janak JF, Wallis JR (1972) Some ecological consequences of a computer model of forest growth. Ecology 60:849–872

Box GEP, Cox DR (1964) An analysis of transformations. J R Stat Soc B 26:1059–1072

Brännström Å, Johansson J, von Festenberg N (2013) The hitchhiker's guide to adaptive dynamics. Games 4:304–328

Bravo-Oviedo A, del Rio M, Calama R, Valentine HT (2013) New approaches to modelling cross-sectional area to height allometry in four mediterranean pine species. Forestry 87(3):399–406

Brouwer R (1962) Distribution of dry matter in the plant. Neth J Agric Sci 10:361–376

Brown GS (1962) The importance of stand density in pruning prescriptions. Emp For Rev 41(3):246–257

Bruning EF (1976) Tree forms in relation to environmental conditions. An ecological viewpoint. In: Cannell MGR, Last FT (eds) Tree physiology and yield improvement. Academic Press, New York, pp 139–156

Bugmann H (2001) A review of forest gap models. Clim Change 51:259–305

Burkhart HE, Tomé M (2012) Modeling trees and stands. Springer, New York

Cajander AK (1949) Finnish forest types and their significance. Acta Forestalia Fennica 56:1–71

Caldwell MM, Dawson TE, Richards JH (1998) Hydraulic lift: consequences of water efflux from the roots of plants. Oecologia 113:151–161

Campolongo F, Cariboni J, Saltelli A (2007) An effective screening design for sensitivity analysis of large models. Environ Model Softw 22:1509–1518

Cannell MGR, Dewar RC (1994) Carbon allocation in trees: a review of concepts for modelling. Adv Ecol Res 25:59–104

Cannell MGR, Thornley JHM (2000) Modelling the components of plant respiration: some guiding principles. Ann Bot 85:45–54

Cao T, Valsta L, Mäkelä A (2010) A comparison of carbon assessment methods for optimizing timber production and carbon sequestration in Scots pine stands. For Ecol Manage 260: 1726–1734

Carlson WC, Harrington CA (1987) Cross-sectional area relationships in root systems of loblolly and shortleaf pine. Can J For Res 17(6):556–558

Chen HYH, Klinka K, Kayahara GJ (1996) Effects of light on growth, crown architecture, and specific leaf area for naturally established *Pinus contorta* var. *latifolia* and *Pseudotsuga menziesii* var. *glauca* saplings. Can J For Res 26:1149–1157

Chiba Y, Fujimori T, Kiyono Y (1988) Another interpretation of the profile diagram and its availability with consideration of the growth process of forest trees. J Jap For Soc 70:245–254

Clark A III , Saucier JR (1989) Influence of initial planting density, geographic location, and species on juvenile wood formation in southern pine. For Prod J 39(7/8):42–48

Clutter JL, Jones EP Jr (1980) Prediction of growth after thinning in old-field slash pine plantations. USDA Forest Service Research Paper SE-217

Comins HN, McMurtrie RE (1993) Long-term response of nutrient-limited forests to CO_2 enrichment; equilibrium behavior of plant-soil models. Ecol Appl 3:666–681

Coops NC, Waring RH, Landsberg JJ (1998) Assessing forest productivity in Australia and New Zealand using a physiologically-based model driven with averaged monthly weather data and satellite-derived estimates of canopy photosynthetic capacity. For Ecol Manage 104:113–127

Cowan IR, Farquhar GD (1977) Stomatal function in relation to leaf metabolism and environment. In: Jennings DH (ed) Integration of activity in the higher plant. Cambridge University Press, Cambridge, pp 471–505

Cramer W, Bondeau A, Woodward FI, Prentice CI, Betts RA, Brovkin V (2001) Global response of terrestrial ecosystem structure and function to CO_2 and climate change: results from six dynamic global vegetation models. Glob Change Biol 7:357–373

Crawley MJ (2007) Plant population dynamics. In: May RM, McLean AR (eds) Theoretical ecology. Oxford University Press, Oxford, pp 62–83

Davidson RL (1969) Effect of root/leaf temperature differentials on root/shoot ratios in some pasture grasses and clover. Ann Bot 33:561–569

de Wit CT (1978) Simulation of assimilation, respiration and transpiration of crops. Pudoc, Wageningen

Dean TJ, Jerez M, Cao QV (2013) A simple stand growth model based on canopy dynamics and biomechanics. For Sci 59(3):335–344

Dewar RC (1996) The correlation between plant growth and intercepted radiation: an interpretation in terms of optimal plant nitrogen content. Ann Bot 78:125–136

Dewar RC (2001) A model of the coupling between respiration, active processes and passive transport. Ann Bot 86:279–286

Dewar RC, Medlyn BE, McMurtrie RE (1999) Acclimation of the respiration/photosynthesis ratio to temperature: insights for a model. Glob Change Biol 5:612–622

Dewar RC, Franklin O, Mäkelä A, McMurtrie RE, Valentine HT (2009) Optimal function explains forest response to global change. BioScience 59:127–139

Dewar RC, Lineweaver CH, Niven RK, Regenauer-Lieb K (2010) Maximum entropy production and plant optimization theories. Philos Trans R Soc B Biol Sci 365:1429 –1435

Dewar R, Mauranen A, Mäkelä A, Hölttä T, Medlyn B, Vesala T (2018) New insights into the covariation of stomatal, mesophyll and hydraulic conductances from optimization models incorporating nonstomatal limitations to photosynthesis. New Phytol 217:571–585

de Kroon H, Hendriks M, van Ruijven J, Ravenek J, Padilla FM, Jongejans E, Visser EJW, Mommer L (2012) Root responses to nutrients and soil biota: drivers of species coexistence and ecosystem productivity. J Ecol 100(1):6–15

de Reffye P, Houllier F, Blaise F, Barthélémy D, Dauzat J, Auclair D (1995) A model simulating above- and belowground tree architecture with agroforestry applications. Agroforestry Syst 30:175–197

Diaz-Balteiro L, Romero C (1998) Modeling timber harvest scheduling problems with multiple criteria: an application in Spain. For Sci 44:47–57

Diaz-Nieto J, Wilby RL (2005) A comparison of statistical downscaling and climate change factor methods: impacts on low flows in the River Thames, United Kingdom. Clim Change 69: 245–268

Dieler J, Pretzsch H (2013) Morphological plasticity of European beech (*Fagus sylvatica* L.) in pure and mixed-species stands. For Ecol Manage 295:97–108

Diéguez-Aranda U, Castedo-Dorado F, Álvarez-González JG, Rodríguez-Soalleiro R (2005) Modelling mortality of Scots pine (*Pinus sylvestris* L.) plantations in the northwest of Spain. Eur J For Res 124(2):143–153

Doetterl S, Berhe AA, Arnold C, Bodé S, Fiener P, Finke P, Fuchslueger L, Griepentrog M, Harden JW, Nadeu E, Schnecker J, Six J, Trumbore S, Van Oost K, Vogel C, Boeckx P (2018) Links among warming, carbon and microbial dynamics mediated by soil mineral weathering. Nat Geosci 11:589–593

Duursma RA, Mäkelä A (2007) Summary models for light interception and light-use efficiency of non-homogeneous canopies. Tree Physiol 27:859–870

Duursma RA, Kolari P, Perämäki M, Pulkkinen M, Mäkelä A, Nikinmaa E, Hari P, Aurela M, Berbigier P, Bernhofer C, Grunwald T, Loustau D, Molder M, Verbeeck H, Vesala T (2009) Contributions of climate, leaf area index and leaf physiology to variation in gross primary production of six coniferous forests across Europe: a model-based analysis. Tree Physiol 29:621–639

Duursma RA, Mäkelä A, Reid DEB, Jokela EJ, Porté A, Roberts SD (2010) Branching networks in gymnosperm trees: implications for metabolic scaling. Funct Ecol 24:723–730

Duursma RA, Falster DS, Valladares F, Sterck FJ, Pearcy RW, Lusk CH, Sendall KM, Nordenstahl M, Houter NC, Atwell BJ, Kelly N, Kelly JWG, Liberloo M, Tissue DT, Medlyn BE, Ellsworth DS (2011) Light interception efficiency explained by two simple variables: a test using a diversity of small- to medium-sized woody plants. New Phytol 193:397–408

Dybzinski R, Farrior C, Wolf A, Reich PB, Pacala SW (2011) Evolutionary stable strategy carbon allocation to foliage, wood, and fine roots in trees competing for light and nitrogen: an analytically tractable, individual-based model and quantitative comparisons to data. Am Nat 177:153–166

Eggers J, Lindhagen A, Lind T, Lamas T, Ohman K (2018) Balancing landscape-level forest management between recreation and wood production. Urban For Urban Green 33:1–11

Exbrayat JF, Bloom AA, Falloon P, Ito A, Smallman TL, Williams M (2018) Reliability ensemble averaging of 21st century projections of terrestrial net primary productivity reduces global and regional uncertainties. Earth Syst Dyn 9:153–165

Fischer R, Bohn F, de Paula MD, Dislich C, Groeneveld J, Gutierrez AG, Kazmierczak M, Knapp N, Lehmann S, Paulick S, Puetz S, Roedig E, Taubert F, Koehler P, Huth A (2016) Lessons learned from applying a forest gap model to understand ecosystem and carbon dynamics of complex tropical forests. Ecol Modell 326:124–133

Foley JA, Prentice IC, Ramankutty N, Levis S, Pollard D, Sitch S, Haxeltine A (1996) An integrated biosphere model of land surface processes, terrestrial carbon balance, and vegetation dynamics. Global Bigeochem Cycles 10:603–628

Fontaine S, Bardoux G, Abbadie L, Mariotti A (2004) Carbon input to soil may decrease soil carbon content. Ecol Lett 7:314–320

Fontes L, Bontemps J, Bugmann H, van Oijen M, Gracia C, Kramer K, Lindner M, Rötzer T, Skovsgaard JP (2010) Models for supporting forest management in a changing environment. For Syst 19(Specia):8–29

Ford ED (1992) The control of tree structure and productivity through the interaction of morphological development and physiological processes. Int J Plant Sci 153:S147–S162

Forrester DI, Guisasola R, Tang X, Albrecht AT, Dong TA, le Maire G (2014) Using a stand-level model to predict light absorption in stands with vertically and horizontally heterogeneous canopies. For Ecosyst 1:1–17

Forrester DI, Ammer C, Annighöfer PJ, Avdagic A, Barbeito I, Bielak K, Brazaitis G, Coll L, del Rìo M, Drössler L, Heym M, Hurt V, Löf M, Matović B, Meloni F, den Ouden J, Pach M, Pereira MG, Ponette Q, Pretzsch H, Skrzyszewski J, Stojanović D, Svoboda M, Ruiz-Peinaido R, Vacchiano G, Verheyen K, Zlatanov T, Bravo-Oviedo A (2017) Predicting the spatial and temporal dynamics of species interactions in *Fagus sylvatica* and *Pinus sylvestris* forests across Europe. For Ecol Manage 405:112–133

Franklin O (2007) Optimal nitrogen allocation controls tree responses to elevated CO2. New Phytol 174:811–822

Franklin O, Ågren GI (2002) Leaf senescence and resorption as mechanisms of maximizing photosynthetic production during canopy development at n limitation. Funct Ecol 16(6): 727–733

Franklin O, McMurtrie RE, Iverson CM, Crous KY, Finzi AC, Tissue DT, Ellsworth DS, Oren R, Norby RJ (2009) Forest fine-root production and nitrogen use under elevated co_2: contrasting responses in evergreen and deciduous trees explained by a common principle. Glob Change Biol 15:132–144

Franklin O, Johansson J, Dewar RC, Dieckmann U, McMurtrie RE, Brännström Å, Dybzinski R (2012) Modeling carbon allocation in trees: a search for principles. Tree Physiol 32(6):648–666

Franklin O, Näsholm T, Högberg P, Högberg MN (2014) Forests trapped in nitrogen limitation: an ecological market perspective on ectomycorrhizal symbiosis. New Phytol 203:657–666
Franklin O, Harrison SP, Dewar R, Farrior CE, Brännström Å, Dieckmann U, Pietsch S, Falster D, Cramer W, Loreau M, Wang H, Mäkelä A, Rebel KT, Meron E, Schymanski SJ, Rovenskaya E, Stocker BD, Zaehle S, Manzoni S, Van Oijen M, Wright IJ, Ciais P, van Bodegom PM, Peñuelas J, Hofhansl F, Terrer C, Soudzilovskaia NA, Midgley G, Prentice CI (2020) Organizing principles for vegetation dynamics. Manuscript accepted for publication in Nature Plants
Freschet GT, Bellingham PJ, Lyver PO, Bonner KI, Wardle DA (2013) Plasticity in above-and belowground resource acquisition traits in response to single and multiple environmental factors in three tree species. Ecol Evol 3(4):1065–1078
Friend AD, Lucht W, Rademacher TT, Keribin R, Betts R, Cadule P, Ciais P, Clark DB, Dankers R, Falloon PD, Ito A, Kahana R, Kleidon A, Lomas MR, Nishina K, Ostberg S, Pavlick R, Peylin P, Schaphoff S, Vuichard N, Warszawski L, Wiltshire A, Woodward FI (2014) Carbon residence time dominates uncertainty in terrestrial vegetation responses to future climate and atmospheric CO_2. Proc Natl Acad Sci U S A 111:3280–3285
Fulford G, Forrester P, Jones A (1997) Modelling with differential and difference equations. Cambridge University Press, Cambridge
García O (1983) A stochastic differential equation model for the height growth of forest stands. Biometrics 39:1059–1072
García O (2005) Unifying sigmoid univariate growth equations. For Biom Modell Inform Sci 1: 63–68
García O (2008) Visualization of a general family of growth functions and probability distributions – the growth-curve explorer. Environ Model Softw 23:1474–1475
García O (2009) A simple and effective forest stand mortality model. Math Comput For Nat Resour Sci (MCFNS) 1(1):1–9
García O (2017) Cohort aggregation modelling for complex forest stands: spruce–aspen mixtures in British Columbia. Ecol Modell 343:109–122
García O (2014) Can plasticity make spatial structure irrelevant in individual-tree models? For Ecosyst 1(1):16
Geritz SAH, Kisdi É, Meszéna G, Metz JAJ (1998) Evolutionarily singular strategies and the adaptive growth and branching of the evolutionary tree. Evol Ecol 12:35–57
Gersani M, Brown JS, O'Brien EE, Maina GM, Abramsky Z (2001) Tragedy of the commons as a result of root competition. J Ecol 89:660–669
Godfrey K (1983) Compartmental models and their applications. Academic Press, New York
Gompertz B (1825) On the nature of the function expressive of the law of human mortality, and on a new mode of determining the value of life contingencies. Philos Trans R Soc Lond 115: 513–585
Gonzalez-Benecke CA, Teskey RO, Martin TA, Jokela EJ, Fox TR, Kane MB, Noormets A (2016) Regional validation and improved parameterization of the 3-PG model for *Pinus taeda* stands. For Ecol Manage 361:237–256
Gove JH, Hollinger DY (2006) Application of a dual unscented Kalman filter for simultaneous state and parameter estimation in problems of surface-atmosphere exchange. J Geophys Res Atmos 111(D8). https://doi.org/10.1029/2005JD006021
Grace JC (1990) Modeling the interception of solar radiant energy and net photosynthesis. In: Dixon RK, Meldahl RS, Ruark GA, Warren WG (eds) Process modeling of forest growth responses to environmental stress. Timber Press, Portland, pp 142–158
Grace JC, Jarvis P, Norman JM (1987) Modelling the interception of solar radiant energy in intensively managed stands. N Z J For Sci 17:193–209
Gray HR (1956) The form and taper of forest-tree stems. Institute paper 32, Imperial Forestry Institute, University of Oxford
Green EJ, MacFarlane DW, Valentine HT, Strawderman WE (1999) Assessing uncertainty in a stand growth model by Bayesian synthesis. For Sci 45(4):528–538
Green EJ, MacFarlane DW, Valentine HT (2000) Bayesian synthesis for quantifying uncertainty in predictions from process models. Tree Physiol 20(5–6):415–419

Greenhill AG (1881) Determination of the greatest height consistent with stability that a vertical pole or mast can be made, and of the greatest height to which a tree of given proportions can grow. Proc Camb Philos Soc 4(2):65–73

Gregoire TG, Valentine HT (2008) Sampling strategies for natural resources and the environment. Chapman & Hall/CRC, Boca Raton

Gregoire TG, Valentine HT, Furnival GM (1995) Sampling methods to estimate foliage and other characteristics of individual trees. Ecology 76:1181–1194

Guillemot J, Martin-StPaul NK, Dufrêne E, François C, Soudani K, Ourcival JM, Delpierre N (2015) The dynamic of the annual carbon allocation to wood in European tree species is consistent with a combined source-sink limitation of growth: implications for modelling. Biogeosciences 12:2773–2790

Guillemot J, Francois C, Hmimina G, Dufrâne E, Martin-StPaul NK, Soudani K, Marie G, Ourcival JM, Delpierre N (2017) Environmental control of carbon allocation matters for modelling forest growth. New Phytol 214:180–193

Härkönen S, Pulkkinen M, Duursma RA, Mäkelä A (2010) Estimating annual GPP, NPP and stem growth in Finland using summary models. For Ecol Manage 259:524–533

Härkönen S, Tokola T, Packalen P, Korhonen L, Mäkelä A (2013) Predicting forest growth based on airborne light detection and ranging data, climate data, and a simplified process-based model. Can J For Res 43:354–375

Härkönen S, Neumann M, Mues V, Berninger F, Bronisz K, Cardellini G, Chirici G, Hasenauer H, Koehl M, Lang M, Merganicova K, Mohren F, Moiseyev A, Moreno A, Mura M, Muys B, Olschofsky K, Del Perugia B, Rorstad P, Solberg B, Thivolle-Cazat A, Trotsiuk V, Mäkelä A (2019) A climate-sensitive forest model for assessing impacts of forest management in Europe. Environ Model Softw 115:128–143

Högberg P, Nordgren A, Buchmann N, Taylor AFS, Ekblad A, Högberg MN, Nyberg G, Ottosson-Löfvenius M, Read DJ (2001) Large-scale forest girdling shows that current photosynthesis drives soil respiration. Nature 411:789–792

Högberg P, Näsholm T, Franklin O, Högberg MN (2017) Tamm review: on the nature of the nitrogen limitation to plant growth in Fennoscandian boreal forests. For Ecol Manage 403: 161–185

Hallé F, Oldeman RAA, Tomlinson PB (1978) Tropical trees and forests. An architectural analysis. Springer, Berlin

Hari P, Kulmala Le (2008) Boreal forest and climate change. Advances in global change research, vol 34. Springer, Berlin

Hari P, Mäkelä A, Korpilahti E, Holmberg M (1986) Optimal control of gas exchange. Tree Physiol 2:169–175

Harmsen K (2000) A modified Mitscherlich equation for rainfed crop production in semi-arid areas: 1. theory. NJAS-Wagen J Life Sci 48(3):237–250

Hartig F, Dyke J, Hickler T, Higgins SI, O'Hara RB, Scheiter S, Huth A (2012) Connecting dynamic vegetation models to data – an inverse perspective. J Biegeogr 39:2240–2252

Helmisaari HS, Derome J, Nöjd P, Kukkola M (2007) Fine root biomass in relation to site and stand characteristics in norway spruce and scots pine stands. Tree Physiol 27:1493–1504

Henttonen H, Mäkinen H, Heiskanen J, Peltoniemi M, Laurèn A, Hordo M (2014) Response of radial increment variation of scots pine to temperature, precipitation and soil water content along a latitudinal gradient across Finland and Estonia. Agric For Meteorol 198–199:294–308

Hilbert DW (1990) Optimization of plant root:shoot ratios and internal nitrogen concentration. Ann Bot 66:91–99

Ho YC, Lee R (1964) A Bayesian approach to problems in stochastic estimation and control. IEEE Trans Automat Contr AC-9:333–339

Hodge A (2004) The plastic plant: root responses to heterogeneous supplies of nutrients. New Phytol 162:9–24

Holmberg M, Aalto T, Akujärvi A, Arslan AN, Bergström I, Böttcher K, Lahtinen I, Mäkelä A, Markkanen T, Minunno F, Peltoniemi M, Rankinen K, Vihervaara P, Forsius M (2019)

Ecosystem services related to carbon cycling – modeling present and future impacts in boreal forests. Front Plant Sci 10:343–351
Honda H (1971) Description of the form of trees by the parameters of the tree-like body: effects of the branching angle and the branch length on the shape of the tree-like body. J Theor Biol 31:331–338
Honda H, Fisher JB (1978) Tree branch angle: maximizing effective leaf area. Science 199: 888–890
Honda H, Tomlinson PB, Fisher JB (1982) Two geometrical models of branching of botanical trees. Ann Bot 49:1–11
Horn HS (1971) The adaptive geometry of trees. Princeton University Press, Princeton
Horn HS (2000) Twigs, trees, and the dynamics of carbon in the landscape. In: Brown JH, West GB (eds) Scaling in biology. Oxford University Press, Oxford, pp 199–220
Hu M, Lehtonen A, Minunno F, Mäkelä A (2020) Age effect on tree structure and biomass allocation in Scots pine (*Pinus sylvestris* L.) and Norway spruce (*Picea abies* [L.] Karst.). Manuscript submitted to Ann For Sci
Huang C, He HS, Hawbaker TJ, Liang Y, Gong P, Wu Z, Zhu Z (2017) A coupled modeling framework for predicting ecosystem carbon dynamics in boreal forests. Environ Model Softw 93:332–343
Hubbard RM, Bond BJ, Ryan MG (1999) Evidence that hydraulic conductance limits photosynthesis in old *Pinus ponderosa* trees. Tree Physiol 19:165–172
Hyink DM (1983) A generalized framework for projecting forest yield and stand structure using diameter distributions. For Sci 29:89–95
Hyvönen R, Ågren GI, Linder S, Persson T, Cotrufo MF, Ekblad A, Freeman M, Grelle A, Janssens IA, Jarvis PG, Kellomäki S, Lindroth A, Loustau D, Lundmark T, Norby RJ, Oren R, Pilegaard K, Ryan MG, Sigurdsson BD, Strömgren M, van Oijen M, Wallin G (2007) The likely impact of elevated [CO_2], nitrogen deposition, increased temperature and management on carbon sequestration in temperate and boreal forest ecosystems: a literature review. New Phytol 173:463–480
Hyytiäinen K, Hari P, Kokkila T, Mäkelä A, Tahvonen O, Taipale J (2004) Connecting a process-based forest growth model to stand-level economic optimization. Can J For Res 34:2060–2073
Ilomäki S, Nikinmaa E, Mäkelä A (2003) Crown rise due to competition drives biomass allocation in silver birch (*Betula pendula* l.). Can J For Res 33:2395–2404
Ingestad T (1980) Growth, nutrition and nitrogen fixation in grey alder at varied rate of nitrogen addition. Physiol Plant 50:353–364
Ingestad T, Ågren GI (1992) Theories and methods on plant nutrition and growth. Physiol Plant 84:177–184
Ingestad T, Aronsson A, Ågren GI (1981) Nutrient flux density model of mineral nutrition in conifer ecosystems. Studia Forestalia Suecica 160:61–72
IPCC (2013) Climate change 2013: the physical science basis. In: Stocker TF, Qin D, Plattner GK, Tignor M, Allen SK, Boschung J, Nauels A, Xia Y, Bex V, Midgley PM (eds) Contribution of working group I to the fifth assessment report of the intergovernmental panel on climate change. Cambridge University Press, Cambridge/New York, p 1535
Ishii H, Kitaoka S, Fujisaki Y, Maruyama T, Koike T (2007) Plasticity of shoot and needle morphology and photosynthesis of two Picea species with different site preferences in northern Japan. Tree Physiol 27:1595–1605
Iwasa Y, Cohen D, León JA (1985) Tree height and crown shape, as results of competitive games. J Theor Biol 112:279–297
Iwasa Y, Andreasen V, Levin S (1987) Aggregation in model-ecosystems. 1. Perfect aggregation. Ecol Modell 37:287–302
Jessen RJ (1955) Determining the fruit count on a tree by randomized branch sampling. Biometrics 11:99–109
Johnson IR, Thornley JHM (1987) A model of shoot:root partitioning with optimal growth. Ann Bot 60:133–142

Jones O, Maillardet R, Robinson A (2009) Introduction to scientific programming and simulation using R. Chapman & Hall/CRC, Boca Raton

Körner C (2003) Carbon limitation in trees. J Ecol 91:4–17

Kalliokoski T, Nygren P, Sievänen R (2008) Coarse root architecture of three boreal tree species growing in mixed stands. Silva Fenn 42:189–210

Kalliokoski T, Mäkinen H, Linkosalo T, Mäkelä A (2016) Evaluation of stand-level hybrid PipeQual model with permanent sample plot data of Norway spruce. Can J For Res 47:234–245

Kalliokoski T, Mäkelä A, Fronzek S, Minunno F, Peltoniemi M (2018) Decomposing sources of uncertainty in climate change projections of boreal forest primary production. Can J For Res 262:192–205

Kalliokoski T, Heinonen T, Holder J, Lehtonen A, Minunno F, Mäkelä A, Packalen T, Peltoniemi M, Pukkala T, Salminen O, Schelhaas MJ, Vauhkonen J, Kanninen M (2019) Scenario analysis of similarities and differences between models of forest development (in Finnish). Finnish Climate Panel, Helsinki

Kalman RE (1960) A new approach to linear filtering and prediction problems. J Basic Eng 82(1):35–45

Kantola A, Mäkelä A (2006) Development of biomass proportions in Norway spruce (*Picea abies* [l.] Karst.). Trees 20:111–121

Kantola A, Mäkinen H, Mäkelä A (2007) Stem form and branchiness of Norway spruce as sawn timber – predicted by a process-based model. For Ecol Manage 241:209–222

Keenan TF, Carbone MS, Reichstein M, Richardson AD (2011) The model–data fusion pitfall: assuming certainty in an uncertain world. Oecologia 167(3):587–597

Keenan TF, Davidson E, Moffat AM, Munger W, Richardson AD (2012) Using model-data fusion to interpret past trends, and quantify uncertainties in future projections, of terrestrial ecosystem carbon cycling. Glob Change Biol 18(8):2555–2569

Kershaw JA, Ducey MJ, Beers TW, Husch B (2017) Forest mensuration. John Wiley & Sons, West Sussex

Ketterings QM, Coe R, van Noordwijk M, Ambagau Y, Palm CA (2001) Reducing uncertainty in the use of allometric biomass equations for predicting above-ground tree biomass in mixed secondary forests. For Ecol Manage 146:199–209

Kikuzawa K (1991) A cost-benefit analysis of leaf habit and leaf longevity of trees and their geographical pattern. Am Nat 138:1250–1263

Kikuzawa K, Lechowicz MJ (2011) Ecology of leaf longevity. Ecological research monographs. Springer, Tokyo/Dordrecht/Heidelberg/London/New York

Kindermann G, Obersteiner M, Sohngen B, Sathaye J, Andrasko K, Rametsteiner E, Schlamadinger B, Wunder S, Beach R (2008) Global cost estimates of reducing carbon emissions through avoided deforestation. For Ecol Manage 105:10303–10307

King DA (1990) The adaptive significance of tree height. Am Nat 135:809–828

King DA (1993) A model analysis of the influence of root and foliage allocation on forest production and competition between trees. Tree Physiol 12:119–135

King D, Loucks OL (1978) The theory of tree bole and branch form. Radiat Environ Biophys 15:141–165

Kira T, Shidei T (1967) Primary production and turnover of organic matter in different forest ecosystems of the Western Pacific. Jap J Ecol 17:70–87

Klir GE (1972) Trends in general systems theory. Wiley & Sons, Inc., New York

Koivisto P (1959) Growth and yield tables (in Finnish). Comm Inst Forestalis Fenniae 51(8):1–49

Kokkila T, Mäkelä A, Franc A (2006) Comparison of distance-dependent and distance-independent stand growth models—is perfect aggregation possible? For Sci 26:623–635

Kozlowski J, Konarzewski M (2004) Is West, Brown and Enquist's model of allometric scaling mathematically correct and biologically relevant? Funct Ecol 8:283–289

Kull O (2002) Acclimation of photosynthesis in canopies: models and limitations. Oecologia 133:267–279

Laasasenaho J, Koivuniemi J (1990) Dependence of some stand characteristics on stand density. Tree Physiol 7:183–187

Landsberg JJ (1986) Physiological ecology of forest production. Academic Press, London
Landsberg JJ (2003) Modeling forest ecosystems: state of the art, challenges, and future directions. Can J For Res 33:385–397
Landsberg JJ, Sands P (2011) Physiological production of forest production: principles, processes and models. Academic Press, London
Landsberg JJ, Waring RH (1997) A generalised model of forest productivity using simplified concepts of radiation use efficiency, carbon balance and partitioning. For Ecol Manage 95: 209–228
Landsberg JJ, Mäkelä A, Sievänen R, Kukkola M (2005) Analysis of biomass accumulation and stem size distributions over long periods in managed stands of *Pinus sylvestris* in Finland using the 3-PG model. Tree Phys 25:781–792
Lang ARG (1991) Application of some of Cauchy's theorems to the estimation of surface areas of leaves, needles and branches of plants, and light transmittance. Agric For Meteorol 55:191–212
Larson PR (1965) Stem form of young *Larix* as influenced by wind and pruning. For Sci 11: 412–423
Lasch P, Badeck FW, Suckow F, Lindner M, Mohr P (2005) Model-based analysis of management alternatives at stand and regional level in Brandenburg (Germany). For Ecol Manage 207:59–74
Lehnebach R, Beyer R, Letort V, Heuret P (2018) The pipe model theory half a century on: a review. Ann Bot 121:773–795
Lehtonen A (2005) Estimating foliage biomass in scots pine (*Pinus sylvestris*) and Norway spruce (*Picea abies*) plots. Tree Physiol 25:803–811
Lehtonen A, Heikkinen J, Petersson H, Tupek B, Liski E, Mäkelä A (2020) Scots pine and Norway spruce foliage biomass in Finland and Sweden–testing traditional models vs. the pipe model theory. Can J For Res 50:146–154. https://doi.org/10.1139/cjfr-2019-0211
Leisch F, R-core (2012) Sweave user manual. R Help files
Leppälammi-Kujansuu J, Aro L, Salemaa M, Hansson K, Kleja DB, Helmisaari HS (2014a) Fine root longevity and carbon input into soil from below- and aboveground litter in climatically contrasting forests. For Ecol Manage 326:79–90
Leppälammi-Kujansuu J, Salemaa M, Kleja DB, Linder S, Helmisaari HS (2014b) Fine root turnover and litter production of Norway spruce in a long-term temperature and nutrient manipulation experiment. Plant Soil 374:73–88
Le Roux X, Lacointe A, Escobar-Gutiérrez A, Le Dizès S (2001) Carbon-based models of individual tree growth: a critical appraisal. Ann For Sci 58(5):469–506
Lind T, Söderberg U (1994) Considering costs and revenues in long-term forecasts of timber yields. Scand J For Res 9:397–404
Lindh M (2016) Evolution of plants. A mathematical perspective. Umea University, Umea
Lindh M, Zhang L, Falster D, Franklin O, Brännström Å (2014) Plant diversity and drought: the role of deep roots. Ecol Modell 290:85–93
Lindner M, Sievänen R, Pretzsch H (1997) Improving the simulation of stand structure in a forest gap model. For Ecol Manage 95:183–195
Liski J, Palosuo T, Peltonimei M, Sievänen R (2005) Carbon and decomposition model Yasso for forest soils. Ecol Modell 189:168–182
Litton CM, Giardina CP (2008) Below-ground carbon flux and partitioning: global patterns and response to temperature. Funct Ecol 22:941–954
Litton CM, Kauffmann JB (2008) Allometric models for predicting aboveground biomass in two widespread woody plants in Hawaii. Biotropica 40:313–320
Litton CJ, Raich JW, Ryan MG (2007) Carbon allocation in forest ecosystems. Glob Change Biol 13:2089–2109
Lloyd J, Taylor JA (1994) On the temperature dependence of soil respiration. Funct Ecol 8(3): 315–323
Lonsdale WM (1990) The self-thinning rule: dead or alive? Ecology 71:1373–1388
Loomis WE (1934) Daily growth of maize. Am J Bot 21:1–6

Ludlow AR, Randle TJ, Grace JC (1990) Developing a process-based growth model for Sitka spruce. In: Dixon RK, Meldahl RS, Ruark GA, Warren WG (eds) Process modeling of forest growth responses to environmental stress. Timber Press, Portland, pp 249–262

Luenberger DG (1979) Introduction to dynamic systems. John Wiley & Sons, New York

Mäkelä A (1985) Differential games in evolutionary theory: height growth strategies of trees. Theor Popul Biol 27:239–267

Mäkelä A (1986) Implications of the pipe model theory on dry matter partitioning and height growth in trees. J Theor Biol 123(1):103–120

Mäkelä A (1997) A carbon balance model of growth and self-pruning in trees based on structural relationships. For Sci 43:239–267

Mäkelä A (1999) Acclimation in dynamic models based on structural relationships. Funct Ecol 13:145–156

Mäkelä A (2002) Derivation of stem taper from the pipe theory in a carbon balance framework. Tree Physiol 22:891–905

Mäkelä A (2003) Process-based modelling of tree and stand growth: towards a herarchical treatment of multiscale processes. Can J For Res 23:398–409

Mäkelä A, Hari P (1986) Stand growth model based on carbon uptake and allocation in individual trees. Ecol Modell 33:205–229

Mäkelä A, Mäkinen H (2003) Generating 3D sawlogs with a process-based growth model. For Ecol Manage 184:337–354

Mäkelä A, Sievänen R (1987) Comparison of two shoot-root partitioning models with respect to substrate utilization and functional balance. Ann Bot 59:129–140

Mäkelä A, Sievänen R (1992) Height-growth strategies in opengrown trees. J Theor Biol 159: 443–467

Mäkelä A, Valentine HT (2001) The ratio of NPP to GPP: evidence of change over the course of stand development. Tree Physiol 21:1015–1030

Mäkelä A, Valentine HT (2006a) Crown ratio influences allometric scaling in trees. Ecology 87(12):2967–2972

Mäkelä A, Valentine HT (2006b) The quarter-power scaling model does not imply size invariant hydraulic resistance in plants. J Theor Biol 243:283–285

Mäkelä A, Vanninen P (1998) Impacts of size and competition on tree form and distribution of aboveground biomass in Scots pine. Can J For Res 28:216–227

Mäkelä A, Berninger F, Hari P (1996) Optimal control of gas exchange during drought: theoretical analysis. Ann Bot 77:461–467

Mäkelä A, Landsberg J, Ek AR, Burk TE, Ter-Mikaelian M, Ågren GI, Oliver CD, Puttonen P (2000) Process-based models for forest ecosystem management: current state of the art and challenges for practical implementation. Tree Physiol 20:289–298

Mäkelä A, Givnish TJ, Berninger F, Buckley TN, Farquhar GD, Hari P (2002) Challenges and opportunities of the optimality approach in plant ecology. Silva Fenn 36:605–614

Mäkelä A, Kolari P, Karimäki J, Nikinmaa E, Perämäki M, Hari P (2006) Modelling five years of weather-driven variation of GPP in a boreal forest. Agric For Meteorol 139:382–398

Mäkelä A, Hynynen J, Hawkins MJ, Reyer C, Soares P, van Oijen M, Tomé M (2012) Using stand-scale forest models for estimating sustainable forest management. For Ecol Manage 285: 164–178

Mäkelä A, Pulkkinen M, Mäkinen H (2016) Bridging empirical and carbon-balance based forest site productivity – significance of below-ground allocation. For Ecol Manage 372:64–77

Mäkelä A, Pulkkinen M, Kolari P, Lagergren F, Berbigier P, Lindroth A, Loustau D, Nikinmaa E, Vesala T, Hari P (2008a) Developing an empirical model of stand GPP with the LUE approach: analysis of eddy covariance data at five contrasting conifer sites in Europe. Glob Change Biol 14:92–108

Mäkelä A, Valentine HT, Helmisaari H (2008b) Optimal co-allocation of carbon and nitrogen in a forest stand at steady state. New Phytol 180:114–123

Mandallaz D (2008) Sampling techniques for forest inventories. Chapman & Hall/CRC, Boca Raton

Mandelbrot B (1983) The fractal geometry of nature. W. H. Freeman, New York
Marklund LG (1988) Biomass functions for pine, spruce and birch in Sweden. Sveriges Lantbruksuniversitet Rapporter-Skog 246:1–73
Maynard-Smith J, Price GR (1973) The logic of animal conflict. Nature 246:15–18
Maynard Smith J (1982) Evolution and the theory of games. Cambridge University Press, Cambridge
McCarthy HR, Oren R, Johnsen KH, Gallet-Budynek A, Pritchard SG, Cook CW, LaDeau SL, Jackson RB, Finzi AC (2010) Re-assessment of plant carbon dynamics at the duke free-air CO_2 enrichment site: interactions of atmospheric [CO_2] with nitrogen and water availability over stand development. New Phytol 185:514–528
McMahon TA (1973) Size and shape in biology. Science 179:1201–1204
McMahon TA, Kronaurer RE (1976) Tree structures: deducing the principle of mechanical design. J Theor Biol 59:443–466
McMurtrie RE (1991) Relationship of forest productivity to nutrient and carbon supply: a modeling analysis. Tree Physiol 9:87–99
McMurtrie R, Wolf L (1983) Above- and below-ground growth of forest stands: a carbon budget model. Ann Bot 52(4):437–448
McMurtrie RE, Norby RJ, Medlyn BE, Dewar RC, Pepper DA, Reich PB, Barton CVM (2008) Why is plant-growth response to elevated CO_2 amplified when water is limiting, but reduced when nitrogen is limiting? a growth-optimization hypothesis. Funct Plant Biol 35:521–534
McRoberts RE, Tomppo EO, Naesset E (2010) Advances and emerging issues in national forest inventories. Scand J For Res 25:368–381
Medlyn BE (2004) A MAESTRO Retrospective. In: Mencuccini M, Grace J, Moncrieff JB, McNaughton K (eds) Forests at the land-atmosphere interface. CABI, Oxfordshire, pp 105–121
Medlyn BE, Duursma RA, Zeppel MJ (2011) Forest productivity under climate change: a checklist for evaluating model studies. Wiley Interdiscip Rev Clim Change 2:332–355
Meehl GA, Covey C, Delworth T, Latif M, McAvaney B, Mitchell JFB, Stouffer RJ, Taylor KE (2007) The WCRP CMIP3 multimodel dataset: a new era in climate change research. Bull Am Meteorol Soc 88:1383–1394
Meinhold R, Singpurwalla N (1983) Understanding the kalman filter. Am Stat 37:123–127
Merganičová K, Merganič J, Lehtonen A, Vacchiano G, Zorana Ostrogović Sever M, Augustynczik A, Grote R, Kyselová I, Mäkelä A, Yousefpour R, Krejza J, Collalti A, Reyer C (2019) Forest carbon allocation modelling under climate change. Tree Physiol 11:11–12
Meyer A, Grote R, Polle A, Butterbach-Bahl K (2010) Simulating mycorrhiza contribution to forest C- and N cycling-the MYCOFON model. Plant Soil 327:493–517
Minunno F, Peltoniemi M, Launiainen S, Aurela M, Lindroth A, Lohila A, Mammarella I, Minkkinen K, Mäkelä A (2016) Calibration and validation of a semi-empirical flux ecosystem model for coniferous forests in the Boreal region. Ecol Modell 341:37–52
Minunno F, Peltoniemi M, Härkönen S, Kalliokoski T, Makinen H, Mäkelä A (2019) Bayesian calibration of a carbon balance model PREBAS using data from permanent growth experiments and national forest inventory. For Ecol Manage 440:208–257
Mitchell KJ (1975) Dynamics and simulated yield of Douglas-fir. For Sci Monogr 17:1–37
Monsi M, Saeki T (1953) Uber den lichtfaktor in den pflanzengesellschaften und seine bedeutung fur die stoffproduktion. Jap J Bot 14:22–52
Morgan J, Cannell MGR (1994) Shape of tree stems: a re-examination of the uniform stress hypothesis. Tree Physiol 14:49–62
Morris MD (1991) Factorial sampling plans for preliminary computational experiments. Technometrics 33:161–174
Munro DD (1974) Forest growth models – a prognosis. In: Fries J (ed) Growth models for tree and stand simulation, Research notes, vol 30. Royal College of Forestry, Department of Forest Yield Research, Stockholm, pp 7–21
Nakicenovic N, Swart R (2000) Special report on emissions scenarios (SRES). Cambridge University Press, Cambridge

Naudts K, Ryder J, McGrath MJ, Otto J, Chen Y, Valade A, Bellasen V, Berhongaray G, Boenisch G, Campioli M, Ghattas J, De Groote T, Haverd V, Kattge J, MacBean N, Maignan F, Merilä P, Penuelas J, Peylin P, Pinty B, Pretzsch H, Schulze ED, Solyga D, Vuichard N, Yan Y, Luyssaert S (2015) A vertically discretised canopy description for ORCHIDEE (SVN r2290) and the modifications to the energy, water and carbon fluxes. Geosci Model Dev 8:2035–2065

Neumann M, Moreno A, Mues V, Härkönen S, Mura M, Bouriaud O, Lang M, Achten WMJ, Thivolle-Cazat A, Bronisz K, Merganic J, Decuyper M, Alberdi I, Astrup R, Mohren F, Hasenauer H (2016) Comparison of carbon estimation methods for european forests. For Ecol Manage 361:397–420

Newnham RM (1965) Stem form and the variation of taper with age and thinning regime. Forestry 38(2):218–224

Niinimäki S, Tahvonen O, Mäkelä A (2012) Applying a process-based model in Norway spruce management. For Ecol Manage 265:102–115

Nikinmaa E, Goulet J, Messier C, Sievänen R, Perttunen J, Lehtonen M (2003) Shoot growth and crown development; the effect of crown position in 3D simulations. Tree Physiol 23:129–136

Nikinmaa E, Hölttä T, Sievänen R (2014) Dynamics of leaf gas exchange, xylem and phloem transport, water potential and carbohydrate concentration in a realistic 3-D model tree crown. Ann Bot 114:653–666

Niklas KJ (1994) Plant allometry. The Univeristy of Chicago Press, Chicago

Niklas KJ (1995) Size-dependent allometry of tree height, diameter and trunk-taper. Ann Bot 75:217–227

Niklas KJ, Kerchner V (1984) Mechanical and photosynthetic constraints on the evolution of plant shape. Paleobiology 10(1):79–101

Nilson T (1999) Inversion of gap frequency data in forest stands. Agric For Meteorol 98/99: 437–448

Norberg RA (1988) Theory of growth geometry of plants and self-thinning of plant populations: geometric similarity, elastic similarity, and different growth modes of plant parts. Am Nat 131:220–256

Norby RJ, Zak DR (2011) Ecological lessons from free-air CO_2 enrichment (face) experiments. Annu Rev Ecol Evol Syst 42:181–203

Norman JM, Welles JM (1983) Radiative transfer in an array of canopies. Agron J 75:481–488

Näsholm T, Ekblad A, Nordin A, Glesler R, Högberg M, Högberg P (1998) Boreal forest plants take up organic nitrogen. Nature 392:914–916

Näsholm T, Högberg P, Franklin O, Metcalfe D, Keel SG, Campbell C, Hurry V, Linder S, Högberg MN (2013) Are ectomycorrhizal fungi alleviating or aggravating nitrogen limitation of tree growth in boreal forests? New Phytol 198(1):214–221

O'Neill RV, Deangelis DL, Waide JB, Allen TFH, Allen GE (1986) A hierarchical concept of ecosystems, vol 23. Princeton University Press, Princeton

O'Neill BC, Kriegler E, Riahi K, Ebi KL, Hallegatte S, Carter TR, Mathur R, van Vuuren DP (2014) A new scenario framework for climate change research: the concept of shared socioeconomic pathways. Clim Change 122:387–400

O'Neill BC, Kriegler E, Ebi KL, Kemp-Benedict E, Riahi K, Rothman DS, van Ruijven BJ, van Vuuren DP, Birkmann J, Kok K, Levy M, Solecki W (2017) The roads ahead: narratives for shared socioeconomic pathways describing world futures in the 21st century. Glob Environ Change 42:169–180

Oker-Blom P, Kellomäki S (1982) Theoretical computations on the role of crown shape in the absorption of light by forest trees. Math Biosci 59:291–311

Oker-Blom P, Pukkala T, Kuuluvainen T (1989) Relationship between radiation interception and photosynthesis in forest canopies: effect of stand structure and latitude. Ecol Modell 49:73–87

Osawa A, Sugita S (1989) The self-thinning rule: another interpretation of Weller's results. Ecology pp 279–283

Osawa A, Ishizuka M, Kanazawa Y (1991) A profile theory of tree growth. For Ecol Manage 41:33–63

Ostonen I, Helmisaari HS, Borken W, Tedersoo L, Kiukumägi M, Bahram M, Lindroos AJ, Nöjd P, Uri V, Merilä P, Asi E, Lõhmus K (2011) Fine root foraging strategies in norway spruce forests across a european climate gradient. Glob Change Biol 17:3620–3632
Pacala SW, Canham CD, Silander JA (1993) Forest models defined by field measurements. 1. The design of a Northeastern forest simulator. Can J For Res 23:1980–1988
Pacala SW, Canham CD, Saponara J, Silander JA, Kobe RK, Ribbens E (1996) Forest models defined by field measurements: estimation, error analysis and dynamics. Ecol Monogr 66:1–43
Panik MJ (2013) Growth curve modeling: theory and applications. John Wiley & Sons, Hoboken
Patrick Bentley L, Stegen JC, Savage VM, Smith DD, von Allmen EI, Sperry JS, Reich PB, Enquist BJ (2013) An empirical assessment of tree branching networks and implications for plant allometric scaling models. Ecol Lett 16:1069–1078
Pearcy RW, Muraoka H, Valladares F (2005) Crown architecture in sun and shade environments: assessing function and trade-offs with a three-dimensional simulation model. New Phytol 166:791–800
Peltoniemi M, Pulkkinen M, Kolari P, Duursma R, Montagnani L, Wharton S, Lagergren F, Takagi K, Verbeeck H, Christensen T, Vesala T, Falk M, Loustau D, Mäkelä A (2012) Does canopy mean N concentration explain differences in light use efficiencies of canopies in 14 contrasting forest sites? Tree Physiol 32:200–218
Peltoniemi M, Pulkkinen M, Aurela M, Pumpanen J, Kolari P, Mäkelä A (2015) A semi-empirical model of boreal forest gross primary production, evapotranspiration, and soil water – calibration and sensitivity analysis. Boreal Environ Res 20:151–171
Penning de Vries FWT (1975) Use of assimilates in higher plants. In: Cooper JP (ed) Photosynthesis and productivity in different environments, Cambridge University Press, Cambridge, pp 459–480
Pienaar LV, Turnbull KJ (1973) The Chapman-Richards generalization of von Bertalanffy's growth model for basal area growth and yield in even-aged stands. For Sci 19:2–22
Pigliucci M (2005) Evolution of phenotypic plasticity: where are we going now? Trends Ecol Evol 20:481–486
Poole D, Raftery AE (2000) Inference for deterministic simulation models: the Bayesian melding approach. J Am Stat Assoc 95(452):1244–1255
Poorter L, Oberbauer SF, Clark DB (1995) Leaf optical properties along a vertical gradient in a tropical rain forest canopy in Costa Rica. Am J Bot 82(10):1257–1263
Poorter H, Niklas KJ, Reich PB, Oleksyn J, Pooti P, Mommer L (2011) Biomass allocation to leaves, stems and roots: meta-analyses of interspecific variation and environmental control. New Phytol 193:30–50
Pregitzer KS (2002) Fine roots of trees–a new perspective. New Phytol 154(2):267–270
Pretzsch H (2009) Forest dynamics, growth and yield. Springer, Berlin
Pretzsch H, Grote R, Reineking B, Rötzer TH, Steifert ST (2008) Models for forest ecosystem management: a European perspective. Ann Bot 101:1065–1087
Pretzsch H, Biber P, Schuetze G, Uhl EO, Roetzer T (2014) Forest stand growth dynamics in Central Europe have accelerated since 1870. Nat Commun 5. https://doi.org/10.1038/ncomms5967
Prusinkiewicz P, Lindenmayer A (1990) The algorithmic beauty of plants. Springer, Berlin
Pruyn ML, Gartner BL, Harmon ME (2002) Within-stem variation of respiration in *Pseudotsuga menziesii* (Douglas-fir) trees. New Phytol 154:359–372
Pruyn ML, Gartner BL, Harmon ME (2005) Storage versus substrate limitation to bole respiratory potential in two coniferous tree species of contrasting sapwood width. J Exp Bot 56:2637–2649
Purves DW, Lichstein JW, Strigul N, Pacala SW (2008) Predicting and understanding forest dynamics using a simple tractable model. Proc Natl Acad Sci U S A 105:17018–17022
Raftery AE, Bao L (2010) Estimating and projecting trends in HIV/AIDS generalized epidemics using incremental mixture importance sampling. Biometrics 66(4):1162–1173
Rajaniemi TK (2003) Evidence for size asymmetry of belowground competition. Basic Appl Ecol 4:239–247
Rantala S (2011) Finnish forestry practice and management. Metsäkustannus, Helsinki

Rasinmaki J, Mäkinen A, Kalliovirta J (2009) SIMO: an adaptable simulation framework for multiscale forest resource data. Comput Electron Agric 66:76–84

Reich PB, Rich RL, Lu X, Wang YP, Oleksyn J (2014) Biogeographic variation in evergreen conifer needle longevity and impacts on boreal forest carbon cycle projections. Proc Natl Acad Sci U S A 111:13703–13708

Reineke LH (1933) Perfecting a stand-density index for even-aged forests. J Agric Res 46(7): 626–637

Repola J (2006) Models for vertical wood density of Scots pine, Norway spruce and birch stems, and their application to determine average wood density. Silva Fenn 40(4):673–685

Repola J (2009) Biomass equations for Scots pine and Norway spruce in Finland. Silva Fenn 43:625–647

Reyer CPO, Flechsig M, Lasch-Bron P, van Oijen M (2016) Integrating parameter uncertainty of a process-based model in assessments of climate change effects on forest productivity. Clim Change 137:395–409

Reynolds JF, Chen J (1996) Modelling whole-plant allocation in relation to carbon and nitrogen supply: coordination versus optimization: opinion. Plant Soil 185(1):65–74

Reynolds JH, Ford ED (2005) Improving competition representation in theoretical models of self-thinning: a critical review. J Ecol 93(2):362–372

Reynolds JF, Hilbert DW, Kemp PR (1993) Scaling ecophysiology from the plant to the ecosystem: a conceptual framework. In: Ehleringerl JR, Field CB (eds) Scaling physiological processes. Leaf to globe. Academic Press, San Diego, pp 127–140

Richards FJ (1959) A flexible growth function for empirical use. J Exp Bot 10:290–300

Richardson AD, ZuDohna H (2003) Predicting root biomass from branching patterns of douglas-fir root systems. Oikos 100(1):96–104

Richardson AD, Williams M, Hollinger DY, Moore DJP, Dail DB, Davidson EA, Scott NA, Evans RS, Hughes H, Lee JT, Rodriques C, Savage K (2010) Estimating parameters of a forest ecosystem C model with measurements of stocks and fluxes as joint constraints. Oecologia 164(1):25–40

Robinson AP, Ek AR (2000) The consequences of hierarchy for modeling in forest ecosystems. Can J For Res 30:1837–1846

Robinson AR, Lermusiaux PFJ (2002) Data assimilation for modeling and predicting coupled physical-biological interactions in the sea. Sea 12:475–536

Rose MR (1978) Cheating in evolutionary games. J Theor Biol 75:21–34

Running SW, Coughlan JC (1988) A general model of forest ecosystem processes for regional applications I. Hydrological balance, canopy gas exchange and primary production processes. Ecol Modell 42:125–154

Running SW, Gower ST (1991) A general model of forest ecosystem processes for regional applications II. Dynamic carbon allocation and nitrogen budgets. Tree Physiol 9:147–160

Ryan MG (1991) A simple method for estimating gross carbon budgets for vegetation in forest ecosystems. Tree Physiol 9:255–266

Ryan MG (1995) Foliar maintenance respiration of subalpine and boreal trees and shrubs in relation to nitrogen content. Plant Cell Environ 18:765–772

Ryan MG, Hubbard RM, Pongracic S, Raison RJ, McMurtrie RE (1996) Foliage, fine-root, woody-tissue and stand respiration in *Pinus radiata* in relation to nitrogen status. Tree Physiol 16: 333–343

Ryan MG, Phillips N, Bond BJ (2006) The hydraulic limitation hypothesis revisited. Plant Cell Environ 29:367–381

Räisänen J, Räty O (2013) Projections of daily mean temperature variability in the future: cross-validation tests with ENSEMBLES regional climate simulations. Clim Dyn 41:1553–1568

Sala A, Hoch G (2009) Height-related growth declines in ponderosa pine are not due to carbon limitation. Plant Cell Environ 32:22–30

Sala A, Woodruff DR, Meinzer FC (2012) Carbon dynamics in trees: feast or famine? Tree Physiol 32:764–775

Schelhaas MJ, Eggers J, Lindner M, Nabuurs GJ, Päivinen R, Schuck A, Verkerk PJ, Van der Werf DC, Zudin S (2007) Model documentation for the European Forest Information Scenario model (EFISCEN 3.1.3). Alterra report 1559 and EFI technical report 26, Wageningen and Joensuu

Schiestl-Aalto P, Kulmala L, Mäkinen H, Nikinmaa E, Mäkelä A (2015) CASSIA – a dynamic model for predicting intra-annual sink demand and interannual growth variation in Scots pine. New Phytol 206:647–659

Schiestl-Aalto P, Ryhti K, Mäkelä A, Peltoniemi M, Bäck J, Kulmala L (2019) Analysis of the NSC storage dynamics in tree organs reveals the allocation to belowground symbionts in the framework of whole tree carbon balance. Front For Glob Change 2. https://doi.org/10.3389/ffgc.2019.00017

Schimel JP, Weintraub MN (2003) The implications of exoenzyme activity on microbial carbon and nitrogen limitation in soil: a theoretical model. Soil Biol Biochem 35:549–563

Schlecht RM, Affleck DLR (2014) Branch aggregation and crown allometry condition the precision of randomized branch sampling estimators of conifer crown mass. Can J For Res 44(5):499–508

Schneider R, Berninger F, Ung CH, Mäkelä A, Swift DE, Zang SY (2011) Within crown variation in the relationship between foliage biomass and sapwood area in jack pine. Tree Physiol 31: 22–29

Shinozaki K, Yoda K, Hozumi K, Kira T (1964a) A quantitative analysis of plant form – the pipe model theory. I. Basic analysis. Jap J Ecol 14:97–105

Shinozaki K, Yoda K, Hozumi K, Kira T (1964b) A quantitative analysis of plant form – the pipe model theory. II. Further evidence of the theory and its application in forest ecology. Jap J Ecol 14:133–139

Shugart HH (1984) A theory of forest dynamics: the ecological implications of forest succession models. Springer, New York

Sievänen R, Nikinmaa E, Nygren P, Ozier-Lafontaine H, Perttunen J, Hakula H (2000) Components of functional-structural tree models. Ann For Sci 57:399–412

Sitch S, Huntingford C, Gedney N, Levy PE, Lomas M, Piao SL, Betts R, Ciais P, Cox P, Friedlingstein P, Jones CD, Prentice IC, Woodward FI (2008) Evaluation of the terrestrial carbon cycle, future plant geography and climate-carbon cycle feedbacks using five Dynamic Global Vegetation Models (DGVMs). Glob Change Biol 14:2015–2039

Skovsgaard JP, Vanclay JK (2008) Forest site productivity: a review of the evolution of dendrometric concepts for even-aged stands. Forestry 81(1):13–31

Snow GRS (1931) Experiments on growth and inhibition. II. New phenomena of inhibition. Proc R Soc Lond 108:305–316

Soetaert K, Petzoldt T (2010) Inverse modelling, sensitivity and Monte Carlo analysis in R using package fme. J Stat Softw 33(3):1–28

Soetaert K, Petzoldt T, Setzer RW (2010) Solving differential equations in R: package desolve. J Stat Softw 33(9):1–25

Soetaert K, Cash J, Mazzia F (2012) Solving differential equations in R. Springer, Heidelberg

Sorrensen-Cothern KA, Ford ED, Sprugel DG (1993) A model of competition incorporating plasticity through modular foliage and crown development. Ecol Monogr 63(3):277–304

Sprugel DG (1990) Components of woody-tissue respiration in young *Abies amabilis* (Dougl.) Forbes trees. Trees 4:88–99

Sprugel DG (2002) When branch autonomy fails: Milton's Law of resource availability and allocation. Tree Physiol 22:1119–1124

Stage AR (1973) Prognosis model for stand development. Usda forest service research paper int-137. Intermountain Forest and Range Experiment Station, Ogden

Starr AW, Ho YC (1969) Nonzero-sum differential games. J Optim Theory Appl 3:184–206

Sterck FJ, Schieving F (2007) 3-D growth patterns of trees: effects of carbon economy, meristem activity, and selection. Ecol Monogr 77:405–420

Stocker BD, Prentice IC, Cornell SE, Davies-Barnard T, Finzi AC, Franklin O, Janssens I, Larmola T, Manzoni T, Näsholm T, Raven JA, Rebel KT, Reed S, Vicca S, Wiltshire A, Zaehle S (2016) Terrestrial nitrogen cycling in Earth system models revisited. New Phytol 210:1165–1168

Stockfors J, Linder S (1998) The effect of nutrition on the seasonal course of needle respiration in Norway spruce stands. Trees 12:130–138

Stout BB (1956) Studies of the root systems of deciduous trees. Harvard Black Rock Forest, Cornwall-on-the-Hudson

Strigul N, Pristinski D, Purves D, Dushoff J, Pacala S (2008) Scaling from trees to forests: tractable macroscopic equations for forest dynamics. Ecol Monogr 78:523–545

Strub MR, Burkhart HE (1975) A class interval-free method for obtaining expected yields from diameter distributions. For Sci 21:67–69

Sulman BN, Phillips RP, Oishi AC, Shevliakova E, Pacal SW (2014) Microbe-driven turnover offsets mineral-mediated storage of soil carbon under elevated CO_2. Nat Clim Change 4: 1099–1102

Terrer C, Vicca S, Hungate BA, Phillips RP, Prentice IC (2016) Mycorrhizal association as a primary control of the CO_2 fertilization effect. Science 353:72–74

Thompson DW (1992) On growth and form. Dover Publications, Inc., New York

Thornley JHM (1972) A model to describe the partitioning of photosynthate during vegetative plant growth. Ann Bot 36:419–430

Thornley JHM (1976) Mathematical models in plant physiology. Academic Press, London

Thornley JHM, France J (2007) Mathematical models in agriculture, 2nd edn. CABI, Wallingford

Thornley JHM, Johnson IR (1990) Plant and crop modelling. Clarendon Press, Oxford

Thornley JHM, Parsons AJ (2014) Allocation of new growth between shoot, root and mycorrhiza in relation to carbon, nitrogen and phosphate supply: teleonomy with maximum growth rate. J Theor Biol 342:1–14

Tomlinson PB (1983) Tree architecture. New approaches help to define the elusive biological property of tree form. Am Sci 71:141–149

Tomppo E, Heikkinen J, Henttonen HM, Ihalainen A, Katila M, Mäkelä H, Tuomainen T, Vainikainen N (2011) Designing and conducting a forest inventory – case: 9th national forest inventory of Finland. Managing forest ecosystems, vol 21. Springer Science & Business Media, p 272

Tuomi M, Thum T, Järvinen H, Fronzek S, Berg B, Harmon ME, Trofymow JA, Sevanto S, Liski J (2009) Leaf litter decomposition – estimates of global variability based on Yasso07 model. Ecol Modell 220:3362–3371

Ťupek B, Mäkipää R, Heikkinen J, Peltoniemi M, Ukonmaanaho L, Hokkanen T, Nöjd P, Nevalainen S, Lindgren M, Lehtonen A (2015) Foliar turnover rates in Finland – comparing estimates from needle-cohort and litterfall-biomass methods. Boreal Environ Res 20:283–304

Valentine HT (1985) Tree-growth models: derivations employing the pipe-model theory. J Theor Biol 117(4):579–585

Valentine HT (1997) Height growth, site index, and carbon metabolism. Silva Fenn 31(3):251–263

Valentine HT, Hilton SJ (1977) Sampling oak foliage by the randomized-branch method. Can J For Res 7:295–298

Valentine HT, Mäkelä A (2005) Bridging process-based and empirical approaches to modeling tree growth. Tree Physiol 25:769–779

Valentine HT, Mäkelä A (2012) Modeling forest stand dynamics from optimal balances of carbon and nitrogen. New Phytol 194:961–971

Valentine HT, Tritton LM, Furnival GM (1984) Subsampling trees for biomass, volume,, or mineral content. For Sci 30:673–681

Valentine HT, Gregoire TG, Burkhart HE, Hollinger DY (1997) A stand-level model of carbon allocation and growth, calibrated for loblolly pine. Can J For Res 27:817–830

Valentine HT, Herman DA, Gove JH, Hollinger DY, Solomon DS (2000) Initializing a model stand for process-based projection. Tree Physiol 20(5–6):393–398

Valentine HT, Green EJ, Mäkelä A, Amateis RL, Mäkinen H, Ducey MJ (2012) Models relating stem growth to crown length dynamics: application to loblolly pine and Norway spruce. Trees 26:469–478

Valentine HT, Amateis RL, Gove JH, Mäkelä A (2013) Crown-rise and crown-length dynamics: application to loblolly pine. Forestry 86:371–375

Valentine HT, Baldwin VC Jr, Gregoire TG, Burkhart HE (1994a) Surrogates for foliar dry matter in loblolly pine. For Sci 40(3):576–585
Valentine HT, Ludlow AR, Furnival GM (1994b) Modeling crown rise in even-aged stands of Sitka spruce or loblolly pine. For Ecol Manage 69:189–197
Vanninen P, Mäkelä A (1999) Fine root biomass of Scots pine stands differing in age and soil fertility in southern Finland. Tree Physiol 19:823–830
Vanninen P, Mäkelä A (2000) Needle and stem wood production in Scots pine (*Pinus sylvestris*) trees of different age, size and competitive status. Tree Physiol 20:527–533
Van Oijen M, Rougier J, Smith R (2005) Bayesian calibration of process-based forest models: bridging the gap between models and data. Tree Physiol 25(7):915–927
Van Oijen M, Schapendonk A, Höglind M (2010) On the relative magnitude of photosynthesis, respiration, growth andcarbon storage in vegetation. Ann Bot 105:793–797
Van Oijen M, Reyer C, Bohn FJ, Cameron DR, Deckmyn G, Flechsig M, Härkönen S, Hartig F, Huth A, Kiviste A, Lasch P, Mäkelä A, Mette T, Minunno F, Rammer W (2013) Bayesian calibration, comparison and averaging of six forest models, using data from Scots pine stands across Europe. For Ecol Manage 289:255–268
Venäläinen A, Tuomenvirta H, Pirinen P, Drebs A (2005) A basic Finnish climate data set 1961–2000 – description and illustrations. Reports of the Finnish Meteorological Institute 5:1–27
Verhulst PF (1838) Notice sur la loi que la population poursuit dans son accroissement. Correspondance Mathématique et physique 10:113–121
Vogt KA, Vogt DJ, Asbjornsen H, Dahlgren RA (1995) Roots, nutrients and their relationship to spatial patterns. Plant Soil 168:113–123
Vuokila Y, Väliaho H (1980) Growth and yield models of planted conifer forests (in Finnish). Publications of the Finnish Forest Research Institute, Helsinki
Wang C (2006) Biomass allometric equations for 10 co-occurring tree species in Chinese temperate forests. Can J For Res 222:9–16
Wang YP, Jarvis PG (1990) Description and validation of an array model – MAESTRO. Agric For Meteorol 51:257–280
Wang YP, Trudinger CM, Enting IG (2009) A review of applications of model-data fusion to studies of terrestrial carbon fluxes at different scales. Agric For Meteor 149(11):1829–1842
Waring RH, Schlesinger WH (1985) Forest ecosystems. Concepts and management. Academic Press, Orlando
Waring RH, Schroeder PE, Oren R (1982) Application of the pipe model theory to predict canopy leaf area. Can J For Res 12:556–560
Waring RH, Coops NC, Landsberg JJ (2010) Improving predictions of forest growth using the 3-PGS model with observations made by remote sensing. For Ecol Manage 259:1722–1729
Warren JM, Brooks JR, Meinzer FC, Eberhart JL (2008) Hydraulic redistribution of water from *Pinus ponderosa* trees to seedlings: evidence for an ectomycorrhizal pathway. New Phytol 178:382–394
Weiner J (1990) Asymmetric competition in plant populations. Tree 5:360–364
Weiner J, Wright DB, Castro S (1997) Symmetry of below-ground competition between *Kochia scoparia* individuals. Oikos 79:85–91
Weiskittel AR, Hann DW, Kershaw JA Jr, Vanclay JK (2011) Forest growth and yield modeling. John Wiley & Sons, West Sussex
Weller DE (1987) A reevaluation of the -3/2 power rule of plant self-thinning. Ecol Monogr 57(1):23–43
Weller DE (1990) Will the real self-thinning rule please stand up? – a reply to Osawa and Sugita. Ecology 71(3)1204–1207
Weller DE (1991) The self-thinning rule: dead or unsupported? – a reply to Lonsdale. Ecology 72(2):747–750
Wertin TM, Teskey RO (2008) Close coupling of whole-plant respiration to net photosynthesis and carbohydrates. Tree Physiol 28:1831–1840
West GB, Brown JH, Enquist BJ (1999) A general model for the structure and allometry of plant vascular systems. Nature 400:664–667

West GB, Brown JH, Enquist BJ (1997a) The fourth dimension of life: fractal geometry and allometric scaling of organisms. Science 284:1677–1679

West GB, Brown JH, Enquist BJ (1997b) A general model for the origin of allometric scaling laws in biology. Science 276:122–126

Westoby M (1981) The place of the self-thinning rule in population dynamics. Am Nat 118(4): 581–587

White HL (1935) The interaction of factors in the growth of Lemna. XII. The interaction of nitrogen and light intensity in relation to root length. Ann Bot 1:649–654

White J (1981) The allometric interpretation of the self-thinning rule. J Theor Biol 89(3):475–500

Williams K, Field CB, Mooney HA (1989) Relationships among leaf construction cost, leaf longevity, and light environment in rain-forest plants of the genus *Piper*. Am Nat 133: 198–211

Williams CJ, LePage BA, Vann DR, Tange T, Ikeda H, Ando M, Kusakabe T, Tsuzuki H, Sweda T (2003) Structure, allometry, and biomass of plantation *Metasequoia glyptostroboides* in Japan. For Ecol Manage 180:287–301

Winsor CP (1932) The Gompertz curve as a growth curve. Proc Natl Acad Sci U S A 18:1–8

Yoda K, Kira T, Ogawa H, Hozumi K (1963) Self-thinning in overcrowded pure stands under cultivation and natural conditions. J Biol Osaka City Univ 14:107–129

Yoder BJ, Ryan MG, Waring RH, Schoettle AW, Kaufmann MR (1994) Evidence of reduced photosynthetic rates in old trees. For Sci 40:513–527

Yousefpour R, Hanewinkel M (2014) Balancing decisions for adaptive and multipurpose conversion of Norway spruce (*Picea abies* L. Karst) monocultures in the Black Forest area of Germany. For Sci 60:73–84

Zeide B (1985) Tolerance and self-tolerance of trees. For Ecol Manage 13(3):149–166

Zeide B (1987) Analysis of the 3/2 power law of self-thinning. For Sci 33(2):517–537

Zeide B (1993) Analysis of growth equations. For Sci 39:594–616

Zeide B (1998) Fractal analysis of foliage distribution in loblolly pine crowns. Can J For Res 28:106–114

Zhao D, Borders B, Wang M, Kane M (2007) Modeling mortality of second-rotation loblolly pine plantations in the Piedmont/Upper Coastal Plain and Lower Coastal Plain of the southern United States. For Ecol Manage 252(1):132–143

Zianis D, Muukkonen P, Mäkipää R, Mencuccini M (2005) Biomass and stem volume equations for tree species in Europe. Silva Fenn Monogr 4:1–63

Zimmermann MH (1978) Hydraulic architecture of some diffuse-porous trees. Can J Bot 56: 2286–2295

Zobel B, Sprague JR (1998) Juvenile wood in forest trees. Springer, Berlin

Zobitz JM, Desai AR, Moore DJP, Chadwick MA (2011) A primer for data assimilation with ecological models using Markov chain Monte Carlo (MCMC). Oecologia 167(3):599–611

Index

A. Mäkelä, H. T. Valentine, *Models of Tree and Stand Dynamics*,
https://doi.org/10.1007/978-3-030-35761-0

MIX
Papier aus verantwortungsvollen Quellen
Paper from responsible sources
FSC® C105338

If you have any concerns about our products,
you can contact us on
ProductSafety@springernature.com

In case Publisher is established outside the EU,
the EU authorized representative is:
Springer Nature Customer Service Center GmbH
Europaplatz 3, 69115 Heidelberg, Germany

Printed by Libri Plureos GmbH
in Hamburg, Germany